LUBRICATION AND RELIABILITY HANDBOOK

LUBRICATION AND RELIABILITY HANDBOOK

Edited by

M. J. NEALE
OBE, BSc(Eng), DIC, FCGI, WhSch, FREng, FIMechE

BUTTERWORTH HEINEMANN

BOSTON OXFORD AUCKLAND JOHANNESBURG MELBOURNE NEW DELHI

Recognizing the importance of preserving what has been written, Butterworth-Heinemann prints its books on acid-free paper whenever possible.

Butterworth-Heinemann supports the efforts of American Forests and the Global ReLeaf program in its campaign for the betterment of trees, forests, and our environment.

Library of Congress Cataloging in Publication Data
Lubrication and reliability handbook/edited by M.J. Neale.
    p. cm.
    ISBN 0 7506 5154 7
    1. Lubrication and lubricants – Handbooks, manuals, etc.   2. Reliability
    (Engineering) – Handbooks, manual, etc.   I Neale, M. J. (Michael John)
    TJ1075.L812 2000
    621.8'9–dc21   00–049378

British Library Cataloguing in Publication Data
A catalogue record for this book is available from the British Library

The publisher offers special discounts on bulk orders of this book.
For information, please contact:
Manager of Special Sales
Butterworth-Heinemann
225 Wildwood Avenue
Woburn, MA 01801-2041
Tel: 781-904-2500
Fax: 781-904-2620

For information on all Butterworth-Heinemann publications available, contact our World Wide Web home page at:
http://www.bh.com

10 9 8 7 6 5 4 3 2 1

Composition by Genesis Typesetting, Rochester, Kent, England
Printed by J. W. Arrowsmith Ltd, Bristol

CONTENTS

INTRODUCTION

This handbook is intended to help engineers in industry with the operation and maintenance of machinery. It gives the information that these engineers need in a form that is instantly accessible and easy to read.

The manufacturers of machinery provide guidance on the operation, lubrication and maintenance required for their particular machines. However, there are, of course, many different machines in an industrial plant or service organisation, supplied by various manufacturers, and there is a need to select as many similar lubricants as possible, and to use related maintenance techniques. This book attempts to bridge the gap which exists between the available data on the various machines, by providing overall guidance on how to co-ordinate the recommendations of the various manufacturers.

The handbook is structured in a number of sections to make it easier to use, and to bring together related subjects, so that the reader when focusing on a particular problem can also refer to related material that is likely to be of interest. The various sections are listed here in this introduction, to provide some overall guidance, additional to that available in the contents list and the index.

Lubricants

This section describes the various types of lubricant that are available with guidance on their overall properties and performance. Detailed information is provided on mineral oils, synthetic oils, greases and solid lubricants, as well as on the various oil additives that are commonly used. Since some machines are now lubricated by their own process fluids information is also given on the viscosity of water, refrigerants and various hydrocarbons and chemicals.

Lubrication of components

The lubrication of machines relates to the lubrication of their various moving components. This section gives guidance on the selection of lubricants to match the needs of the components under a range of operating conditions. The components covered are plain and rolling bearings, gears, roller chains, wire ropes, flexible couplings and slides.

Lubrication systems

The next subject requiring review is the optimum method of feeding the lubricant to the various machines and their components. This can range from manual greasing to automated centralising greasing systems, and from splash, wick and ring oil feeding to pressurised mist systems and full size oil circulation systems. Detailed guidance is also given on the selection and design of circulation system components such as oil tanks, pumps, filters and coolers as well as the interconnecting piping systems and the necessary instrumentation and warning devices.

Machine operation

The machine manufacturers and/or process designers will usually provide the necessary guidance on machine operating conditions. The operating engineers can however benefit from additional guidance on running in procedures, and on lubricant related operating problems, such as potential lubricant deterioration due to high or low temperatures, and the effect of contaminant process gases and liquids. Information is provided on these areas, together with data on fire or health hazards from lubricants.

Machine maintenance

To keep the machines in a plant or fleet operating effectively, requires good maintenance procedures. The handbook reviews the suitability of the various maintenance methods for various types of machines and gives guidance on their selection. Condition based maintenance is covered in detail with the various methods by which the condition of a machine can be monitored while it is in operation, so that future essential maintenance can be planned. Such methods include temperature measurement, vibration analysis, wear debris analysis, and lubricant tests, as well as methods of assessing the operating performance of machine components.

Component failure

When a failure does occur on one of the working components of a machine, such as a bearing, gear, seal or coupling, it is useful to have guidance on understanding the causes of the failure from the appearance of the failed component. This section therefore includes a large number of photographs of machine components showing the typical surface appearance associated with the various failure modes.

Component repair

Finally, after a failure has occurred it is useful to have guidance on how a worn surface can be rebuilt or refaced, or how a bearing or friction surface can be relined.

This handbook is based on experience from around the world, over many years, of the investigation of problems with machines of all kinds, and of dealing with these by practical and economical solutions. It is hoped that it will be helpful to the many engineers involved in machine operation and maintenance of all kinds of machinery and plant.

CONTRIBUTORS

Section	Author
Selection of lubricant type	A. R. Lansdown MSc, PhD, FRIC, FInstPet
Mineral oils	T. I. Fowle BSc(Hons), ACGI, CEng, FIMechE
Synthetic oils	A. R. Lansdown MSc, PhD, FRIC, FInstPet
Greases	N. Robinson & A. R. Lansdown MSc, PhD, FRIC, FInstPet
Solid lubricants and coatings	J. K. Lancaster PhD, DSc, FInstP
Other liquids	D. T. Jamieson FRIC
Plain bearing lubrication	J. C. Bell BSc, PhD
Rolling bearing lubrication	E. L. Padmore CEng, MIMechE
Gear and roller chain lubrication	J. Bathgate BSc, CEng, MIMechE
Wire rope lubrication	D. M. Sharp
Lubrication of flexible couplings	J. D. Summers-Smith BSc, PhD, CEng, FIMechE
Slide lubrication	M. J. Neale OBE, BSc(Eng), DIC, FCGI, WhSch, FEng, FIMechE
Lubricant selection	R. S. Burton
Selection of lubrication systems	W. J. J. Crump BSc, ACGI, FInstP
Total loss grease systems	P. L. Langborne BA, CEng, MIMechE
Total loss oil and fluid grease systems	P. G. F. Seldon CEng, MIMechE
Mist systems	R. E. Knight BSc, FCGI
Dip splash systems	J. Bathgate BSc, CEng, MIMechE
Circulation systems	D. R. Parkinson FInstPet
Design of oil tanks	A. G. R. Thomson BSc(Eng), CEng, AFRAeS
Selection of oil pumps	A. J. Twidale
Selection of filters and centrifuges	R. H. Lowres CEng, MIMechE, MIProdE, MIMarE, MSAE, MBIM
Selection of heaters and coolers	J. H. Gilbertson CEng, MIMechE, AMIMarE
A guide to piping design	P. D. Swales BSc, PhD, CEng, MIMechE
Selection of warning and protection devices	A. J. Twidale
Commissioning lubrication systems	N. R. W. Morris
Running-in procedures	W. C. Pike BSc, ACGI, CEng, MIMechE
Industrial plant environmental data	R. L. G. Keith BSc
High pressure and vacuum	A. R. Lansdown MSc, PhD, FRIC, FInstPet J. D. Summers-Smith BSc, PhD, CEng, FIMechE
High and low temperatures	M. J. Todd MA
Chemical effects	H. H. Anderson BSc(Hons), CEng, FIMechE
Maintenance methods	M. J. Neale OBE, BSc(Eng), DIC, FCGI, WhSch, FEng, FIMechE
Condition monitoring	M. J. Neale OBE, BSc(Eng), DIC, FCGI, WhSch, FEng, FIMechE
Operating temperature limits	J. D. Summers-Smith BSc, PhD, CEng, FIMechE
Vibration analysis	M. J. Neale OBE, BSc(Eng), DIC, FCGI, WhSch, FEng, FIMechE
Wear debris analysis	M. H. Jones BSc(Hons), CEng, MIMechE, MInstNDT M. J. Neale OBE, BSc(Eng), DIC, FCGI, WhSch, FEng, FIMechE
Lubricant change periods and tests	J. D. Summers-Smith BSc, PhD, CEng, FIMechE
Lubricant biological deterioration	E. C. Hill MSc, FInstPet
Component performance analysis	M. J. Neale OBE, BSc(Eng), DIC, FCGI, WhSch, FEng, FIMechE
Allowable wear limits	H. H. Heath FIMechE

CONTRIBUTORS

Section	Author
Failure patterns and failure analysis	J. D. Summers-Smith BSc, PhD, CEng, FIMechE
	M. J. Neale OBE, BSc(Eng), DIC, FCGI, WhSch, FEng, FIMechE
Plain bearing failures	P. T. Holingan BSc(Tech), FIM
Rolling bearing failures	W. J. J. Crump BSc, ACGI, FInstP
Gear failures	T. I. Fowle BSc(Hons), ACGI, CEng, FIMechE
	H. J. Watson BSc(Eng), CEng, MIMechE
Piston and ring failures	M. J. Neale OBE, BSc(Eng), DIC, FCGI, WhSch, FEng, FIMechE
Seal failures	B. S. Nau BSc, PhD, ARCS, CEng, FIMechE, MemASME
Brake and clutch failures	T. P. Newcombe DSc, CEng, FIMechE, FInstP
	R. T. Spurr BSc, PhD
Wire rope failures	S. Maw MA, CEng, MIMechE
Fretting of surfaces	R. B. Waterhouse MA, PhD, FIM
Wear mechanisms	K. H. R. Wright PhD, FInstP
Repair of worn surfaces	G. R. Bell BSc, ARSM, CEng, FIM, FWeldI, FRIC
Wear resistant materials	H. Hocke CEng, MIMechE, FIPlantE, MIMH, FIL
	M. Bartle CEng, MIM, DipIM, MIIM, AMWeldI
Repair of plain bearings	P. T. Holligan BSc(Tech), FIM
Repair of friction surfaces	T. P. Newcomb DSc, CEng, FIMechE, FInstP
	R. T. Spurr BSc, PhD
Viscosity of lubricants	H. Naylor BSc, PhD, CEng, FIMechE
Surface hardness	M. J. Neale OBE, BSc(Eng), DIC, FCGI, WhSch, FEng, FIMechE
Surface finish and shape	R. E. Reason DSc, ARCS, FRS
Shape tolerances of components	J. J. Crabtree BSc(Tech)Hons
S.I. units and conversion factors	M. J. Neale OBE, BSc(Eng), DIC, FCGI, WhSch, FEng, FIMechE

Table 1.1 Importance of lubricant properties in relation to bearing type

Lubricant property \ Type of component	Plain journal bearing	Rolling bearing	Closed gears	Open gears, ropes, chains, etc.	Clock and instrument pivots	Hinges, slides, latches, etc.
1. Boundary lubricating properties	+	+ +	+ + +	+ +	+ +	+
2. Cooling	+ +	+ +	+ + +	−	−	−
3. Friction or torque	+	+ +	+ +	−	+ +	+
4. Ability to remain in bearing	+	+ +	−	+	+ + +	+
5. Ability to seal out contaminants	−	+ +	−	+	−	+
6. Temperature range	+	+ +	+ +	+	−	+
7. Protection against corrosion	+	+ +	−	+ +	−	+
8. Volatility	+	+	−	+ +	+ +	+

Note: The relative importance of each lubricant property in a particular class of component is indicated on a scale from + + + = highly important to − = quite unimportant.

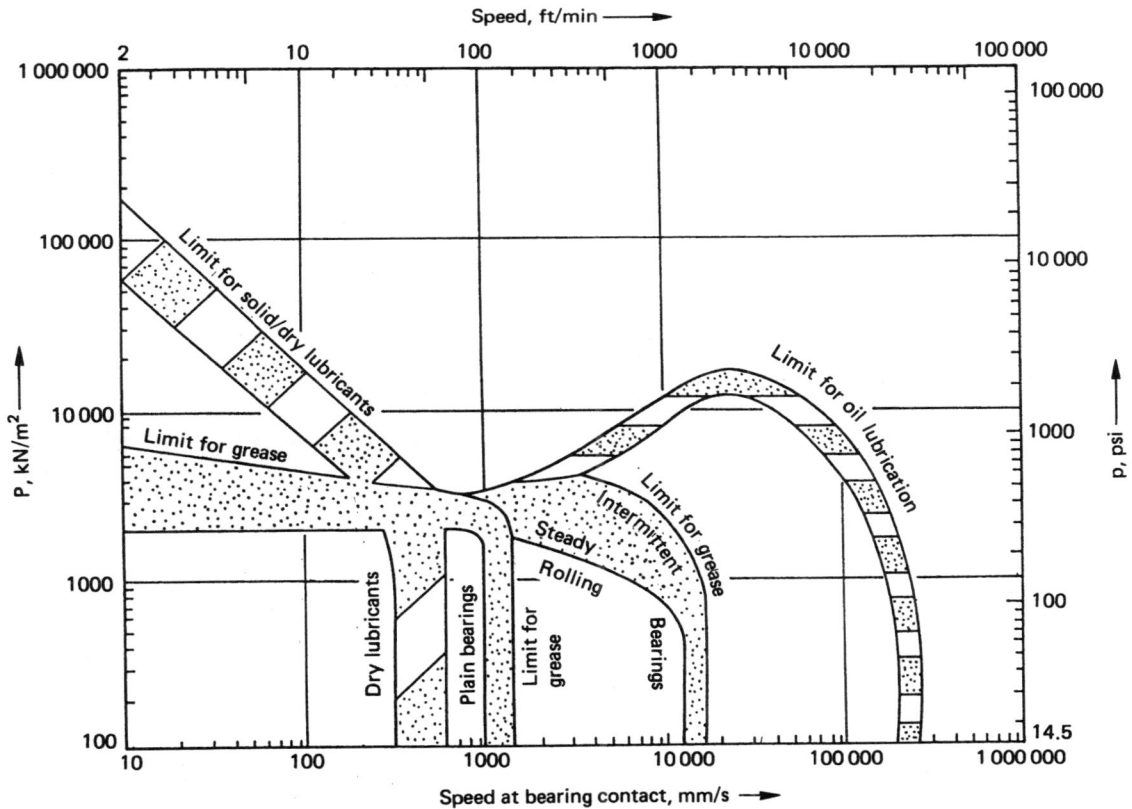

Figure 1.1 Speed/load limitations for different types of lubricant

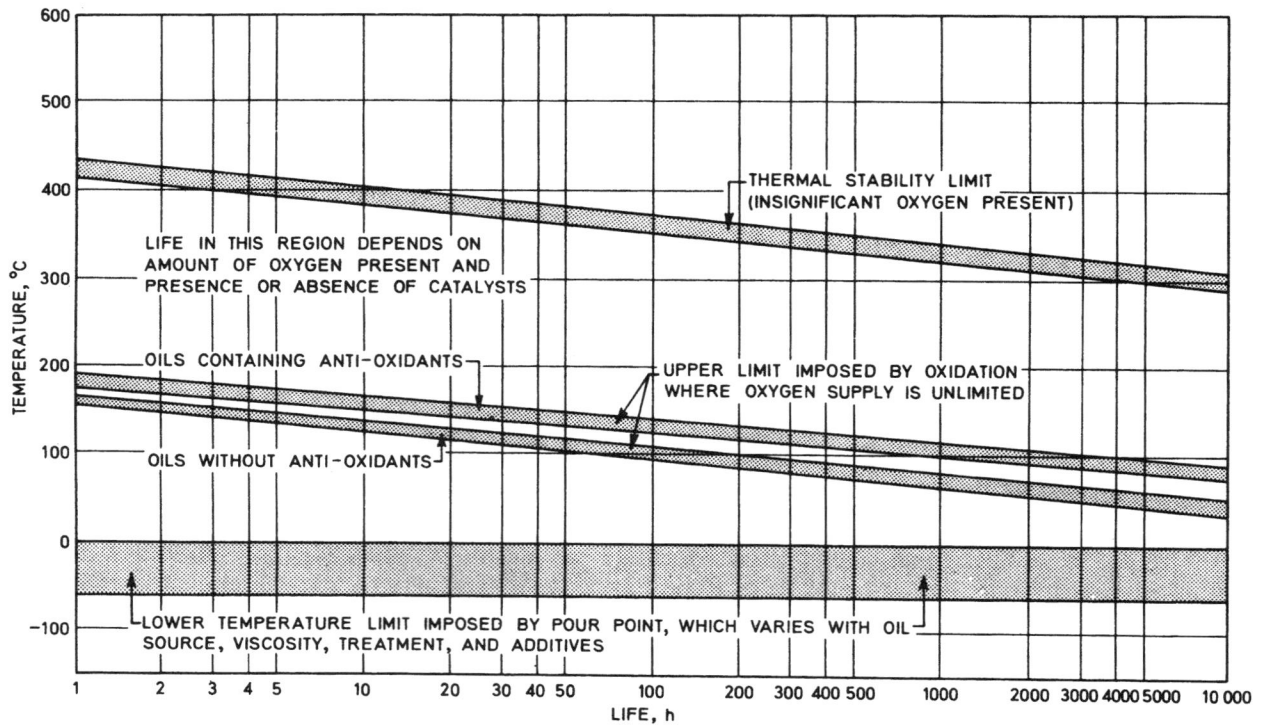

Figure 1.2 Temperature limits for mineral oils

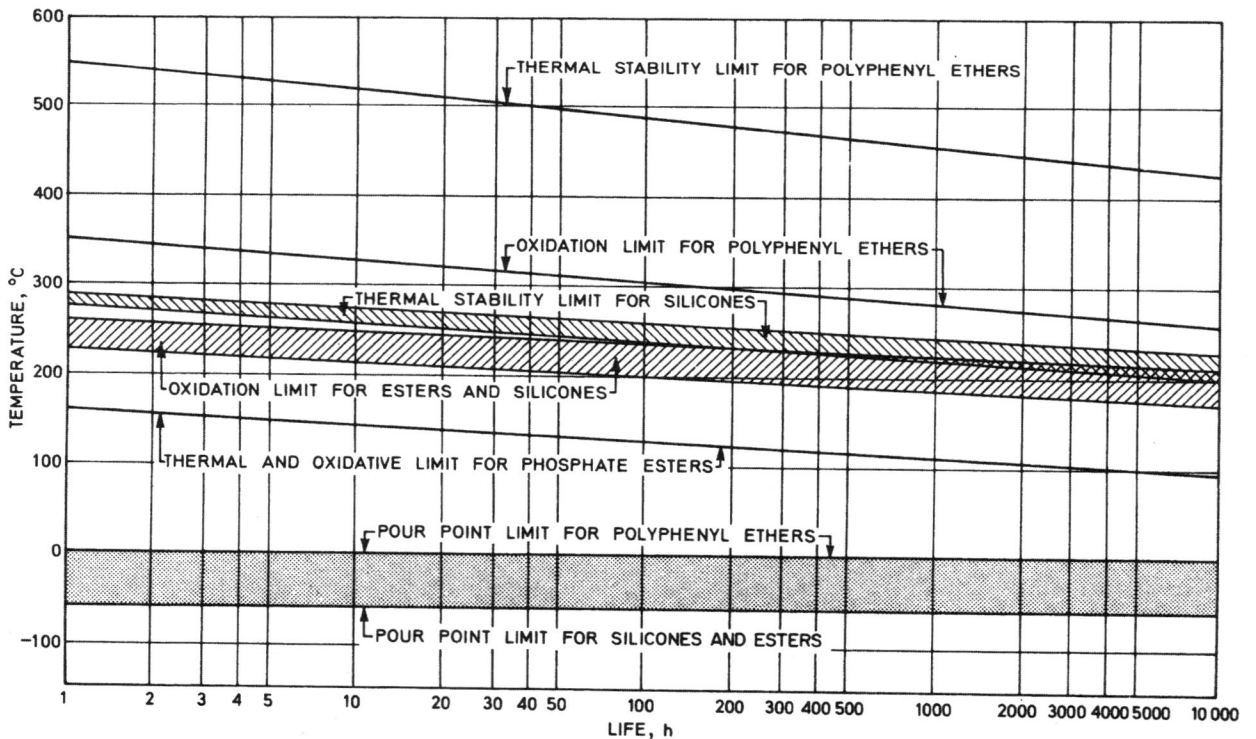

Figure 1.3 Temperature limits for some synthetic oils

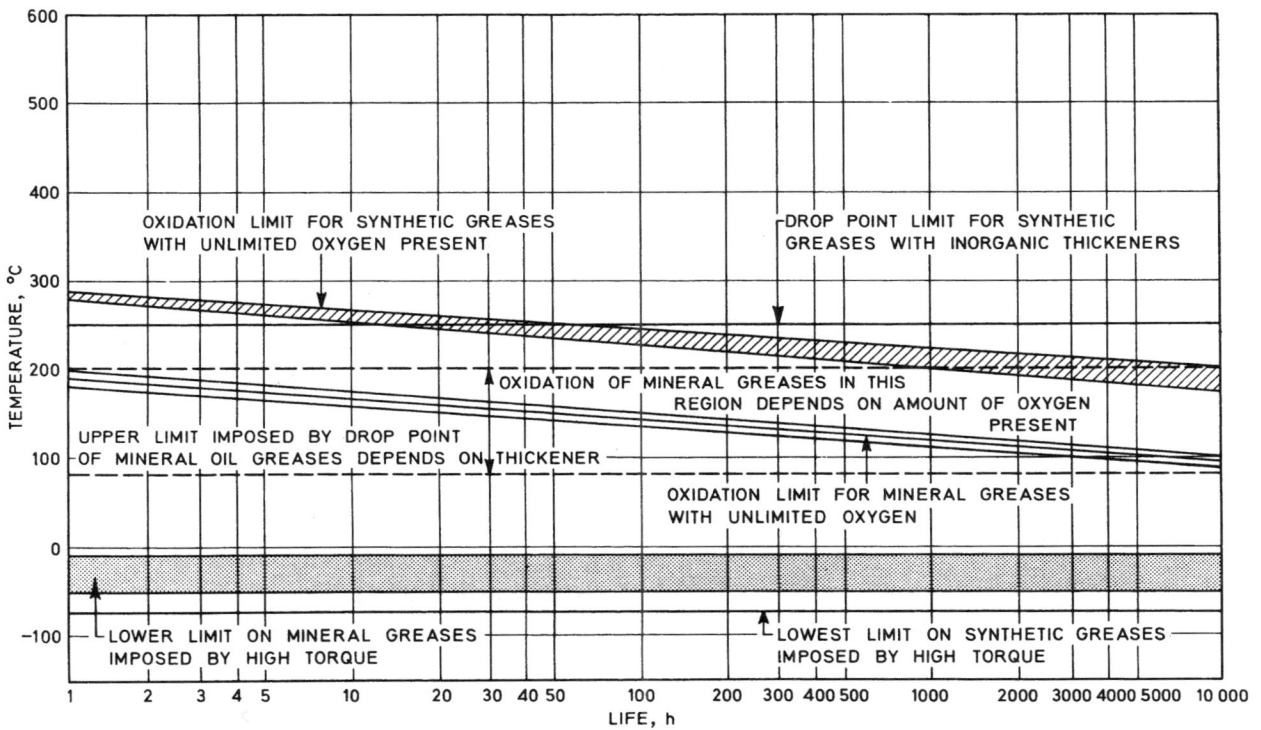

Figure 1.4 Temperature limits for greases. *In many cases the grease life will be controlled by volatility or migration. This cannot be depicted simply, as it varies with pressure and the degree of ventilation, but in general the limits may be slightly below the oxidation limits*

The effective viscosity of a lubricant in a bearing may be different from the quoted viscosity measured by a standard test method, and the difference depends on the shear rate in the bearing.

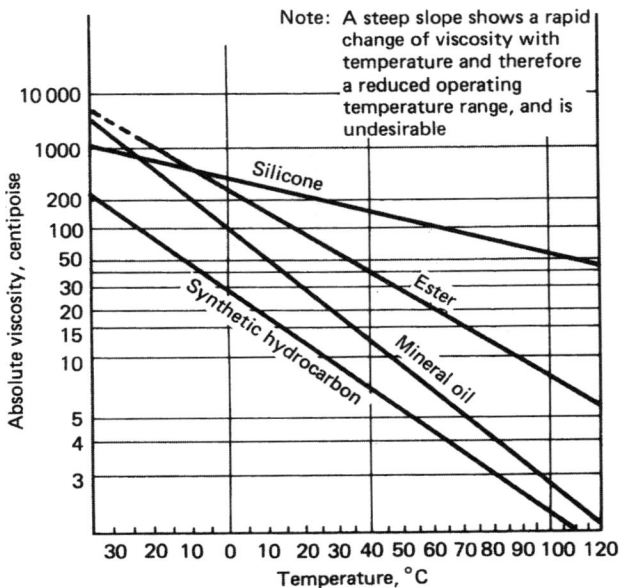

Figure 1.5 Viscosity/temperature characteristics of various oils

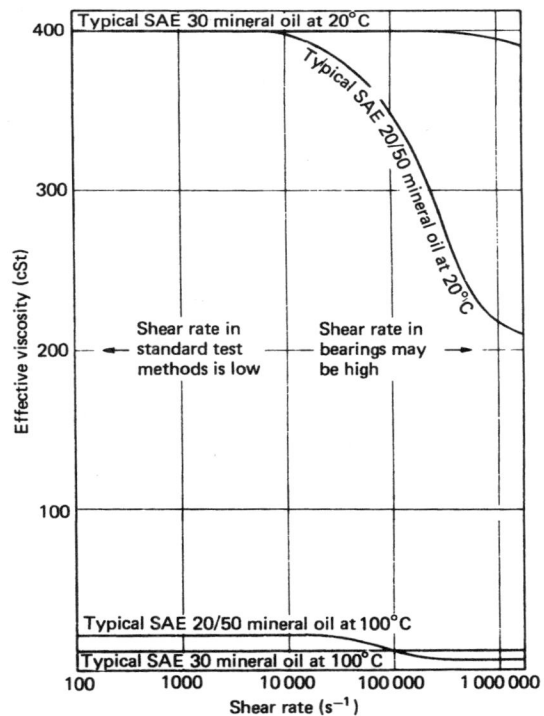

Figure 1.6 Variation of viscosity with shear rate

CLASSIFICATION

Mineral oils are basically hydrocarbons, but all contain thousands of different types of varying structure, molecular weight and volatility, as well as minor but important amounts of hydrocarbon derivatives containing one or more of the elements nitrogen, oxygen and sulphur. They are classified in various ways as follows.

Types of crude petroleum

Paraffinic   Contains significant amounts of waxy hydrocarbons and has 'wax' pour point (see below) but little or no asphaltic matter. Their naphthenes have long side-chains.

Naphthenic   Contains asphaltic matter in least volatile fractions, but little or no wax. Their naphthenes have short side-chains. Has 'viscosity' pour point.

Mixed base   Contains both waxy and asphaltic materials. Their naphthenes have moderate to long sidechains. Has 'wax' pour point.

Viscosity index

Lubricating oils are also commonly classified by their change in kinematic viscosity with temperature, i.e. by their kinematic viscosity index or KVI. Formerly, KVIs ranged between 0 and 100 only, the higher figures representing lower degrees of viscosity change with temperature, but nowadays oils may be obtained with KVIs outside these limits. They are generally grouped into high, medium and low, as in Table 2.1.

Table 2.1 Classification by viscosity index

Group	Kinematic viscosity index
Low viscosity index (LVI)	Below 35
Medium viscosity index (MVI)	35–80
High viscosity index (HVI)	80–110
Very high viscosity index (VHVI)	Over 110

It should be noted, however, that in Table 2.5 viscosity index has been determined from dynamic viscosities by the method of Roelands, Blok and Vlugter,[1] since this is a more fundamental system and allows truer comparison between mineral oils. Except for low viscosity oils, when DVIs are higher than KVIs, there is little difference between KVI and DVI for mineral oils.

Traditional use

Dating from before viscosity could be measured accurately, mineral oils were roughly classified into viscosity grades by their typical uses as follows:

Spindle oils   Low viscosity oils (e.g. below about $0.01\,\mathrm{Ns/m^2}$ at 60°C,) suitable for the lubrication of high-speed bearings such as textile spindles.

Light machine oils   Medium viscosity oils (e.g. 0.01–$0.02\,\mathrm{Ns/m^2}$) at 60°C, suitable for machinery running at moderate speeds.

Heavy machine oils   Higher viscosity oils (e.g. 0.02–$0.10\,\mathrm{Ns/m^2}$) at 60°C, suitable for slow-moving machinery.

Cylinder oils   Suitable for the lubrication of steam engine cylinder; viscosities from 0.12 to $0.3\,\mathrm{Ns/m^2}$ at 60°C.

Hydrocarbon types

The various hydrocarbon types are classified as follows:

(*a*) Chemically saturated (i.e. no double valence bonds) straight and branched chain. (Paraffins or alkanes.)

(*b*) Saturated 5- and 6-membered rings with attached side-chains of various lengths up to 20 carbon atoms long. (Naphthenes.)

(*c*) As (*b*) but also containing 1, 2 or more 6-membered unsaturated ring groups, i.e. containing double valence bonds, e.g. mono-aromatics, di-aromatics, polynuclear aromatics, respectively.

A typical paraffinic lubricating oil may have these hydrocarbon types in the proportions given in Table 2.2.

Table 2.2 Hydrocarbon types in Venezuelan 95 VI solvent extracted and dewaxed distillate

Hydrocarbon types		% Volume
Saturates (KVI = 105)	Paraffins	15
	Naphthenes	60
Aromatics	Mono-aromatics	18
	Di-aromatics	6
	Poly-aromatics	1

The VI of the saturates has a predominant influence on the VI of the oil. In paraffinic oils the VI of the saturates may be 105–120 and 60–80 in naphthenic oils.

Structural group analyses

This is a useful way of accurately characterising mineral oils and of obtaining a general picture of their structure which is particularly relevant to physical properties, e.g. increase of viscosity with pressure. From certain other physical properties the statistical distribution of carbon atoms in aromatic groups ($\%\ C_A$), in naphthenic groups ($\%\ C_N$), in paraffinic groups ($\%\ C_P$), and the total number (R_T) of naphthenic and aromatic rings (R_N and R_A) joined together. Table 2.3 presents examples on a number of typical oils.

Table 2.3 Typical structural group analyses (*courtesy:* Institution of Mechanical Engineers)

Oil type	Specific gravity at 15.6°C	Viscosity Ns/m² at 100°C	Mean molecular weight	$\%$ C_A	$\%$ C_N	$\%$ C_P	R_A	R_N	R_T
LVI spindle oil	0.926	0.0027	280	22	32	46	0.8	1.4	2.2
LVI heavy machine oil	0.943	0.0074	370	23	26	51	1.1	1.6	2.7
MVI light machine oil	0.882	0.0039	385	4	37	59	0.2	2.1	2.3
MVI heavy machine oil	0.910	0.0075	440	8	37	54	0.4	2.7	3.1
HVI light machine oil	0.871	0.0043	405	6	26	68	0.3	1.4	1.7
HVI heavy machine oil	0.883	0.0091	520	7	23	70	0.4	1.8	2.2
HVI cylinder oil	0.899	0.0268	685	8	22	70	0.7	2.3	3.0
Medicinal white oil	0.890	0.0065	445	0	42	58	0	2.8	2.8

REFINING

Distillation

Lubricants are produced from crude petroleum by distillation according to the outline scheme given in Figure 2.1.

The second distillation is carried out under vacuum to avoid subjecting the oil to temperatures over about 370°C, which would rapidly crack the oil.

The vacuum residues of naphthenic crudes are bitumens. These are not usually classified as lubricants but are used as such on some plain bearings subject to high temperatures and as blending components in oils and greases to form very viscous lubricants for open gears, etc.

Refining processes

The distillates and residues are used to a minor extent as such, but generally they are treated or refined both before and after vacuum distillation to fit them for the more stringent requirements. The principal processes listed in Table 2.4 are selected to suit the type of crude oil and the properties required.

Elimination of aromatics increases the VI of an oil. A lightly refined naphthenic oil may be LVI but MVI if highly refined. Similarly a lightly refined mixed-base oil may be MVI but HVI if highly refined. Elimination of aromatics also reduces nitrogen, oxygen and sulphur contents.

The distillates and residues may be used alone or blended together. Additionally, minor amounts of fatty oils or of special oil-soluble chemicals (additives) are blended in to form additive engine oils, cutting oils, gear oils, hydraulic oils, turbine oils, and so on, with superior properties to plain oils, as discussed below. The tolerance in blend viscosity for commercial branded oils is typically ±4% but official standards usually have wider limits, e.g. ±10% for ISO 3448.

PHYSICAL PROPERTIES

Viscosity-temperature

Figure 2.4 illustrates the variation of viscosity with temperature for a series of oils with kinematic viscosity

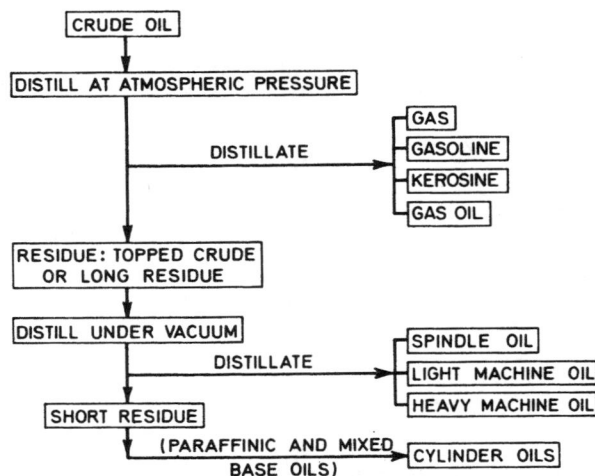

Figure 2.1 (*courtesy:* Institution of Mechanical Engineers)

Table 2.4 Refining processes (*Courtesy:* Institution of Mechanical Engineers)

Process	Purpose
De-waxing	Removes waxy materials from paraffinic and mixed-base oils to prevent early solidification when the oil is cooled to low temperatures, i.e. to reduce pour point
De-asphalting	Removes asphaltic matter, particularly from mixed-base short residues, which would separate out at high and low temperatures and block oil-ways
Solvent extraction	Removes more highly aromatic materials, chiefly the polyaromatics, in order to improve oxidation stability
Hydrotreating	Reduces sulphur content, and according to severity, reduces aromatic content by conversion to naphthenes
Acid treatment	Now mainly used as additional to other treatments to produce special qualities such as transformer oils, white oils and medicinal oils
Earth treatment	Mainly to obtain rapid separation of oil from water, i.e. good demulsibility

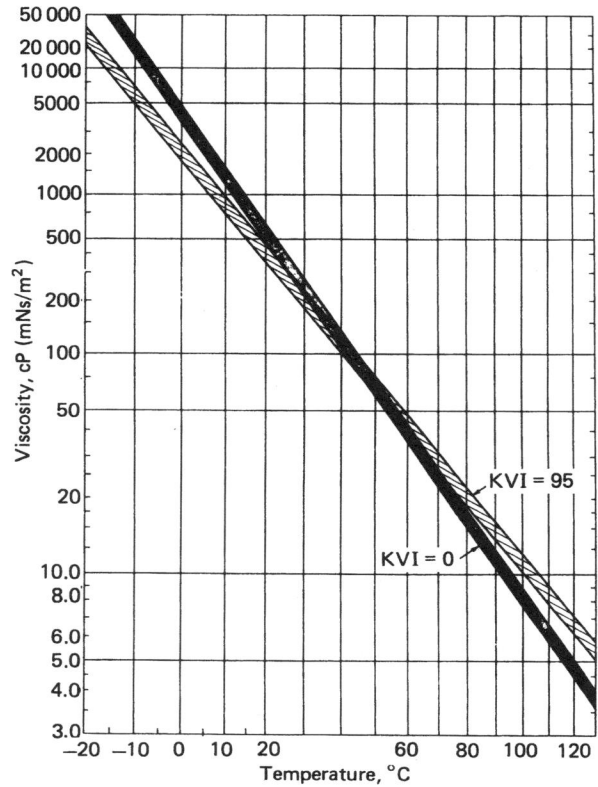

Figure 2.2 150 grade ISO 3448 oils of 0 and 95 KVI

index of 95 (dynamic viscosity index 93). Figure 2.2 shows the difference between 150 Grade ISO 3448 oils with KVIs of 0 and 95.

Viscosity-pressure

The viscosity of oils increases significantly under pressure. Naphthenic oils are more affected than paraffinic but, very roughly, both double their viscosity for every 35 MN/m^2 increase of pressure. Figure 2.3 gives an impression of the variation in viscosity of an SAE 20 W ISO 3448 or medium machine oil, HVI type, with both temperature and pressure.

In elastohydrodynamic (ehl) formulae it is usually assumed that the viscosity increases exponentially with pressure. Though in fact considerable deviations from an exponential increase may occur at high pressures, the assumption is valid up to pressures which control ehl behaviour, i.e. about 35 MN/m^2. Typical pressure viscosity coefficients are given in Table 2.5, together with other physical properties.

Pour point

De-waxed paraffinic oils still contain 1% or so of waxy hydrocarbons, whereas naphthenic oils only have traces of them. At about 0°C, according to the degree of dewaxing, the waxes in paraffinic oils crystallise out of solution and at about –10°C the crystals grow to the extent that the remaining oil can no longer flow. This temperature, or close to it, when determined under specified conditions is known as the pour point. Naph-

thenic oils, in contrast, simply become so viscous with decreasing temperature that they fail to flow, although no wax crystal structure develops. Paraffinic oils are therefore said to have 'wax' pour points while naphthenic oils are said to have 'viscosity' pour points.

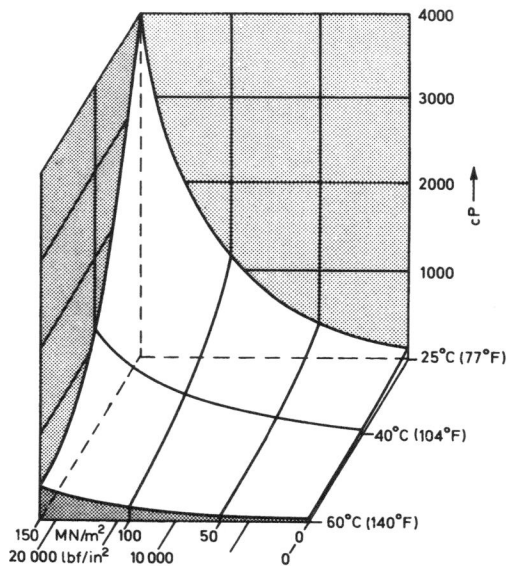

Figure 2.3 Variation of viscosity with temperature and pressure of an SAE 20 W (HVI) oil (*Courtesy:* Institution of Mechanical Engineers)

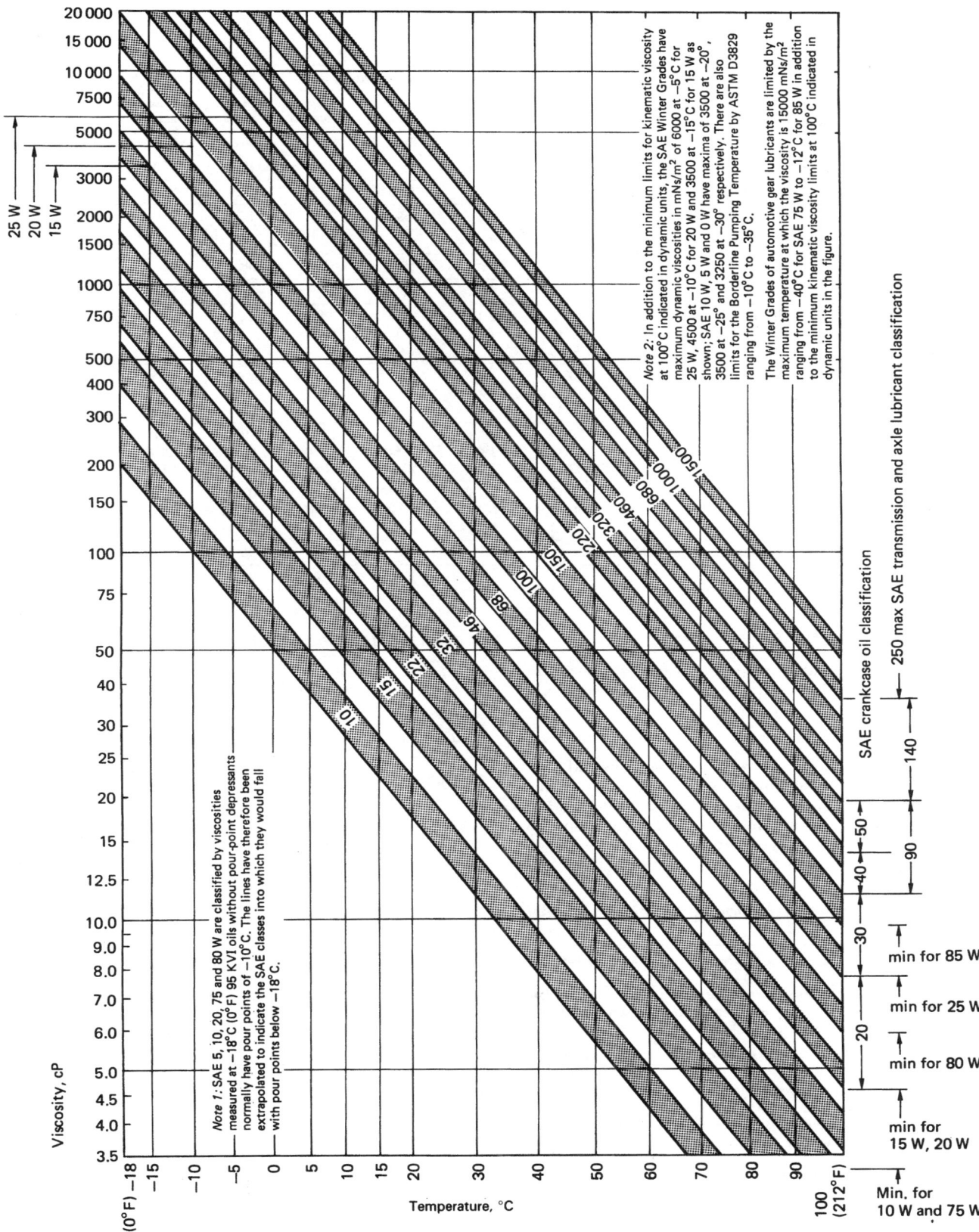

Note 1: SAE 5, 10, 20, 75 and 80 W are classified by viscosities measured at −18°C. (0°F) 95 KVI oils without pour-point depressants normally have pour points of −10°C. The lines have therefore been extrapolated to indicate the SAE classes into which they would fall with pour points below −18°C.

Note 2: In addition to the minimum limits for kinematic viscosity at 100°C indicated in dynamic units, the SAE Winter Grades have maximum dynamic viscosities in mNs/m² of 6000 at −5°C for 25 W, 4500 at −10°C for 20 W and 3500 at −15°C for 15 W as shown; SAE 10 W, 5 W and 0 W have maxima of 3500 at −20°, 3500 at −25° and 3250 at −30° respectively. There are also limits for the Borderline Pumping Temperature by ASTM D3829 ranging from −10°C to −35°C.

The Winter Grades of automotive gear lubricants are limited by the maximum temperature at which the viscosity is 15000 mNs/m² ranging from −40°C for SAE 75 W to −12°C for 85 W in addition to the minimum kinematic viscosity limits at 100°C indicated in dynamic units in the figure.

Figure 2.4 ISO 3448 ranges for 95 KVI oils with approximate SAE limits

Thermal properties

Table 2.5 Typical physical properties of highly refined mineral oils (*Courtesy:* Institution of Mechanical Engineers)

		Naphthenic oils			*Paraffinic oils*		
		Spindle	*Light machine*	*Heavy machine*	*Light machine*	*Heavy machine*	*Cylinder*
Density (kg/m^3) at	25°C	862	880	897	862	875	891
Viscosity (mNs/m^2) at	30°C	18.6	45.0	171	42.0	153	810
	60°C	6.3	12.0	31	13.5	34	135
	100°C	2.4	3.9	7.5	4.3	9.1	27
Dynamic viscosity index		92	68	38	109	96	96
Kinematic viscosity index		45	45	43	98	95	95
Pour point, °C		−43	−40	−29	−9	−9	−9
Pressure—viscosity coefficient ($m^2/N \times 10^8$) at	30°C	2.1	2.6	2.8	2.2	2.4	3.4
	60°C	1.6	2.0	2.3	1.9	2.1	2.8
	100°C	1.3	1.6	1.8	1.4	1.6	2.2
Isentropic secant bulk modulus at 35 MN/m^2 and	30°C	—	—	—	198	206	—
	60°C	—	—	—	172	177	—
	100°C	—	—	—	141	149	—
Thermal capacity (J/kg °C) at	30°C	1880	1860	1850	1960	1910	1880
	60°C	1990	1960	1910	2020	2010	1990
	100°C	2120	2100	2080	2170	2150	2120
Thermal conductivity (Wm/m^2 °C) at	30°C	0.132	0.130	0.128	0.133	0.131	0.128
	60°C	0.131	0.128	0.126	0.131	0.129	0.126
	100°C	0.127	0.125	0.123	0.127	0.126	0.123
Temperature (°C) for vapour pressure of 0.001 mmHg		35	60	95	95	110	125
Flash point, open, °C		163	175	210	227	257	300

DETERIORATION

Lubricating oils can become unfit for further service by: oxidation, thermal decomposition, and contamination.

Oxidation

Mineral oils are very stable relative to fatty oils and pure hydrocarbons. This stability is ascribed to the combination of saturated and unsaturated hydrocarbons and to certain of the hydrocarbon derivatives, i.e. compounds containing oxygen, nitrogen and sulphur atoms – the so-called 'natural inhibitors'.

Factors influencing oxidation

Temperature	Rate doubles for every 8–10°C temperature rise.
Oxygen access	Degree of agitation of the oil with air.
Catalysis	Particularly iron and copper in finely divided or soluble form.
Top-up rate	Replenishment of inhibition (natural or added).
Oil type	Proportions and type of aromatics and especially on the compounds containing nitrogen, oxygen, sulphur.

Table 2.6 Effects of oxidation and methods of test

Types of product produced by oxidation	*Factors involved*	*Methods of test*
Organic acids which are liable to corrode cadmium, lead and zinc and thereby to promote the formation of emulsions Lightly polymerised materials which increase the viscosity of the oil Moderately polymerised materials which become insoluble in the oil, especially when cold. When dispersed these also promote emulsification and increase of viscosity. When settled out they clog filter screens and block oil-ways Highly polymerised coke-like materials formed locally on very hot surfaces where they may remain	The relative proportions of the various types of products depend on the conditions of oxidation and the type of oil The degree of oxidation which can be tolerated depends on the lubrication system: more can be tolerated in simple easily cleaned bath systems without sensitive metals, less in complex circulation systems	Total acid number or neutralisation value, which assesses the concentration of organic acids, and is therefore an indication of the concentration of the usually more deleterious polymerised materials, is the most convenient and precise test to carry out. Limits vary between 0.2 mg KOH/g and 4.0 or more With many additive oils proof of the continued effective presence of the necessary additives, e.g. anti-oxidant, is more important

Thermal decomposition

Mineral oils are also relatively stable to thermal decomposition in the absence of oxygen, but at temperatures over about 330°C, dependent on time, mineral oils will decompose into fragments, some of which polymerise to form hard insoluble products.

Table 2.7 Thermal decomposition products

Product	Effect
Light hydrocarbons	Flash point is reduced; viscosity is reduced
Carbonaceous residues	Hard deposits on heater surfaces reduce flow rates and accentuate overheating

Some additives are more liable to thermal decomposition than the base oils, e.g. extreme pressure additives; and surface temperature may have to be limited to temperatures as low as 130°C.

Contamination

Contamination is probably the most common reason for changing an oil. Contaminants may be classified as shown in Table 2.8.

Table 2.8 Contaminants

Type	Example
Gaseous	Air, ammonia
Liquid	Water, oil of another type or viscosity grade or both
Solid	Fuel soot, road dust, fly ash, wear products

Where appropriate, oils are formulated to cope with likely contaminants, for example turbine oils are designed to separate water and air rapidly, diesel engine oils are designed to suspend fuel soot in harmless finely divided form and to neutralise acids formed from combustion of the fuel.

Solid contaminants may be controlled by appropriate filtering or centrifuging or both. Limits depend on the abrasiveness of the contaminant and the sensitivity of the system.

Oil life

Summarising the data given under the headings Oxidation and Thermal decomposition, above, Figure 2.5 gives an indication of the time/temperature limits imposed by thermal and oxidation stability on the life of a well-refined HVI paraffinic oil.

ADDITIVE OILS

Plain mineral oils are used in many units and systems for the lubrication of bearings, gears and other mechanisms where their oxidation stability, operating temperature range, ability to prevent wear, etc., are adequate. Nowadays, however, the requirements are often greater than plain oils are able to provide, and special chemicals or additives are 'added' to many oils to improve their properties. The functions required of these 'additives' gives them their common names listed in Table 2.9.

Table 2.9 Types of additives

Main type	Function and sub-types
Acid neutralisers	Neutralise contaminating strong acids formed, for example, by combustion of high sulphur fuels or, less often, by decomposition of active EP additives
Anti-foam	Reduces surface foam
Anti-oxidants	Reduce oxidation. Various types are: oxidation inhibitors, retarders; anti-catalyst metal deactivators, metal passivators
Anti-rust	Reduces rusting of ferrous surfaces swept by oil
Anti-wear agents	Reduce wear and prevent scuffing of rubbing surfaces under steady load operating conditions; the nature of the film is uncertain
Corrosion inhibitors	Type (a) reduces corrosion of lead; type (b) reduces corrosion of cuprous metals
Detergents	Reduce or prevent deposits formed at high temperatures, e.g. in ic engines
Dispersants	Prevent deposition of sludge by dispersing a finely divided suspension of the insoluble material formed at low temperature
Emulsifier	Forms emulsions; either water-in-oil or oil-in-water according to type
Extreme pressure	Prevents scuffing of rubbing surfaces under severe operating conditions, e.g. heavy shock load, by formation of a mainly inorganic surface film
Oiliness	Reduces friction under boundary lubrication conditions; increases load-carrying capacity where limited by temperature rise by formation of mainly organic surface films
Pour point depressant	Reduces pour point of paraffinic oils
Tackiness	Reduces loss of oil by gravity, e.g. from vertical sliding surfaces, or by centrifugal force
Viscosity index improvers	Reduce the decrease in viscosity due to increase of temperature

Table 2.10 Types of additive oil required for various types of machinery

Type of machinery	Usual base oil type	Usual additives	Special requirements
Food processing	Medicinal white oil	None	Safety in case of ingestion
Oil hydraulic	Paraffinic down to about −20°C, naphthenic below	Anti-oxidant Anti-rust Anti-wear Pour point depressant VI improver Anti-foam	Minimum viscosity change with temperature; minimum wear of steel/steel
Steam and gas turbines	Paraffinic or naphthenic distillates	Anti-oxidant Anti-rust	Ready separation from water, good oxidation stability
Steam engine cylinders	Unrefined or re-fined residual or high-viscosity distillates	None or fatty oil	Maintenance of oil film on hot surfaces; re-sistance to washing away by wet steam
Air compressor cylinders	Paraffinic or naphthenic distillates	Anti-oxidant Anti-rust	Low deposit formation tendency
Gears (steel/steel)	Paraffinic or naphthenic	Anti-wear, EP Anti-oxidant Anti-foam Pour point depressant	Protections against abrasion and scuffing
Gears (steel/bronze)	Paraffinic	Oiliness Anti-oxidant	Reduce friction, temperature rise, wear and oxidation
Machine tool slideways	Paraffinic or naphthenic	Oiliness; tackiness	Maintains smooth sliding at very low speeds. Keeps film on vertical surfaces
Hermetically sealed refrigerators	Naphthenic	None	Good thermal stability, miscibility with re-frigerant, low flow point
Diesel engines	Paraffinic or naphthenic	Detergent Dispersant Anti-oxidant Acid neutraliser Anti-foam Anti-wear Corrosion inhibitor	Vary with type of engine thus affecting additive combination

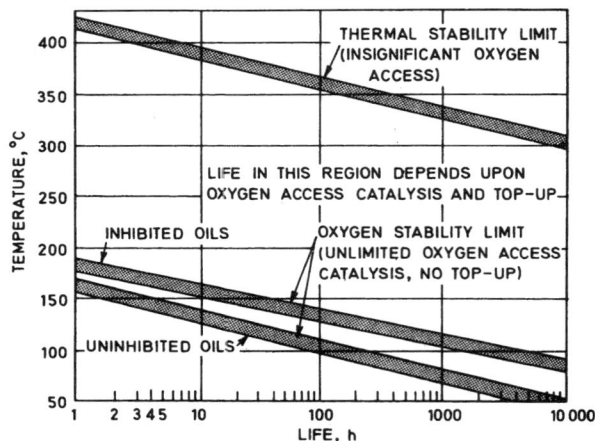

Figure 2.5 Approximate life of well-refined mineral oils (*Courtesy:* Institution of Mechanical Engineers)

Selection of additive combinations

Additives and oils are combined in various ways to provide the performance required. It must be emphas-ised, however, that indiscriminate mixing can produce undesired interactions, e.g. neutralisation of the effect of other additives, corrosivity and the formation of insol-uble materials.

Indeed, some additives may be included in a blend simply to overcome problems caused by other additives. The more properties that are required of a lubricant, and the more additives that have to be used to achieve the result, the greater the amount of testing that has to be carried out to ensure satisfactory performance.

Application data for a variety of synthetic oils are given in the table below. The list is not complete, but most readily available synthetic oils are included.

Table 3.1

Property \ Fluid	Di-ester	Inhibited Esters	Typical Phosphate Ester	Typical Methyl Silicone	Typical Phenyl Methyl Silicone	Chlorinated Phenyl Methyl Silicone	Polyglycol (inhibited)	Perfluorinated Polyether
Maximum temperature in absence of oxygen (°C)	250	300	110	220	320	305	260	370
Maximum temperature in presence of oxygen (°C)	210	240	110	180	250	230	200	310
Maximum temperature due to decrease in viscosity (°C)	150	180	100	200	250	280	200	300
Minimum temperature due to increase in viscosity (°C)	−35	−65	−55	−50	−30	−65	−20	−60
Density (g/ml)	0.91	1.01	1.12	0.97	1.06	1.04	1.02	1.88
Viscosity index	145	140	0	200	175	195	160	100–300
Flash point (°C)	230	255	200	310	290	270	180	
Spontaneous ignition temperature	Low	Medium	Very high	High	High	Very high	Medium	Very high
Thermal conductivity (W/M °C)	0.15	0.14	0.13	0.16	0.15	0.15	0.15	
Thermal capacity (J/kg°C)	2000	1700	1600	1550	1550	1550	2000	
Bulk modulus	Medium	Medium	Medium	Very low	Low	Low	Medium	Low
Boundary lubrication	Good	Good	Very good	Fair but poor for steel on steel	Fair but poor for steel on steel	Good	Very good	Poor
Toxicity	Slight	Slight	Some toxicity	Non-toxic	Non-toxic	Non-toxic	Believed to be low	Low
Suitable rubbers	Nitrile, silicone	Silicone	Butyl, EPR	Neoprene, viton	Neoprene, viton	Viton, fluoro-silicone	Nitrile	Many
Effect on plastics	May act as plasticisers		Powerful solvent	Slight, but may leach out plasticisers	Slight, but may leach out plasticisers	Slight, but may leach out plasticisers	Generally mild	Mild
Resistance to attack by water	Good	Good	Fair	Very good	Very good	Good	Good	Very good
Resistance to chemicals	Attacked by alkali	Attacked by alkali	Attacked by many chemicals	Attacked by strong alkali	Attacked by strong alkali	Attacked by alkali	Attacked by oxidants	Very good
Effect on metals	Slightly corrosive to non-ferrous metals	Corrosive to some non-ferrous metals when hot	Enhance corrosion in presence of water	Non-corrosive	Non-corrosive	Corrosive in presence of water to ferrous metals	Non-corrosive	Removes oxide films at elevated temperatures
Cost (relative to mineral oil)	4	6	6	15	25	40	4	500

The data are generalisations, and no account has been taken of the availability and property variations of different viscosity grades in each chemical type.

Table 3.1 *continued*

Property \\ Fluid	Chlorinated Diphenyl	Silicate Ester or Disiloxane	Polyphenyl Ether	Fluorocarbon	Mineral Oil (for comparison)	Remarks
Maximum temperature in absence of oxygen (°C)	315	300	450	300	200	For esters this temperature will be higher in the absence of metals
Maximum temperature in presence of oxygen (°C)	145	200	320	300	150	This limit is arbitrary. It will be higher if oxygen concentration is low and life is short
Maximum temperature due to decrease in viscosity (°C)	100	240	150	140	200	With external pressurisation or low loads this limit will be higher
Minimum temperature due to increase in viscosity (°C)	−10	−60	0	−50	0 to −50	This limit depends on the power available to overcome the effect of increased viscosity
Density (g/ml)	1.42	1.02	1.19	1.95	0.88	
Viscosity index	−200 to +25	150	−60	−25	0 to 140	A high viscosity index is desirable
Flash point (°C)	180	170	275	None	150 to 200	Above this temperature the vapour of the fluid may be ignited by an open flame
Spontaneous ignition temperature	Very high	Medium	High	Very high	Low	Above this temperature the fluid may ignite without any flame being present
Thermal conductivity (W/m °C)	0.12	0.15	0.14	0.13	0.13	A high thermal conductivity and high thermal capacity are desirable for effective cooling
Thermal capacity (J/kg °C)	1200	1700	1750	1350	2000	
Bulk modulus	Medium	Low	Medium	Low	Fairly high	There are four different values of bulk modulus for each fluid but the relative qualities are consistent
Boundary lubrication	Very good	Fair	Fair	Very good	Good	This refers primarily to anti-wear properties when some metal contact is occurring
Toxicity	Irritant vapour when hot	Slight	Believed to be low	Non-toxic unless overheated	Slight	Specialist advice should always be taken on toxic hazards
Suitable rubbers	Viton	Viton nitrile, fluoro-silicone	(None for very high temperatures)	Silicone	Nitrile	
Effect on plastics	Powerful solvent	Generally mild	Polyimides satisfactory	Some softening when hot	Generally slight	
Resistance to attack by water	Excellent	Poor	Very good	Excellent	Excellent	This refers to breakdown of the fluid itself and not the effect of water on the system
Resistance to chemicals	Very resistant	Generally poor	Resistant	Resistant but attacked by alkali and amines	Very resistant	
Effect on metals	Some corrosion of copper alloys	Non-corrosive	Non-corrosive	Non-corrosive, but unsafe with aluminium and magnesium	Non-corrosive when pure	
Cost (relative to mineral oil)	10	8	100	300	1	These are rough approximations, and vary with quality and supply position

A grease may be defined as solid to semi-fluid lubricant consisting of a dispersion of a thickening agent in a lubricating fluid. The thickening agent may consist of e.g. a soap, a clay or a dyestuff. The lubricating fluid is usually a mineral oil, a diester or a silicone.

Tables 4.1, 4.2 and 4.3 illustrate some of the properties of greases containing these three types of fluid. All values and remarks are for greases typical of their class, some proprietary grades may give better or worse performance in some or even all respects.

TYPES OF GREASE

Table 4.1 Grease containing mineral oils

Soap base or thickener	Min. drop pt. °C (°F) ASTM D 556 IP 132	Min. usable temp. °C (°F)	Max. usable temp. °C (°F)	Rust protection	Available with extreme pressure additive	Use	Cost	Official specification
Lime (calcium)	90 (190)	−20 (0)	60 (140)	—	Yes	General purpose	Low	BS 3223
Lime (calcium) heat stable	99.5 (210) or 140 (280) *	−20 (0) −55 (−65) *	80 (175) *	(Some-times)	Yes	General purpose and rolling bearings	Low	BS 3223 DEF STAN 91–17 (LG 280) DEF STAN 91–27 (XG 279) MIL–G–10924B
Sodium (conventional)	205 (400)	0 (32)	150–175 (300–350)	Yes	No	Glands, seals low–medium speed rolling bearings	Low	—
Sodium and calcium (mixed)	150 (300)	−40 (−40)	120–150 (250–300)	Yes	No	High-speed rolling bearings	Medium	DEF 2261A (XG 271) MIL–L–7711A
Lithium	175 (350)	−40 (−40)	150 (300)	Yes	Yes	All rolling bearings	Medium	DEF STAN 91–12 (XG 271) DEF STAN 91–28 (XG 274) MIL–L–7711A
Aluminium Complex	200 (390)	−40 (−40)	160 (320)	Yes	Mild EP	Rolling bearings	Medium/ high	—
Lithium Complex	235 (450)	−40 (−40)	175 (350)	Yes	Mild EP	Rolling bearings	Medium	—
Clay	None	−30 (−20)	205 (400)*	Poor	Yes	Sliding friction	Medium	—

* Depending on conditions of service.

Although mineral oil viscosity and other characteristics of the fluid have been omitted from this table, these play a very large and often complicated part in grease performance. Certain bearing manufacturers demand certain viscosities and other characteristics of the mineral oil, which should be observed. Apart from these requirements, the finished characteristics of the grease, as a whole, should be regarded as the most important factor.

Table 4.2 Grease containing esters

Soap base or thickener	Min. drop pt. °C (°F) ASTM D 566 IP 132	Min. usable temp. °C (°F)	Max. usable temp. °C (°F)	Rust protection	Available with extreme pressure additive	Use	Cost	Official specification
Lithium	175 (350)	−75 (−100)	120 (250)	Yes	Yes	Rolling bearings	High	DTD 5598 (XG 287) MIL G–23827A
Clay	None	−55 (−65)	Not applicable†	Yes	Yes	Rolling bearings	High	DTD 5598 (XG 287) MIL–G–23827A
Dye	260 (500)	−40 (−40)	175 (350)†	Yes	No	Rolling bearings	Very high	DTD 5579 (XG 292) MIL–G–25760A

† Upper limit will depend on type of ester used and conditions of service.

Table 4.3 Grease containing silicones

Soap base or thickener	Min. drop pt. °C (°F)	Min. usable temp. °C (°F)	Max. usable temp. °C (°F)	Rust protection	Available with extreme pressure additive	Use	Cost	Official specification
Lithium	175 (350)	−55 (−65)	205 (400)	Yes	No	Rolling bearings	High	—
Dye	260 (500)	−75 (−100)	260 (500)	Yes	Yes	Rolling bearings/ miniature	Very high	DTD 5585 (XG 300) MIL–G–25013D
Silicone soap	205 (400)	−55 (−65)	260 (500)	Yes	No	Miniature bearings	Extremely high	—

Table 4.4 NLGI consistency range for greases

Description	NLGI no.	Penetration range (ASTM D 217–IP 50)	Types generally available	Some common uses
Semi-fluid	000	445–475	Not dye	Centralised systems
Semi-fluid	00	400–430	Not dye	Total loss systems
Very soft	0	355–385	Not dye	
Soft	1	310–340	All	Rolling bearings
Medium soft	2	265–295	All	General purpose
Medium	3	220–250	All	
Stiff	4	175–205	Na or Ca only	Plain bearings so
Very stiff	5	130–160	Na or Ca only	called 'Block' or
Very stiff	6	85–115	Na or Ca only	'Brick' greases

CONSISTENCY

The consistency of grease depends on, amongst other things, the percentage of soap, or thickener in the grease. It is obtained by measuring in tenths of a millimetre, the depth to which a standard cone sinks into the grease in five seconds at a temperature of 25°C (77°F) (ASTM D 217-IP 50). These are called 'units', a *non* dimensional value which *strictly should not be regarded as* tenths of a millimetre. It is called Penetration.

Penetration has been classified by the National Lubricating Grease Institute (NLGI) into a series of single numbers which cover a very wide range of consistencies. This classification does not take into account the nature of the grease, nor does it give any indication of its quality or use.

The commonest consistencies used in rolling bearings are in the NLGI 2 or 3 ranges but, since modern grease manufacturing technology has greatly improved stability of rolling bearing greases, the tendency is to use softer greases. In centralised lubrication systems, it is unusual to use a grease stiffer than NLGI 2 and often a grease as soft as an NLGI 0 may be found best. The extremes (000, 00, 0 and 4, 5, 6) are rarely, if ever, used in normal rolling bearings (other than 0 in centralised systems), but these softer greases are often used for gear lubrication applications.

GREASE SELECTION

When choosing a grease consideration must be given to circumstances and nature of use. The first decision is always the consistency range. This is a function of the method of application (e.g. centralised, single shot, etc.). This will in general dictate within one or two NLGI ranges, the grade required. Normally, however, an NLGI 2 will be found to be most universally acceptable and suitable for all but a few applications.

The question of operating temperature range comes next. Care should be taken that the operating range is known with a reasonable degree of accuracy. It is quite common to overestimate the upper limit: for example, if a piece of equipment is near or alongside an oven, it will not necessarily be at that oven temperature – it may be higher due to actual temperature-rise of bearing itself, or lower due to cooling effects by convection, radiation, etc.

Likewise, in very low-temperature conditions, the ambient temperature often has little effect after start-up due to internal heat generation of the bearing. It is always advisable, if possible, to measure the temperature by a thermocouple or similar device. A measured temperature, even if it is not the true bearing temperature, will be a much better guide than a guess. By using Tables 4.1, 4.2 and 4.3 above, the soap and fluid can be readily decided.

Normally, more than one type of grease will be found suitable. Unless it is for use in a rolling bearing or a heavily-loaded plain bearing the choice will then depend more or less on price, but logistically it may be advisable to use a more expensive grease if this is already in use for a different purpose. For a rolling bearing application, speed and size are the main considerations; the following Table 4.5 is intended as a guide only for normal ambient temperature.

If the bearing is heavily loaded for its size, i.e. approaching the maker's recommended maximum, or is subject to shock loading, it is important to use a good extreme-pressure grease. Likewise a heavily-loaded plain bearing will demand a good EP grease.

In general it is advisable always to have good anti-rust properties in the grease, but since most commercial greases available incorporate either additives for the purpose or are in themselves good rust inhibitors, this is not usually a major problem.

Table 4.5 Selection of greases for rolling bearings

	SPEED				
	Very slow *(under 500 r.p.m.)*	*Slow* *(500 r.p.m.)*	*Average* *(1000 r.p.m.)*	*Fast* *(2000 r.p.m.)*	*Very fast* *(over 2000 r.p.m.)*
BEARING SIZE Micro (1–5 mm)	.... A specially selected ultra-clean grease required				
Miniature (under 10 mm)	.... Normally aviation greases are used, e.g. XG 287 type				
Small (20–40 mm) Medium (65 mm)	Calcium (LG 280 type)	Calcium or lithium (XG 271 type)	Lithium (XG 271 type)	(Lithium (XG 271 type)) (Lithium (XG 274 type))	Soda–calcium oil type
Large (100 mm)	Lithium/oil (XG 274 or XG 271 type)	Lithium (XG 274 type)	Lithium (XG 274 type)	Lithium (XG 274 type)	—
Very large (200 mm or more)	Calcium or lithium	Lithium (XG 274 type)	Lithium (XG 274 type)	Lithium or soda–calcium (XG 274 type)	—

Note: for definition of these 'types' refer to Table 4.1

Table 4.6 Uses of greases containing fillers

Filler	*Graphite*	*Molybdenum disulphide*	*Metallic powder**

GREASE
Calcium
	Plain bearings
	Sliding friction and/or localised high
	  temperatures
		Not commonly used
		  see lithium
			High temperature thread lubricant
			Electrically conducting†

Lithium	As for calcium but not commonly used	Sliding friction Anti fretting Knuckle or ball joint Reciprocal motion Rolling bearings Care must be taken with localised high temperatures	Not commonly used
Clay	As for calcium	As for calcium	As for calcium

  * Metal powder is normally zinc dust, copper flake or finely-divided lead.
  † Useful where a current must be passed through rolling bearing from axle to outer race.

Note: filled greases should not normally be used in rolling bearings. Finally, it should be remembered that all of the above remarks refer to typical greases of their type. It is quite possible by special manufacturing techniques, use of specialised additives, etc., to reverse many of the properties either by accident or design. The grease supplier will always advise on any of these abnormal properties exhibited by his grease.

SOLID LUBRICANTS
REQUIRED WHEN
FLUIDS ARE:

Undesirable
- { Contaminate product — Food machinery, electrical contacts
- { Maintenance difficult — Inaccessibility, storage problems

Ineffective
- { Hostile environments — Corrosive gases, dirt and dust
- High temperatures — Metalworking, missiles
- Cryogenic temperatures — Missiles, refrigeration plant
- Radiation — Reactors, space
- Space/vacuum — Satellites, X-ray equipment
- Fretting conditions — } General, often used with oils
- Extreme pressures — }

A   TYPES OF SOLID LUBRICANT

Materials are required which form a coherent film of low shear strength between two sliding surfaces.

Lamellar solids		
	T_{max} (°C)	*Special features*
MoS_2	350	Stable to > 1150 °C in vacuum
WS_2	400	Oxidative stability > MoS_2
Graphite	500	Ineffective in vacuum/dry gases
TaS_2	550	Low electrical resistivity
CaF_2/BaF_2	1000	Ineffective below 300 °C

Polymers		
UHMWPE	100	Exceptionally low wear
FEP	210 }	Chemically inert; useful at
PTFE	275 }	cryogenic temperatures
Polyimides	300	Friction > PTFE
Polyurethanes	100 }	Useful for abrasion resistance but
Nylon 11	150 }	friction relatively high.

Soft metals	
Pb, Au, Ag, Sn, In	Useful in vacuum.

Oxides	
MoO_3, PbO/SiO_2, B_2O_3/PbS .	Effective only at high temperatures

Miscellaneous	
$AsSbS_4$, $Sb(SbS_4)$, $Ce_2(MoS_4)$	Oil & grease additives.
$Zn_2P_2O_7/Ca(OH)_2$	White lubricant additive for grease.

Plasma-sprayed coatings	
Ag/Ni-Cr/CaF_2/Glass	Wide temperature range lubricant formulations,
Ag/Cr_2C_3/Ni-Al/BaF_2-CaF_2	20–1000 °C.

B   METHODS OF USE

General

Powder	– Rubbed on to surfaces to form a 'burnished film', 0.1–10 μm thick. See subsection C.
Dispersion with resin in volatile fluids	– Sprayed on to surfaces and cured to form a 'bonded coating', 5–25 μm thick. See subsection D.
Dispersion in non-volatile fluids	– Directly as a lubricating medium, or as an additive to oils and greases. See subsection E.

Specialised

As lubricating additives to metal, carbon and polymer bearing materials.

As proprietary coatings produced by vacuum deposition, plasma spraying, particle impingement, or electrophoresis.

C   BURNISHED FILMS

Effects of operational variables

Results obtained from laboratory tests with a ball sliding on a film-covered disc. Applicable to MoS_2, WS_2 and related materials, but not to PTFE and graphite.

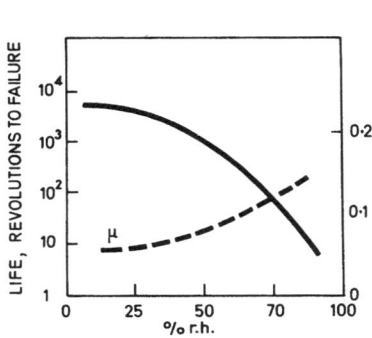

RELATIVE HUMIDITY

Acidity may develop at high r.h. Possibility of corrosion or loss of film adhesion.

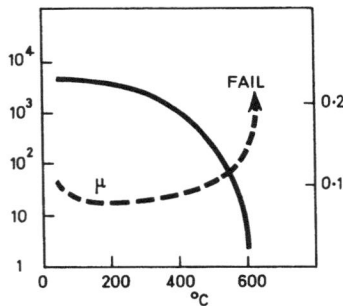

TEMPERATURE

Life limited by oxidation, but MoO_3 is not abrasive. Temperature lowers humidity.

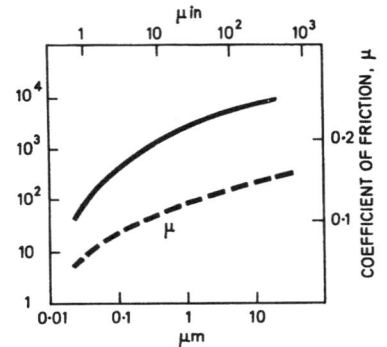

FILM THICKNESS

Difficult to produce films $>10\ \mu m$ thick except at high r.h.

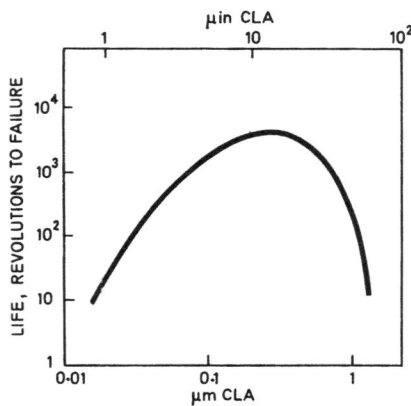

SURFACE ROUGHNESS

Optimum depends on method of surface preparation. Abrasion, 0.5 μm; grit-blast, 0.75 μm; grinding, 1.0 μm; turning, 1.25 μm.

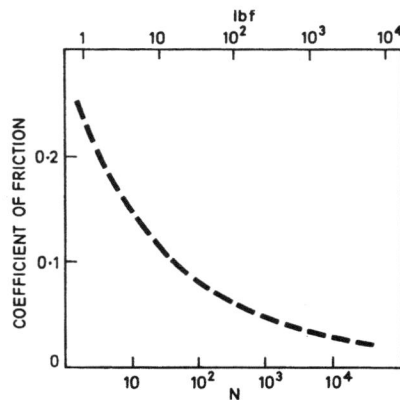

LOAD

As load increases, a greater proportion is supported by the substrate. A similar trend occurs as the substrate hardness increases.

No well-defined trend exists between film life and substrate hardness. Molybdenum is usually an excellent substrate for MoS_2 films. Generally similar trends with film thickness and load also apply to soft metal films.

D   BONDED COATINGS

MoS_2 resin coatings show performance trends broadly similar to those for burnished films but there is less dependence of wear life upon relative humidity.

Both the coefficient of friction and the wear rate of the coating vary with time.

Laboratory testing is frequently used to rate different coatings for particular applications. The most common tests are:

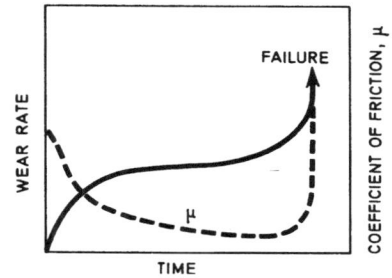

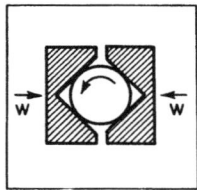

Falex (adopted for specification tests)

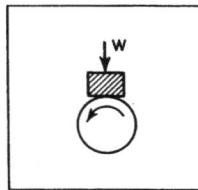

Pad-on-ring (Timken and LFW-1

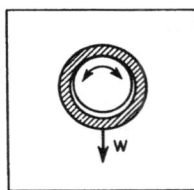

Oscillating plain bearing

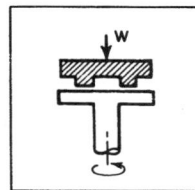

Thrust washer

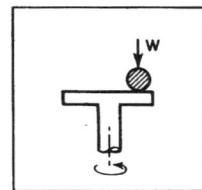

Ball on flat (may be 3 balls)

It is essential to coat the moving surface. Coating both surfaces usually increases the wear life, but by much less than 100% (≃30% for plain bearings, ≃1% for Falex tests). Considerable variations in wear life are often found in replicate tests (and service conditions).

Performance of MoS_2 bonded coatings at elevated temperatures is greatly dependent on the type of resin binder and on the presence of additives in the formulation. Typical additives include graphite, soft metals (Au, Pb, Ag), lead phosphite, antimony trioxide, and sulphides of other metals.

General characteristics of MoS_2 films with different binders

Type	Curing temp. for 1 hr (°C)	Max. temp. (°C)	Adhesion to substrate	Relative wear life *
Acrylic	20	65	Fair	7
Cellulose	20	65	Fair	25
Alkyd	120	95	Good	1
Phenolic	150	150	Good	75
Epoxy	200	200	Excellent	18
Silicone	250	300	Fair	7
Silicates	20	450	Fair	50
Vitreous	300–600	550	Good	50

* Based on simplified laboratory tests.

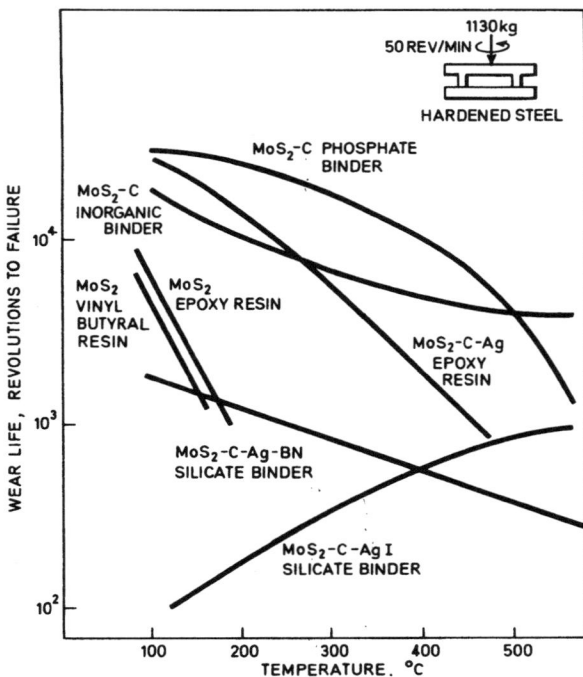

Points to note in design

1 Wide variety of types available; supplier's advice should always be sought.
2 Watch effect of cure temperature on substrate.
3 Use acrylic binders on rubbers, cellulose on wood and plastics.
4 Substrate pretreatment essential.
5 Fluids usually deleterious to life.

Preparation of coatings

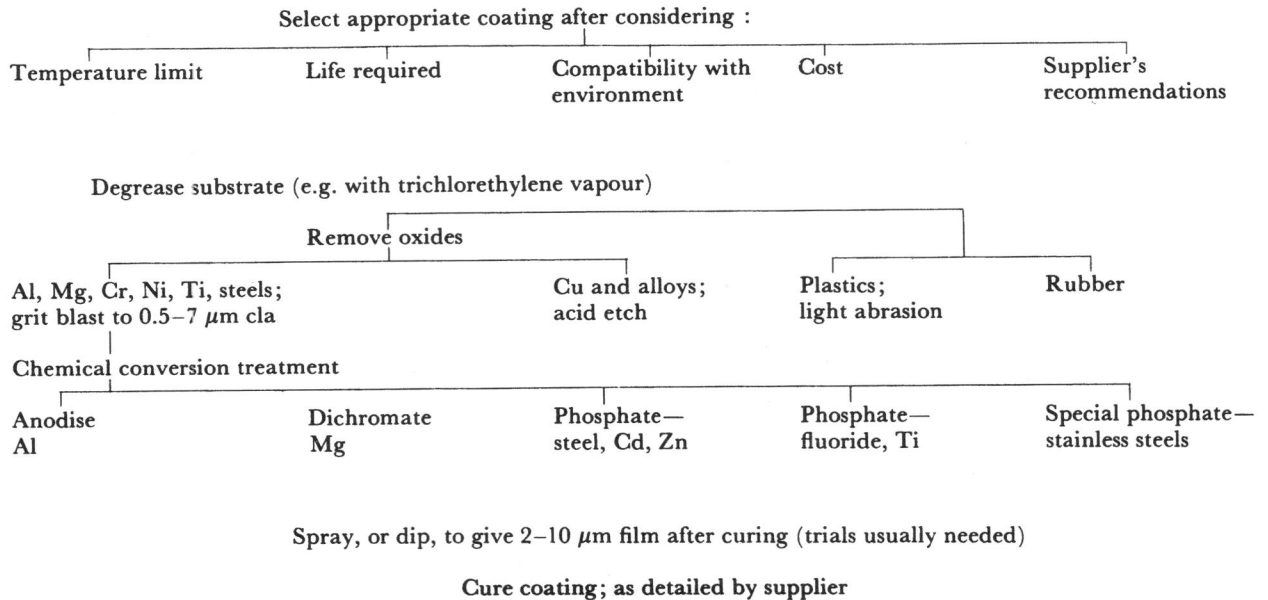

Select appropriate coating after considering :

Temperature limit	Life required	Compatibility with environment	Cost	Supplier's recommendations

Degrease substrate (e.g. with trichlorethylene vapour)

Remove oxides

Al, Mg, Cr, Ni, Ti, steels; grit blast to 0.5–7 μm cla	Cu and alloys; acid etch	Plastics; light abrasion	Rubber

Chemical conversion treatment

Anodise Al	Dichromate Mg	Phosphate— steel, Cd, Zn	Phosphate— fluoride, Ti	Special phosphate— stainless steels

Spray, or dip, to give 2–10 μm film after curing (trials usually needed)

Cure coating; as detailed by supplier

Specifications for solid film bonded coatings

US-MIL-L23398	Lubricant, solid film, air-drying
UK-DEF-STAN 91–19/1 US-MIL-L-8937 }	Lubricant, solid film, heat-curing
US-MIL-L-46010	Lubricant, solid film, heat cured, corrosion inhibited
US-MIL-L-81329	Lubricant, solid film, extreme environment

Other requirements

Satisfactory appearance

Limits on { Curing time/temperature
            Film thickness

Adhesion – tape test

Thermal stability – resistance to flaking/cracking at temperature extremes

Fluid compatibility – no softening/peeling after immersion

Performance { Wear life
               Load carrying capacity

Storage stability of dispersion

Corrosion – anodised aluminium or phosphated steel

E DISPERSIONS

Graphite, MoS_2 and PTFE dispersions are available in a wide variety of fluids: water, alcohol, toluene, white spirit, mineral oils, etc.

In addition to uses for bonded coatings, other applications include:

Additives to oils and greases	Improving lubricant performance Running-in Assembly problems Wire-rope and chain lubrication Grease thickeners	Compatibility with other additives may pose problems
Parting and anti-stick agents	Moulds for plastics Die-casting moulds Saucepans, etc. Cutting-tool treatments Impregnation of grinding wheels	Use PTFE if product discoloration is a problem
Anti-seize compounds	High-temperature thread lubricants Metalworking generally	Formulations generally proprietary and complex

Specifications for solid lubricant dispersions in oils and greases

Paste

UK-DTD-392B
US-MIL-T-5544 } Anti-seize compound, high temperatures (50% graphite in petrolatum)

UK-DTD-5617   Anti-seize compound, MoS_2 (50% MoS_2 in mineral oil)

US-MIL-A-13881   Anti-seize compound, mica base (40% mica in mineral oil)

US-MIL-L-25681C   Lubricant, MoS_2, silicone (50% MoS_2 – anti-seize compound)

Grease

US-MIL-G-23549A   Grease, general purpose (5% MoS_2, mineral oil base)

UK-DTD-5527A
US-MIL-G-21164C } Grease, MoS_2, low and high temperature (5% MoS_2, synthetic oil base)

US-MIL-G-81827   Grease, MoS_2, high load, wide temperature range (5% MoS_2)

UK-DEF-STAN 91–18/1   Grease, graphite, medium (5% in mineral oil base)

UK-DEF-STAN 91–8/1   Grease, graphite (40% in mineral oil base)

Oil

UK-DEF-STAN 91–30/1
US-MIL-L-3572 } Lubricating oil, colloidal graphite (10% in mineral oil)

There is a wide variety of liquids with many different uses and which may interact with tribological components. In these cases, the most important property of the liquid is usually its viscosity. Viscosity values are therefore presented for some common liquids and for some of the more important process fluids.

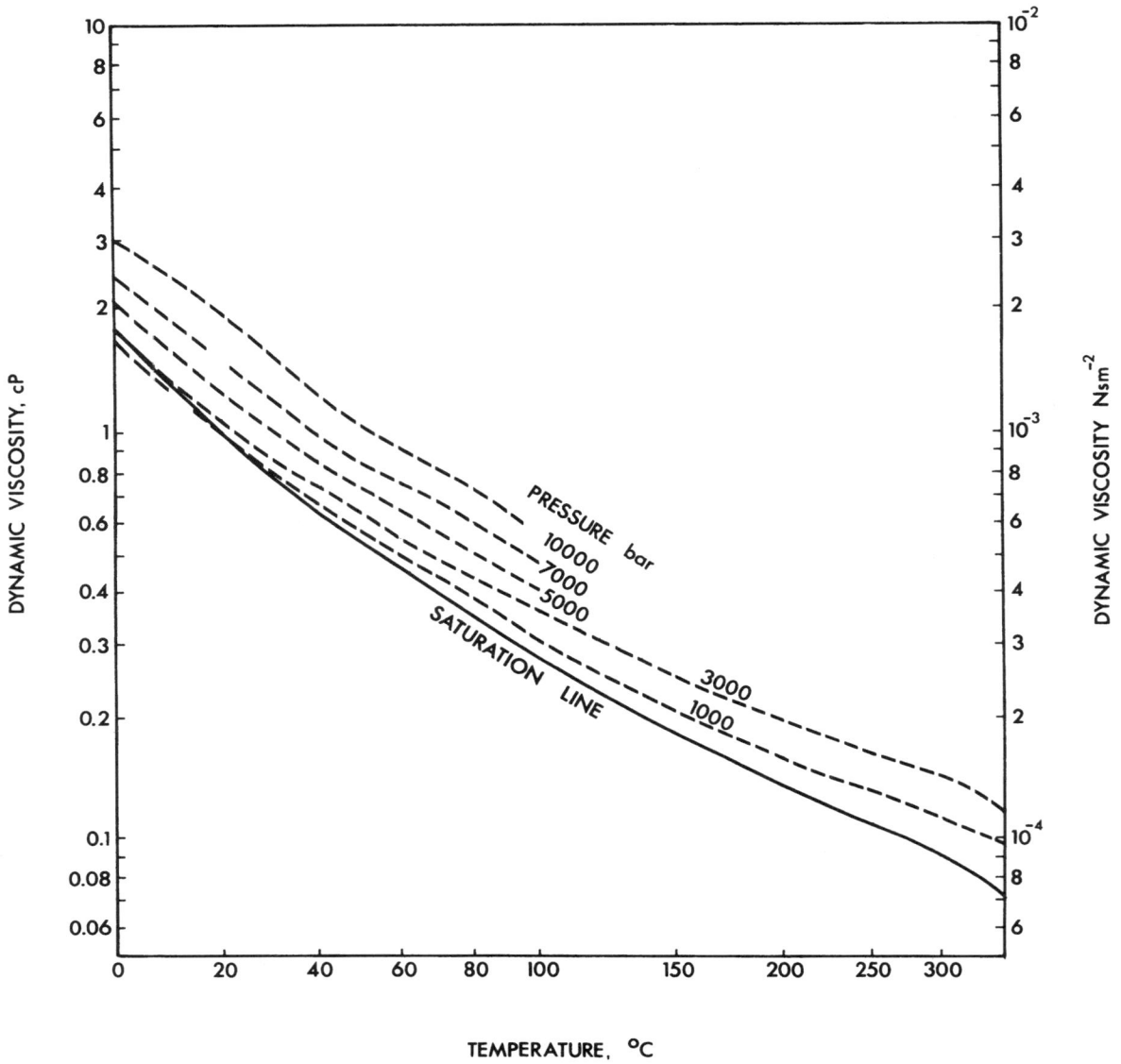

Figure 6.1 The viscosity of water at various temperatures and pressures

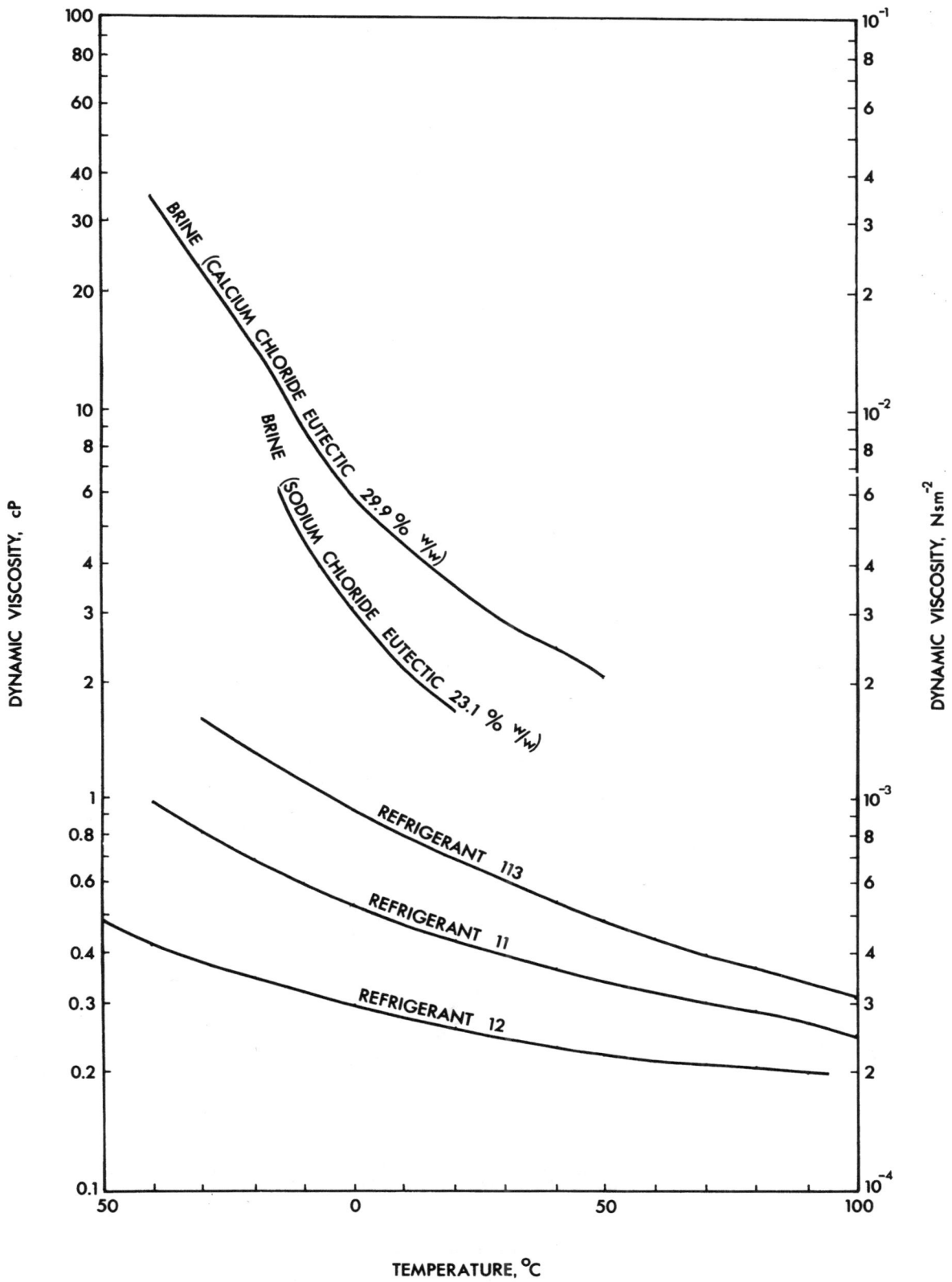

Figure 6.2 *The viscosity of various refrigerant liquids*

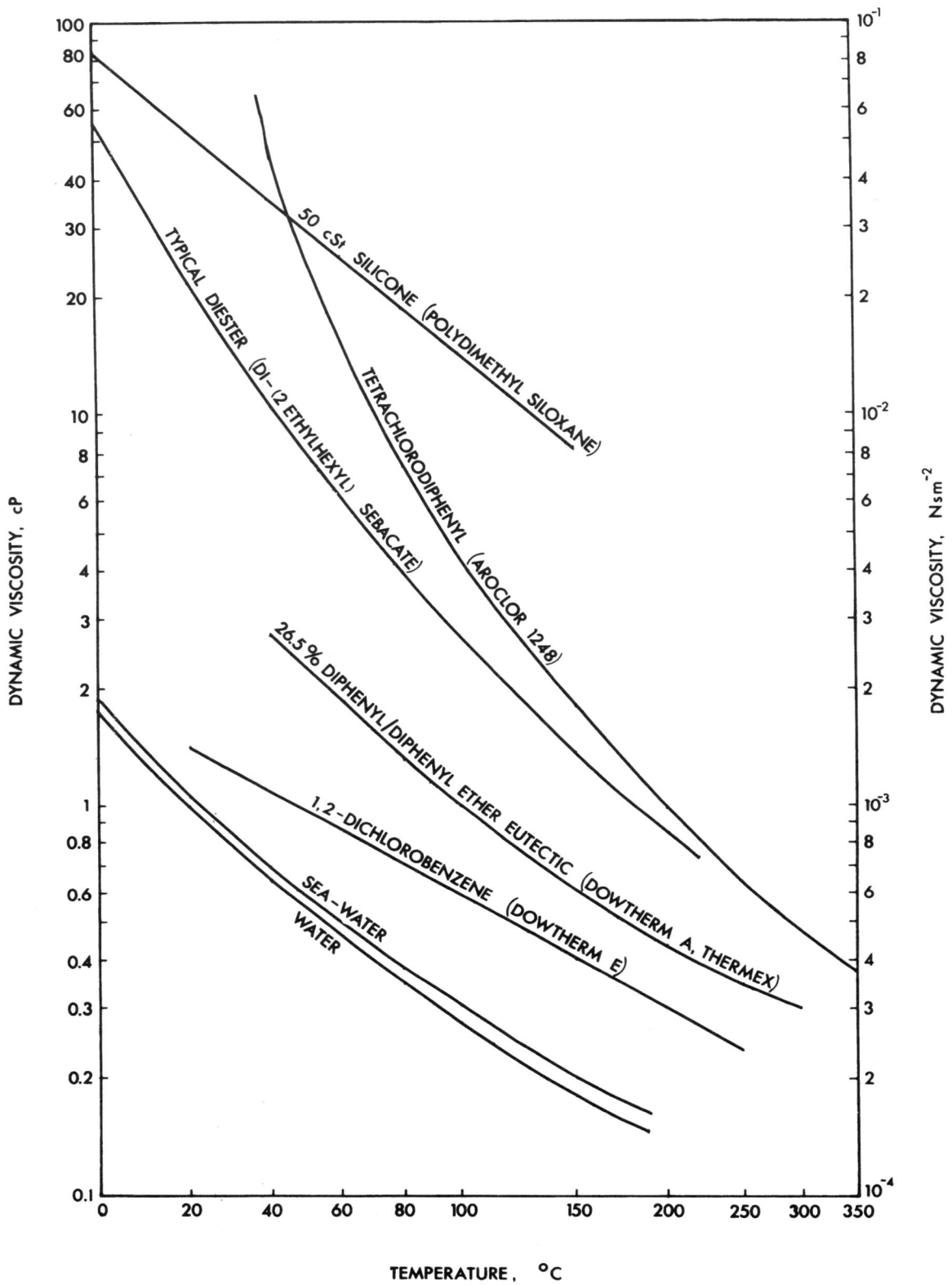

Figure 6.3 *The viscosity of various heat transfer fluids*

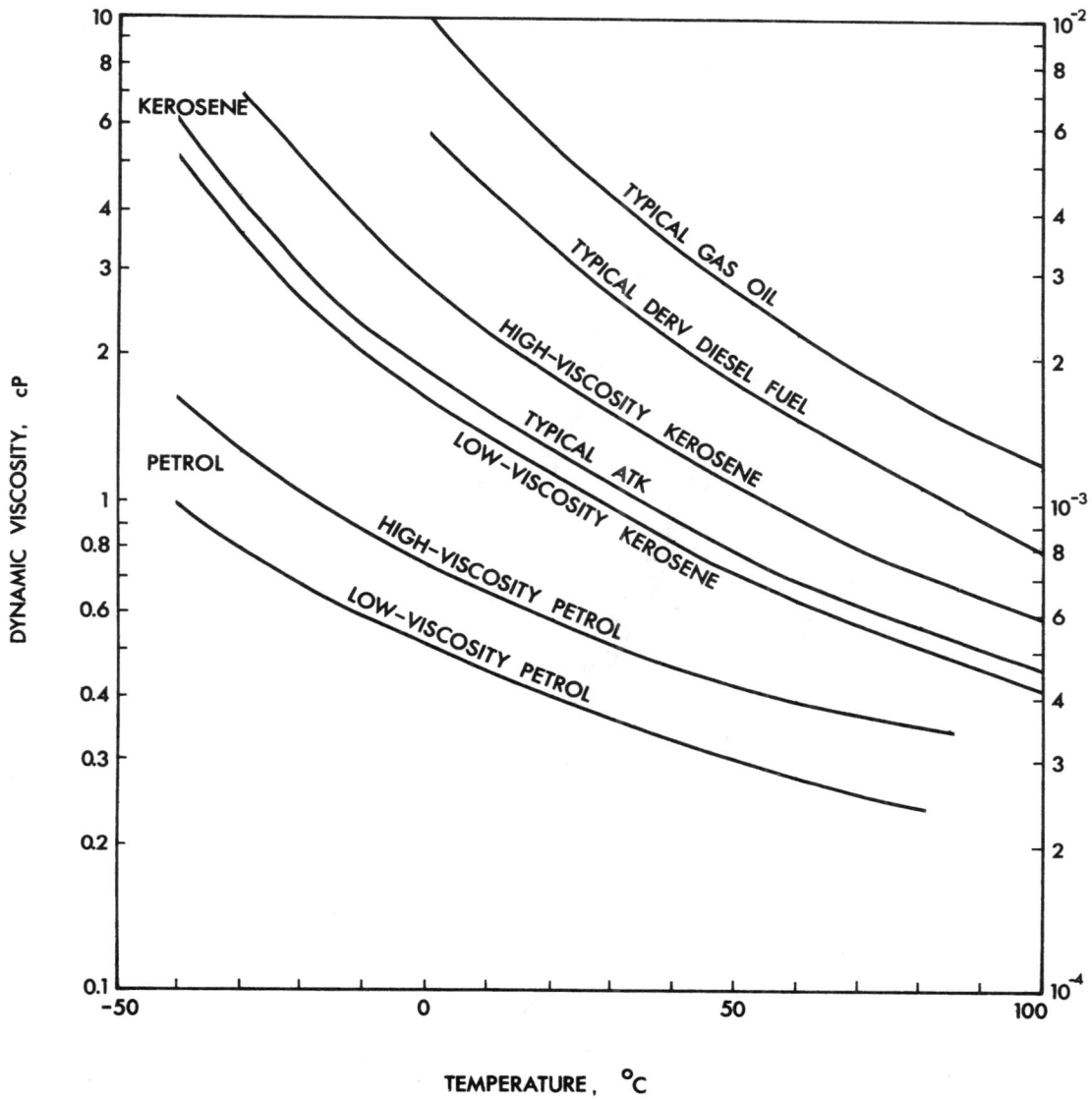

Figure 6.4 The viscosity of various light petroleum products

Petroleum products are variable in composition and so only typical values or ranges of values are given.

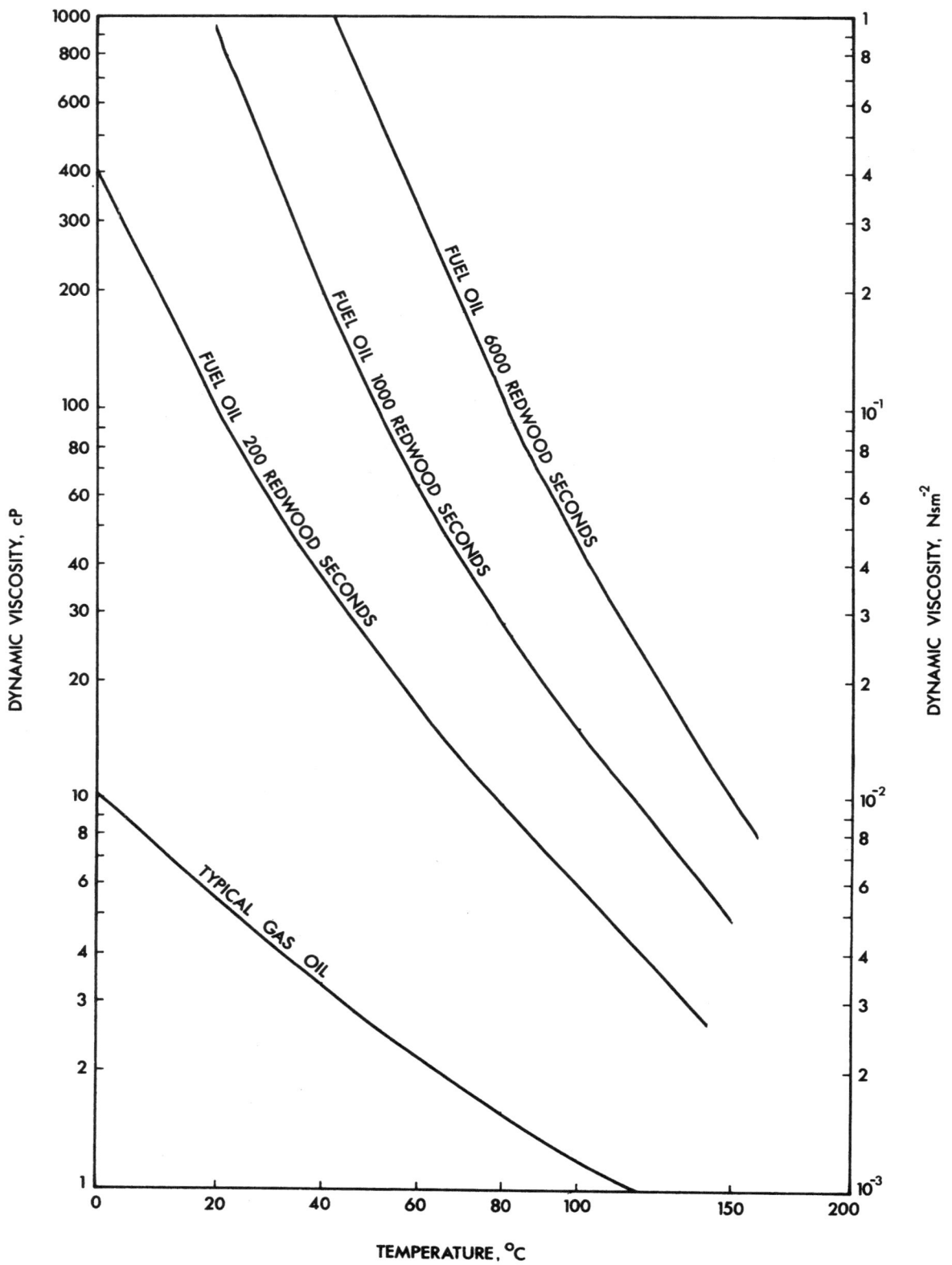

Figure 6.5 The viscosity of various heavy petroleum products

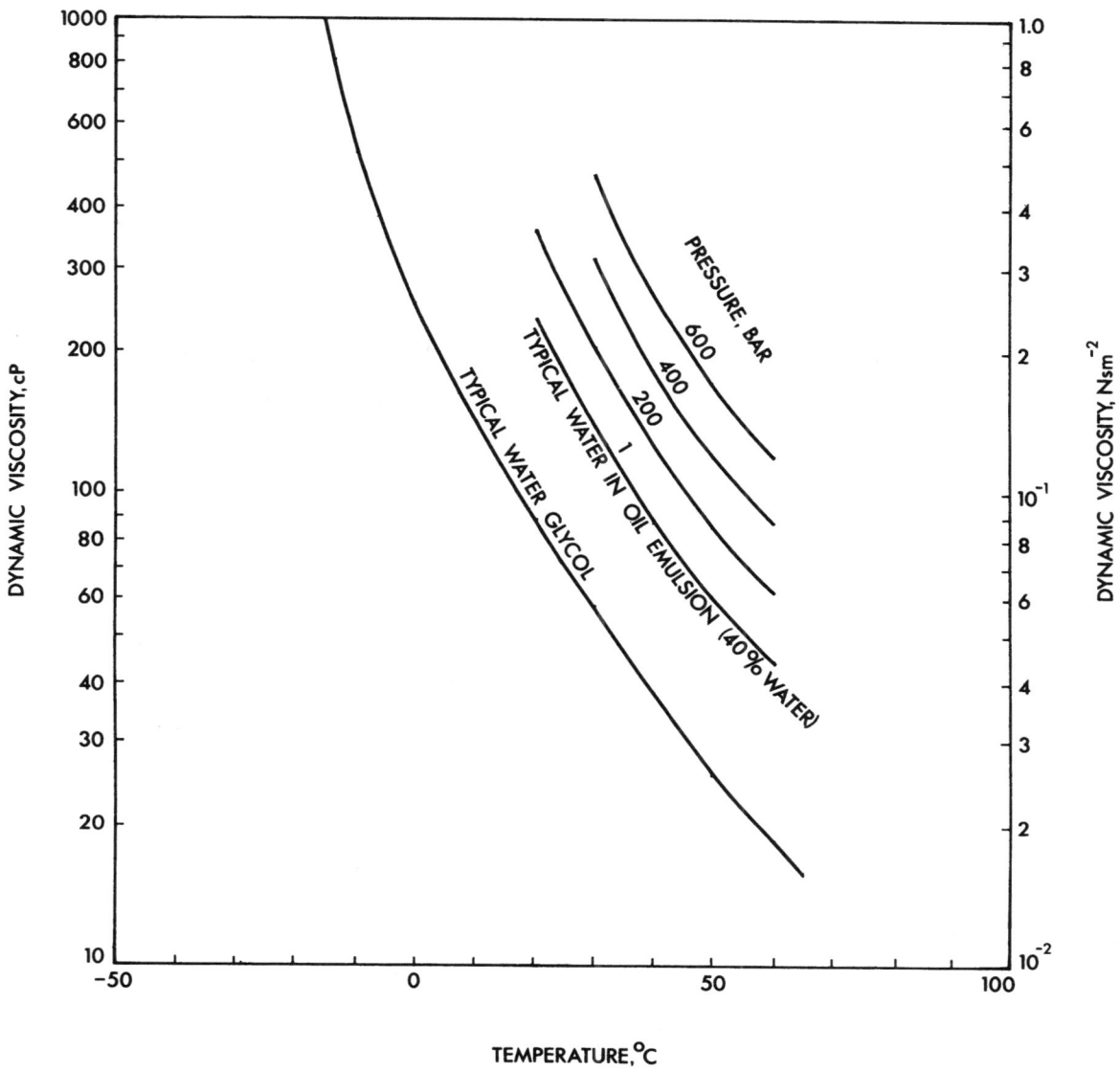

Figure 6.6 The viscosity of various water-based mixtures

For all practical purposes the above fluids may be classed as Newtonian but other fluids, such as water-in-oil emulsions, are non-Newtonian. The viscosity values given for the typical 40% water-in-oil emulsion are for very low shear rates. For this emulsion the viscosity will decrease by 10% at shear rates of about $3000\,s^{-1}$ and by 20% at shear rates of about $10\,000\,s^{-1}$.

Mineral oils and greases are the most suitable lubricants for plain bearings in most applications. Synthetic oils may be required if system temperatures are very high. Water and process fluids can also be used as lubricants in certain applications. The general characteristics of these main classes of lubricants are summarised in Table 7.1.

Table 7.1 Choice of lubricant

Lubricant	Operating range	Remarks
Mineral oils	All conditions of load and speed	Wide range of viscosities available. Potential corrosion problems with certain additive oils (e.g. extreme pressure) (see Table 7.9)
Synthetic oils	All conditions if suitable viscosity available	Good high and low temperature properties. Costly
Greases	Use restricted to operating speeds below 1 to 2 m/s	Good where sealing against dirt and moisture necessary and where motion is intermittent
Process fluids	Depends on properties of fluid	May be necessary to avoid contamination of food products, chemicals, etc. Special attention to design and selection of bearing materials

Table 7.2 Methods of liquid lubricant supply

Method of Supply	Main characteristics	Examples
Hand oiling	Non automatic, irregular. Low initial cost. High maintenance cost	Low-speed, cheap journal bearings
Drip and wick feed	Non automatic, adjustable. Moderately efficient. Cheap	Journals in some machine tools, axles
Ring and collar feed	Automatic, reliable. Efficient, fairly cheap. Mainly horizontal bearings	Journals in pumps, blowers, large electric motors
Bath and splash lubri-cation	Automatic, reliable, efficient. Oil-tight housing required High initial cost	Thrust bearings, bath only. Engines, process machinery, general
Pressure feed	Automatic. Positive and adjustable. Reliable and efficient. High initial cost	High-speed and heavily loaded journal and thrust bearings in machine tools, engines and compressors

Notes

Pressure oil feed: This is usually necessary when the heat dissipation of the bearing housing and its surroundings is not sufficient to restrict its temperature rise to 20°C or less

Journal bearings: Oil must be introduced by means of oil grooves in the bearing housing. Some common arrangements are shown in Figure 7.3

Thrust bearings: These must be lubricated by oil bath or by pressure feed from the centre of the bearing

Cleanliness: Cleanliness of the oil supply is essential for satisfactory performance and long life

The most important property of a lubricant for plain bearings is its viscosity. If the viscosity is too low the bearing will have inadequate load-carrying capacity, whilst if the viscosity is too high the power loss and the operating temperature will be unnecessarily high. Figure 7.1 gives a guide to the value of the minimum allowable viscosity for a range of speeds and loads. It should be noted that these values apply for a fluid at the mean bearing temperature. The viscosity of mineral oils falls with increasing temperature. The viscosity/temperature characteristics of typical mineral oils are shown in Figure 7.2. The most widely used methods of supplying lubricating oils to plain bearings are listed in Table 7.2

The lubricating properties of greases are determined to a large extent by the viscosity of the base oil and the type of thickener used in their manufacture. The section of this handbook on greases summarises the properties of the various types.

Additive oils are not required for plain bearing lubrication but other requirements of the system may demand their use. Additives and certain contaminants may create potential corrosion problems. Tables 7.3 and 7.4 give a guide to additive and bearing material requirements, with examples of situations in which problems can arise.

Table 7.3 Principal additives and contaminants

Problem	Occurs in	Requirements
Oxidation of lubricant	IC engines Steam turbines Compressors High-speed gearboxes	Antioxidant additives
Scuffing	Gearboxes Cam mechanisms	Extreme-pressure additive
Deposit formation	IC engines Compressors	Dispersant additives
Excessive wear of lubricated surfaces	General	Antiwear additives
Water contamination	IC engines Steam turbines Compressors	Good demulsification properties. Turbine-quality oils may be required
Dirt particle contamination	IC engines Industrial plant	Dispersant additives
Weak organic acid contamination	IC engines	Acid neutraliser
Strong mineral acid contamination	Diesel engines Process fluids	Acid neutraliser
Rusting	IC engines Turbines Industrial plant General	Rust inhibitor

Plain journal bearings

Surface speed, $u = \pi d n$, ms^{-1}

Mean pressure, $\bar{p} = \dfrac{W}{ld}$, kNm^{-2}

where n = shaft speed, s^{-1}
l = bearing width, m
d = shaft diameter, m
W = load, kN

Minimum allowable viscosity $\eta_{min.}$, cP, may be read directly

Plain thrust bearings

Surface speed, $u = \pi D n$, ms^{-1}

Mean pressure, $\bar{p} = \dfrac{0.4W}{lD}$, kNm^{-2}

where n = shaft speed, s^{-1}
l = width of bearing ring, m
D = mean pad diameter, m
W = thrust load, kN

Minimum allowable viscosity $\eta_{thrust} = \eta_{min.}\left(\dfrac{D}{l}\right)$

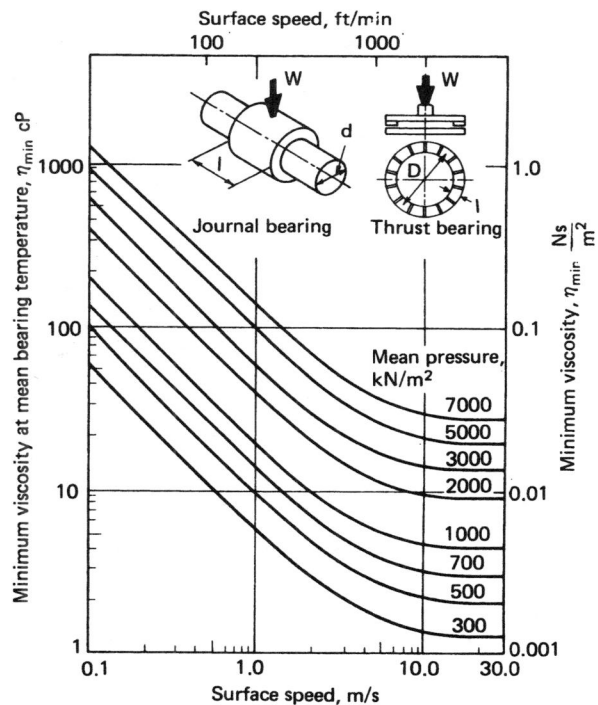

Figure 7.1 Lubricant viscosity for plain bearings

Table 7.4 Resistance to corrosion of bearing metals

	Maximum operating temperature, °C	Additive or contaminant				
		Extreme-pressure additive	Antioxidant	Weak organic acids	Strong mineral acids	Synthetic oil
Lead-base white metal	130	Good	Good	Moderate/poor	Fair	Good
Tin-base white metal	130	Good	Good	Excellent	Very good	Good
Copper–lead (without overlay)	170	Good	Good	Poor	Fair	Good
Lead–bronze (without overlay)	180	Good with good quality bronze	Good	Poor	Moderate	Good
Aluminium–tin alloy	170	Good	Good	Good	Fair	Good
Silver	180	Sulphur-containing additives must not be used	Good	Good—except for sulphur	Moderate	Good
Phosphor–bronze	220	Depends on quality of bronze. Sulphurised additives can intensify corrosion	Good	Fair	Fair	Good
Copper–lead or lead–bronze with suitable overlay	170	Good	Good	Good	Moderate	Good

Note: corrosion of bearing metals is a complex subject. The above offers a general guide. Special care is required with extreme-pressure lubricants; if in doubt refer to bearing or lubricants supplier.

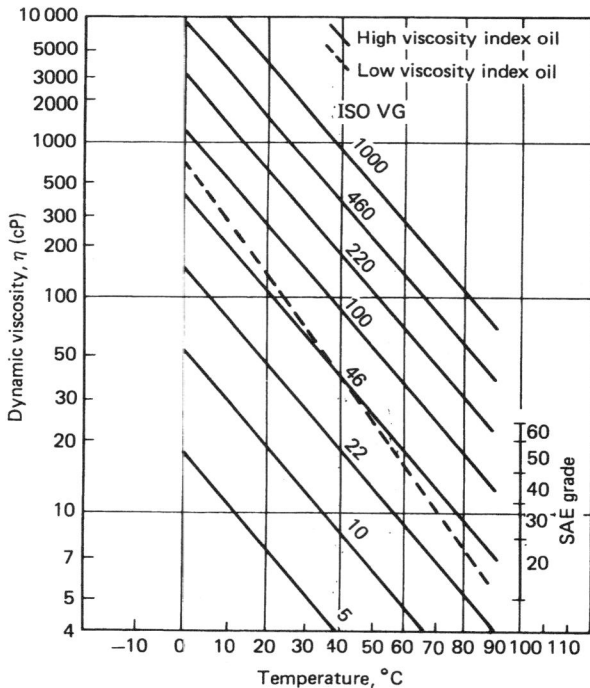

Figure 7.2 Typical viscosity/temperature characteristics of mineral oils

Bearing temperature

Lubricant supply rate should be sufficient to restrict the temperature rise through the bearing to less than 20°C. A working estimate of the mean bearing temperature, $\theta_{bearing}$, is given by

$$\theta_{bearing} = \theta_{supply} + 20, \text{ °C}$$

Dynamic and Kinematic Viscosity

Dynamic Viscosity, η (cP)
   = Density × Kinematic Viscosity (cSt)

Viscosity classification grades are usually expressed in terms of Kinematic Viscosities.

Rotation	Housing	Direction of load	
		Fixed	Variable
Unidirectional	Unit	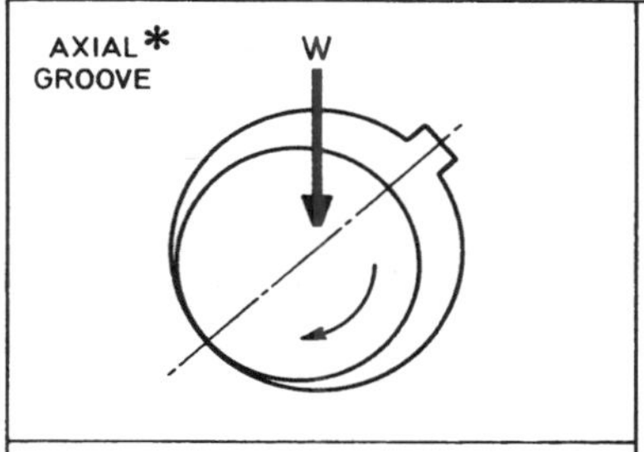 AXIAL* GROOVE W	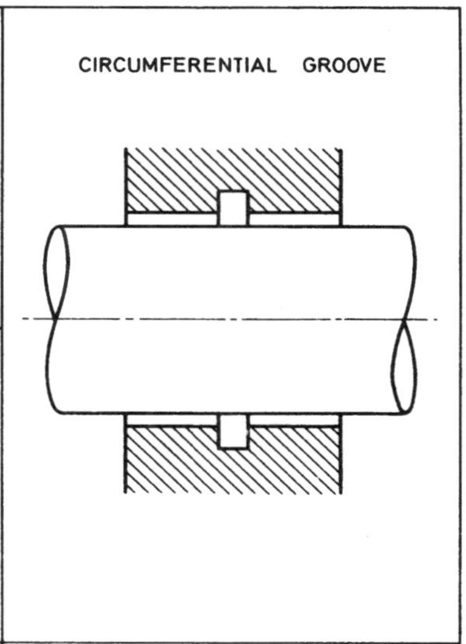 CIRCUMFERENTIAL GROOVE
	Split	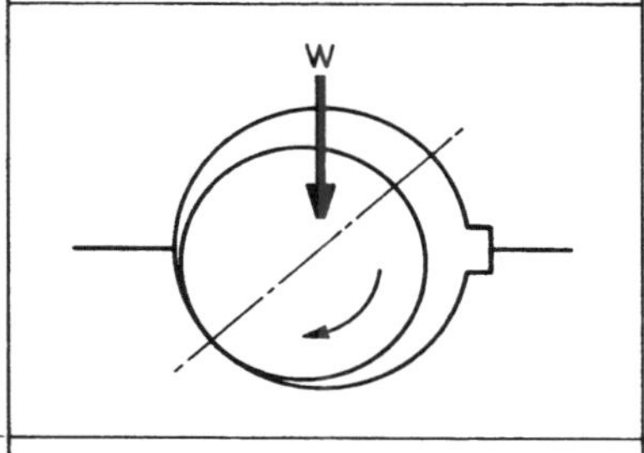 W	
Reversible	Unit/split	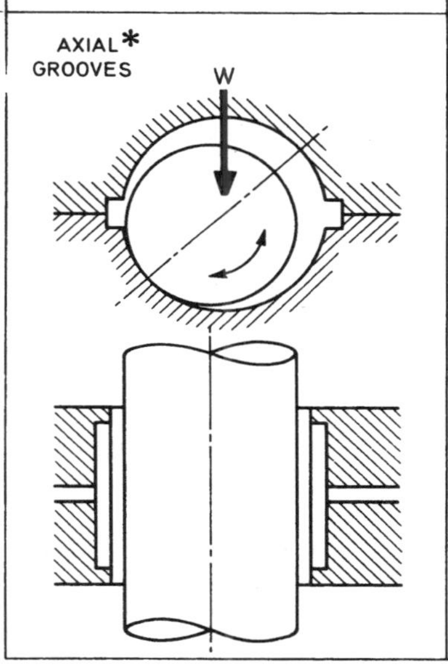 AXIAL* GROOVES W	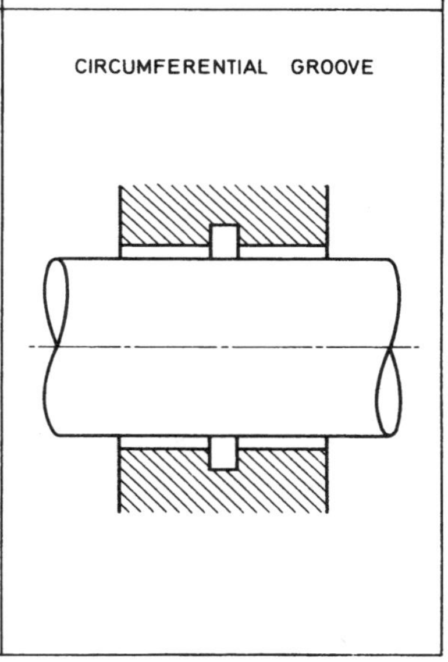 CIRCUMFERENTIAL GROOVE

* At moderate speeds oil holes may be substituted if l/d does not exceed 1.

Note: the load-carrying capacity of bearings with circumferential grooves is somewhat lower than with axial grooves owing to the effect of the groove on pressure generation.

Figure 7.3 Oil grooves in journal bearings

SELECTION OF THE LUBRICANT

Table 8.1 General guide for choosing between grease and oil lubrication

Factor affecting the choice	Use grease	Use oil
Temperature	Up to 120°C—with special greases or short relubrication intervals up to 200/220°C	Up to bulk oil temperature of 90°C or bearing temperature of 200°C—these temperatures may be exceeded with special oils
Speed factor*	Up to *dn* factors of 300 000/350 000 (depending on design)	Up to *dn* factors of 450 000/500 000 (depending on type of bearing)
Load	Low to moderate	All loads up to maximum
Bearing design	Not for asymmetrical spherical roller thrust bearings	All types
Housing design	Relatively simple	More complex seals and feeding devices necessary
Long periods without attention	Yes, depends on operating conditions, especially temperature	No
Central oil supply for other machine elements	No—cannot transfer heat efficiently or operate hydraulic systems	Yes
Lowest torque	When properly packed can be lower than oil on which the grease is based	For lowest torques use a circulating system with scavenge pumps or oil mist
Dirty conditions	Yes—proper design prevents entry of contaminants	Yes, if circulating system with filtration

* *dn* factor (bearing bore (mm) × speed (rev/min)).

Note: for large bearings (> 65 mm bore) use nd_m (d_m is the arithmetic mean of outer diameter and bore (mm)).

GREASE LUBRICATION

Grease selection

The principal factors governing the selection of greases for rolling bearings are speed, temperature, load, environment and method of application. Guides to the selection of a suitable grease taking account of the above factors are given in Tables 8.2 and 8.3.

The appropriate maximum speeds for grease lubrication of a given bearing type are given in Figure 8.1. The life required from the grease is also obviously important and Figure 8.2 gives a guide to the variation of grease operating life with percentage speed rating and temperature for a high-quality lithium hydroxystearate grease as derived from Figure 8.1. (These greases give the highest speed ratings.)

When shock loading and/or high operating temperatures tend to shake the grease out of the covers into the bearing, a grease of a harder consistency should be chosen, e.g. a no. 3 grease instead of a no. 2 grease.

Note: it should be recognised that the curves in Figures 8.1 and 8.2 can only be a guide. Considerable variations in life are possible depending on precise details of the application, e.g. vibration, air flow across the bearing, clearances, etc.

Table 8.2 The effect of the method of application on the choice of a suitable grade of grease

System	NLGI grade no.
Air pressure	0 to 2 depending on type
Pressure-guns or mechanical lubricators	Up to 3
Compression cups	Up to 5
Centralised lubrication	2 or below
(a) Systems with separate metering valves	Normally 1 or 2
(b) Spring return systems	1
(c) Systems with multi-delivery pumps	3

Table 8.3 The effect of environmental conditions on the choice of a suitable type of grease

Type of grease	NLGI grade no.	Speed maximum (percentage recommended maximum for grease)	Environment	Typical service temperature				Base oil viscosity (approximate values)	Comments
				Maximum		Minimum			
				°C	°F	°C	°F		
Lithium	2	{100 / 75	Wet or dry	100 / 135	210 / 275}	−25	−13}	Up to 140 cSt at 100°F	Multi-purpose, not advised at max. speeds or max. temperatures for bearings above 65 mm bore or on vertical shafts
Lithium	3	{100 / 75	Wet or dry	100 / 135	210 / 275}	−25	−13}		For max. speeds recommended where vibration loads occur at high speeds
Lithium EP	1	75	Wet or dry	90	195	−15	5	14.5 cSt at 210°F	Recommended for roll-neck bearings and heavily-loaded taper-roller bearings
Lithium EP	2	{100 / 75	Wet or dry	70 / 90	160} / 195}	−15	5		
Calcium (conventional)	1, 2 and 3	50	Wet or dry	60	140	−10	14	140 cSt at 100°F	
Calcium EP	1 and 2	50	Wet or dry	60	140	−5	25	14.5 cSt at 210°F	
Sodium (conventional)	3	75/100	Dry	80	175	−30	−22	30 cSt at 100°F	Sometimes contains 20% calcium
Clay		50	Wet or dry	200	390	10	50	550 cSt at 100°F	
Clay		100	Wet or dry	135	275	−30	−22	Up to 140 cSt at 100°F	
Clay		100	Wet or dry	120	248	−55	−67	12 cSt at 100°F	Based on synthetic esters
Silicone/lithium		75	Wet or dry	200	390	−40	−40	150 cSt at 25°C	Not advised for conditions where sliding occurs at high speed and load

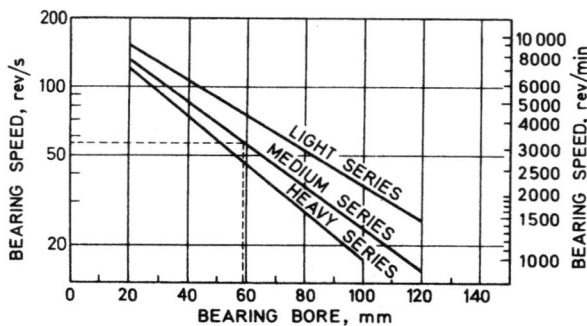

Figure 8.1 Approximate maximum speeds for grease lubrication. (Basic diagram for calculating bearing speed ratings)

Bearing type	Multiply bearing speed from Figure 8.1 by this factor to get the maximum speed for each type of bearing
Cage centred on inner race	As Figure 8.1
Pressed cages centred on rolling elements	1.5–1.75
Machined cages centred on rolling elements	1.75–2.0
Machined cages centred on outer race	1.25–2.0
Taper- and spherical- roller bearings	0.5
Bearings mounted in adjacent pairs	0.75
Bearings on vertical shafts	0.75
Bearings with rotating outer races and fixed inner races	0.5

(Ball bearings and cylindrical roller bearings)

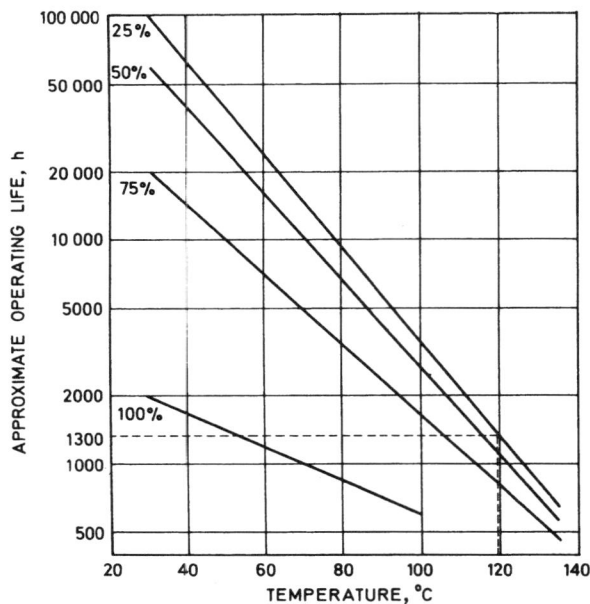

Figure 8.2 Variation of operating life of a high-quality grade 3 lithium hydroxystearate grease with speed and temperature

Calculation of relubrication interval

The relubrication period for ball and roller bearings may be estimated using Figures 8.1 and 8.2. The following is an example in terms of a typical application:

Required to know:	Approximate relubrication period for the following:
Bearing type:	Medium series bearing 60 mm bore.
Cage:	Pressed cage centred on balls.
Speed:	950 rev/min.
Temperature:	120°C [The bearing temperature (not merely the local ambient temperature) i.e. either measured or estimated as closely as possible.]
Position:	Vertical shaft.
Grease:	Lithium grade 3.
Duty:	Continuous.
From Figure 8.1:	60 mm bore position on the lower edge of the graph intersects the medium series curve at approximately 3100 rev/min.

Factor for pressed cages on balls is about 1.5. Thus $3100 \times 1.5 = 4650$ rev/min.

Factor for vertical mounting is 0.75. Thus $4650 \times 0.75 = 3488$ rev/min.

This is the maximum speed rating (100%).

Now actual speed = 950 rev/min; therefore

$$\text{percentage of maximum} = \frac{950}{3488} \times 100 = 27\%$$

(say 25% approximately).

In Figure 8.2 the 120°C vertical line intersects the 25% speed rating curve for the grade 3 lithium grease at approximately 1300 hours, which is the required answer.

Method of lubrication

Rolling bearings may be lubricated with grease by a lubrication system as described in other sections of the handbook or may be packed with grease on assembly.

Packing ball and roller bearings with grease

(a) The grease should not occupy more than one-half to three-quarters of the total available free space in the covers with the bearing packed full.
(b) One or more bearings mounted horizontally – completely fill bearings and space between, if more than one, but fill only two-thirds to three-quarters of space in covers.
(c) Vertically-mounted bearings – completely fill bearing but fill only half of top cover and three-quarters of bottom cover.
(d) Low/medium speed bearings in dirty environments – completely fill bearing and covers.

Relubrication of ball and roller bearings

Relubrication may be carried out in two ways, depending on the circumstances:

(a) Replenishment, by which is meant the addition of fresh grease to the original charge.
(b) Repacking, which normally signifies that the bearing is dismounted and all grease removed and discarded, the bearing then being cleaned and refilled with fresh grease. An alternative, if design permits, is to flush the bearing with fresh grease *in situ*. (Grease relief valves have been developed for this purpose.)

The quantity required per shot is an arbitrary amount. Requirement is only that sufficient grease is injected to disturb the charge in the bearing and to displace same through the seals, or grease relief valves.

A guide can be obtained from

$$W = \frac{D \times w}{200}$$

where W is quantity (g)
         D is outside diameter (mm)
and w is width (mm)

If grease relief valves are not fitted, the replenishment charge should not exceed 5% of the original charge. After grease has been added to a bearing, the housing vent plug (if fitted) should be left out for a few minutes after start-up in order to allow excess grease to escape. A better method, if conditions allow, is to push some of the static grease in the cover back into the bearing to redistribute the grease throughout the assembly. This method is likely to be unsatisfactory when operating temperatures exceed about 100°C.

OIL LUBRICATION

Oil viscosity selection

Generally, when speeds are moderate, the following minimum viscosities at the operating temperatures are recommended:

	cSt
Ball and cylindrical-roller bearings	12
Spherical-roller bearings	20
Spherical-roller thrust bearings	32

The oils will generally be HVI or MVI types containing rust and oxidation inhibitors. Oils containing extreme pressure (EP) additives are normally only necessary for bearings where there is appreciable sliding, e.g. taper-roller or spherical-roller bearings, operating under heavy or shock loads, or if required for associated components, e.g. gears. The nomogram, Figure 8.3, shows how to select more precisely the viscosity needed for known bore and speed when the operating temperatures can be estimated. If the operating temperature is not known or cannot be estimated then the manufacturer's advice should be sought.

To use Figure 8.3, starting with the right-hand portion of the graph for the appropriate bearing bore and speed, determine the viscosity required for the oil at the working temperature. The point of intersection of the horizontal line, which represents this oil viscosity, and the vertical line from the working temperature shows the grade of oil to be selected. If the point of intersection lies between two oils, the thicker oil should be chosen.

Examples:

Bearing bore $d = 60$ mm, speed $n = 5000$ rev/min (viscosity at working temperature = 6.8 cSt), with working temperature = 65°C. *Select oil S 14 (14 cSt at 50°C approx.)*

Bearing bore $d = 340$ mm, speed $n = 500$ rev/min (viscosity at working temperature = 13.2 cSt), with working temperature = 80°C. *Select oil S 38 (38 cSt at 50°C approx.)*

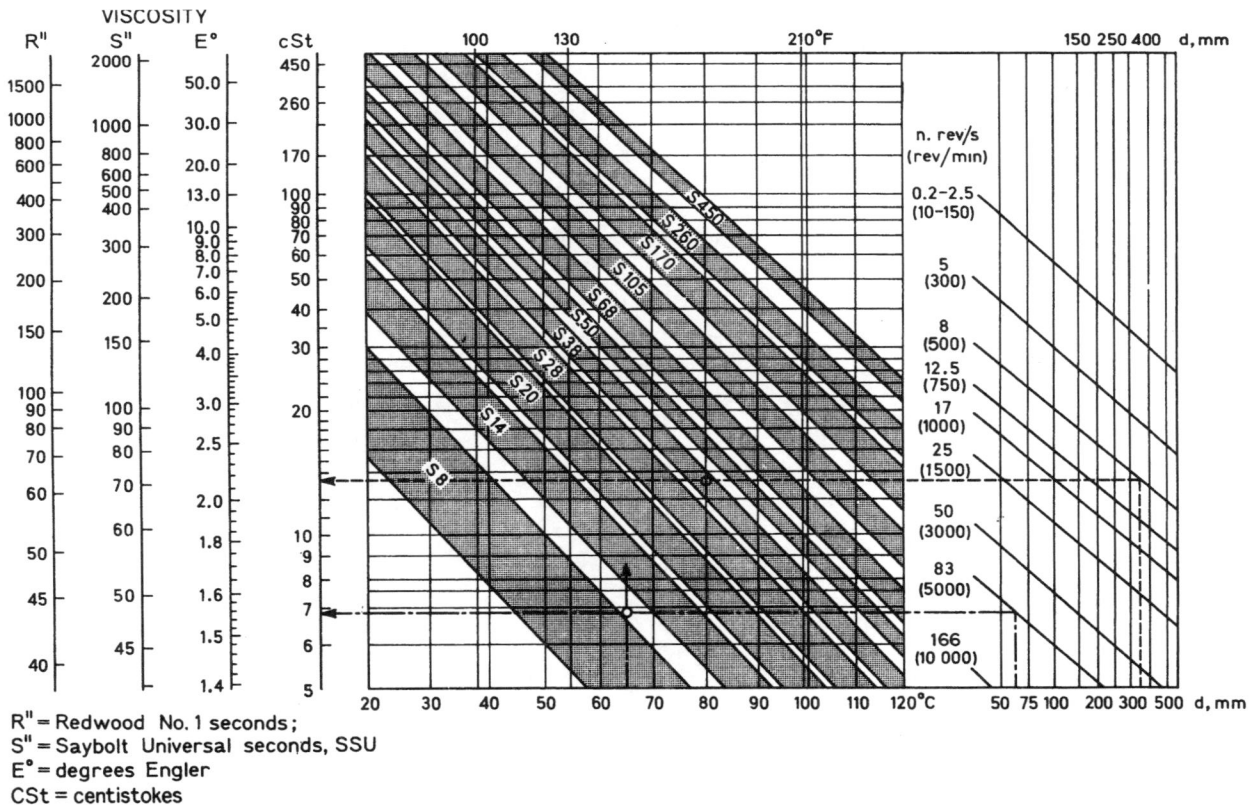

R" = Redwood No. 1 seconds;
S" = Saybolt Universal seconds, SSU
E° = degrees Engler
CSt = centistokes

Figure 8.3 Graph for the selection of oil for roller bearings (Permission of the Skefko Ball Bearing Co. Ltd). *The graph has been compiled for a viscosity index of 85, which represents a mean value of the variation of the viscosity of the lubricating oil with temperature. Differences for 95 VI oils are negligible*

Application of oil to rolling bearings

System	Conditions	Oil levels/oil flow rates	Comments
Bath/splash	Generally used where speeds are low A limit in *dn* value of 100 000 is sometimes quoted, but higher values can be accommodated if churning is not a problem	Bearings on horizontal and vertical shafts, immerse half lowest rolling element Multi-row bearings on vertical shafts, fully immerse bottom row of elements	
Oil flingers, drip feed lubricators, etc.	Normally as for bath/splash	Flow rate dictated by particular application; ensure flow is sufficient to allow operation of bearing below desired or recommended maximum temperature—generally between 70°C and 90°C	Allows use of lower oil level if temperature-rise is too high with bath/splash
Pressure circulating	No real limit to *dn* value Use oil mist where speeds are very high	As a guide, use:* 0.6 cm^3/min cm^2 of projected area of bearing (o.d. × width)	The oil flow rate has generally to be decided by consideration of the operating temperature
Oil mist	No real limit to *dn* value Almost invariably used for small bore bearings above 50 000 rev/min, but also used at lower speeds	As a guide, use:* 0.1 to 0.3 × bearing bore (cm/2.54) × no. of rows—cm^3/hour Larger amounts are required for pre-loaded units, up to 0.6 × bearing bore (cm/2.54) × no. of rows—cm^3/hour	In some cases oil-mist lubrication may be combined with an oil bath, the latter acting as a reserve supply which is particularly valuable when high-speed bearings start to run

* It must be emphasised that values obtained will be very approximate and that the manufacturer's advice should be sought on the performance of equipment of a particular type.

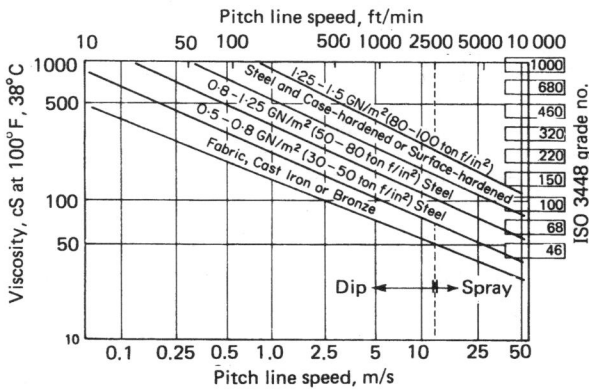

Figure 9.1 Selection of oil for industrial enclosed gear units

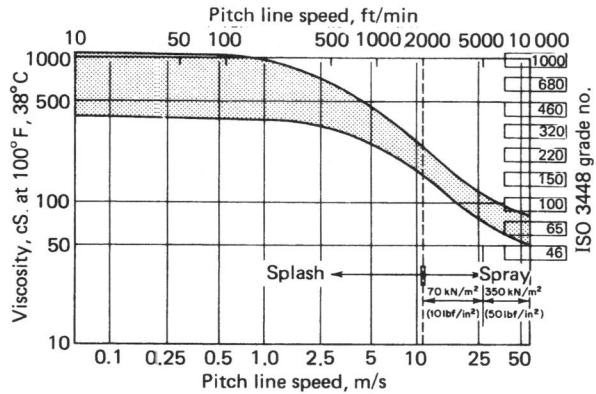

Figure 9.2 Selection of oil for industrial enclosed worm gears

Figure 9.1 is a general guide only. It is based on the criterion: $Sc\, HV/(Vp + 100)$

where Sc = Surface stress factor

$$= \frac{\text{Load/inch line of contact}}{\text{Relative radius of curvature}}$$

and HV = Vickers hardness for the softer member of the gear pair

Vp = Pitch line velocity, ft/min

The chart applies to gears operating in an ambient temperature between 10°C and 25°C. Below 10°C use one grade lower. Above 25°C use one grade higher. Special oils are required for very low and very high temperatures and the manufacturer should be consulted.

With shock loads, or highly-loaded low-speed gears, or gears with a variable speed/load duty cycle, EP oils may be used. Mild EPs such as lead naphthanate should not be used above 80°C (170°F) running temperature. Full hypoid EP oils may attack non ferrous metals. Best EP for normal industrial purposes is low percentage of good quality sulphur/phosphorus or other carefully inhibited additive.

Suitable lubricants for worm gears are plain mineral oils of a viscosity indicated in Figure 9.2. It is also common practice, but usually unnecessary, to use fatty additive or leaded oils. Such oils may be useful for heavily-loaded, slow-running gears but must not be used above 80°C (170°F) running temperature as rapid oxidation may occur, resulting in acidic products which will attack the bronze wheel and copper or brass bearing-cages.

Worm gears do not usually exceed a pitch line velocity of 2000 ft/min, but if they do, spray lubrication is essential. The sprayed oil must span the face width of the worm.

Quantity	Speed m/s	10	15	20	25
$75 \times 10^{-6} \times C^*$ m³/sec	Pressure kN/m²	100	170	270	340

C^* = centre distance in metres.

Quantity	Speed ft/min	2000	3000	4000	5000
$Q = C/4$ gpm	Pressure lbf/in²	15	25	40	50

Where C = centre distance, inches.

Recent developments in heavily loaded worm gear lubrication include synthetic fluids which:

(a) have a wider operating temperature range
(b) reduce tooth friction losses
(c) have a higher viscosity index and thus maintain an oil film at higher temperatures than mineral oils
(d) have a greatly enhanced thermal and oxidation stability, hence the life is longer

Even more recent developments include the formulation of certain soft synthetic greases which are used in 'lubricated-for-life' worm units. Synthetic lubricants must not be mixed with other lubricants.

Spray lubrication

Quantity	Speed m/sec	10	25	50	100	150
$85 \times 10^{-6} \times kW$ m³/sec	Pressure kN/m²	10	100	140	180	210

Quantity	Speed ft/min	2500	5000	10 000	20 000	30 000
$0.0085 \times hp$ gpm	Pressure lbf/in²	10	15	20	25	30

AUTOMOTIVE LUBRICANTS

SAE classification of transmission and axle lubricants

SAE visco-sity No.	Centistokes				Redwood seconds			
	0°F – 18°C		210°F 99°C		0°F – 18°C		210°F 99°C	
	Min.	Max.	Min.	Max.	Min.	Max.	Min	Max.
75	—	3250	—	—	—	13 100	—	—
80	3250*	21 700	—	—	13 100	87 600	—	—
90	—	—	14	25	—	—	66	107
140	—	—	25	43	—	—	107	179
250	—	—	41	—	—	—	179	—

Note: * The min. viscosity at 0°F may be waived if the viscosity is not less than 7 cSt at 210°F.

These values are approximate and are given for information only.

Selection of lubricants for transmissions and axles

Almost invariably dip-splash.

The modern tendency is towards universal multi-purpose oil.

	Manual gear boxes	Automatic gear boxes	Rear axles (hypoids)	Rear axles (spiral bevel and worms)
Cars	SAE 80 (EP) or multi-purpose	Automatic transmission fluid (ATF)	Highly active SAE 90 EP or multi-purpose	—
Heavy vehicles	SAE 80 EP or SAE 90 (EP) or multi-purpose	SAE 80 EP for semi-auto-matics ATF fluid for autos	—	SAE 140 or multi-pur-pose

Notes: (1) Above only to be used where supplier's recommenda-tions are not available.
(2) Above are suitable for normal conditions. In cold conditions (< 0°C) use one SAE grade less. In hot conditions (> 40°C) use one SAE grade higher.
(3) In most cases (except hypoids), straight oils are accept-able. The above EPs are given for safety if supplier's recommendations are not known.
(4) Some synthetic (polyglycol) oils are very successful with worm gears. They must not be mixed with any other oils. ATF fluids must not be mixed with others.
(5) Change periods: (only if manufacturer's recommenda-tions not known). Rear axles—do not change. Top up as required. All manual and automatic gearboxes—change after 20 000 miles. Before that top up as required.

ROLLER CHAINS

Type of lubricant: Viscosity grade no. 150 (ISO 3448). For slow-moving chains on heavy equipment, bituminous viscous lubricant or grease can be used. Conditions of operation determine method of application and top-ping-up or change periods. Refer to manufacturer for guidance under unusual conditions.

Method of application	Speed limitation m/s(ft/min)	Quantity	Comments
Dip	< 10 (<2000)	—	1.5 x ROLLER PITCH OIL LEVEL
Slow drip	0–3 (0–600)	5–10 drops/ min.	APPLIED TO IN-GOING SIDE OF CHAIN
Fast drip	3–7.5 (600–1500)	> 20	
Spray	> 7·5 (>1500)	Depends upon speed and other conditions	SUMP PUMP DRIVEN SEPARATELY OR FROM CHAIN SPROCKET

OPEN GEARS

Applies to large, slow-running gears without oil-tight housings.

Requirements of lubricant	Types of lubricant	Methods of application
Must form protective film	Generally bituminous and tacky. Sometimes cut back by volatile diluent Can use grease or heavy EP oils	Hand, brush, paddle Dip-shallow pool Drip-automatic Spray-continuous or intermittent
Must not be squeezed out		
Must not be thrown off		
Must be suitable for prevailing ambient conditions		

Temperature °C	Viscosity of open gear lubricant cS at 38°C(100°F)		
	Spray		Drip
	Mild EP oil	Residual compound	Mild EP oil
−10 to 15	—	200–650	—
5 to 35	100–120	650–2000	100–120
25 to 50	180–200	650–2000	180–200

THE ADVANTAGES OF LUBRICATION

Increased fatigue life

Correct lubricants will facilitate individual wire adjustment to equalise stress distribution under bending conditions. An improvement of up to 300% can be expected from a correctly lubricated rope compared with a similar unlubricated rope.

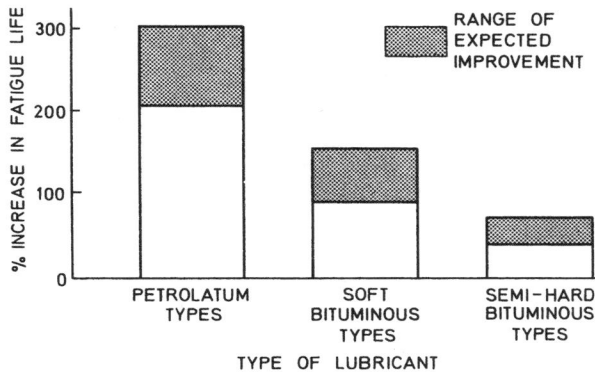

Figure 10.1 Percentage increases in fatigue life of lubricated rope over unlubricated rope

Increased corrosion resistance

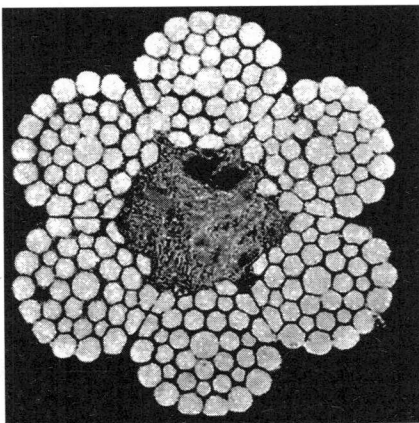

Figure 10.2 Typical effect of severe internal corrosion. Moisture has caused the breakdown of the fibre core and then attacked the wires at the strand/core interface

Figure 10.3 Typical severe corrosion pitting associated with 'wash off' of lubricant by mine water

Increased abrasion resistance

Figure 10.4 Typical abrasion condition which can be limited by the correct service dressing

LUBRICATION DURING MANUFACTURE

The Main Core Fibre cores should be given a suitable dressing during their manufacture. This is more effective than subsequent immersion of the completed core in heated grease.

Independent wire rope cores are lubricated in a similar way to the strands.

The Strands The helical form taken by the individual wires results in a series of spiral tubes in the finished strand. These tubes must be filled with lubricant if the product is to resist corrosive attack. The lubricant is always applied at the spinning point during the stranding operation.

The Rope A number of strands, from three to fifty, will form the final rope construction, again resulting in voids which must be filled with lubricant. The lubricant may be applied during manufacture at the point where the strands are closed to form the rope, or subsequently by immersion through a bath if a heavy surface thickness is required.

Dependent on the application the rope will perform, the lubricant chosen for the stranding and closing process will be either a petrolatum or bituminous based compound. For certain applications the manufacturer may use special techniques for applying the lubricant.

Irrespective of the lubrication carried out during rope manufacture, increased rope performance is closely associated with adequate and correct lubrication of the rope in service.

LUBRICATION OF WIRE ROPES IN SERVICE

	(1)	(2)	(3)	(4)	(5)
Operating conditions	Ropes working in industrial or marine environments	Ropes subject to heavy wear	Ropes working over sheaves where (1) and (2) are not critical	As (3) but for friction drive applications	Standing ropes not subject to bending
Predominant Cause of rope deterioration	Corrosion	Abrasion	Fatigue	Fatigue—corrosion	Corrosion
Typical applications	Cranes and derricks working on ships, on docksides, or in polluted atmospheres	Mine haulage, excavator draglines, scrapers and slushers	Cranes and grabs, jib suspension ropes, piling, percussion and drilling	Lift suspension, compensating and governor ropes, mine hoist ropes on friction winders	Pendant ropes for cranes and excavators. Guys for masts and chimneys
Dressing requirements	Good penetration to rope interior. Ability to displace moisture. Internal and external corrosion protection. Resistance to 'wash off'. Resistance to emulsification	Good antiwear properties. Good adhesion to rope. Resistance to removal by mechanical forces	Good penetration to rope interior. Good lubrication properties. Resistance to 'fling off'	Non slip property. Good penetration to rope interior. Ability to displace moisture. Internal and external corrosion protection	Good corrosion protection. Resistance to 'wash off'. Resistance to surface cracking
Type of lubricant	Usually a formulation containing solvent leaving a thick (0.1 mm) soft grease film	Usually a very viscous oil or soft grease containing MoS_2 or graphite. Tackiness additives can be of advantage	Usually a good general purpose lubricating oil of about SAE 30 viscosity	Usually a solvent-dispersed temporary corrosion preventative leaving a thin, semi-hard film	Usually a relatively thick, bituminous compound with solvent added to assist application
Application technique	Manual or mechanical	Manual or mechanical	Mechanical	Normally by hand	Normally by hand
*Frequency of applications**	Monthly	Weekly	10/20 cycles per day	Monthly	Six monthly/ 2 years

* The periods indicated are for the general case. The frequency of operation, the environmental conditions and the economics of service dressing will more correctly dictate the period required.

APPLICATION TECHNIQUES

Ideally the lubricant should be applied close to the point where the strands of the rope tend to open when passing over a sheave or drum.

The lubricant may be applied manually or mechanically.

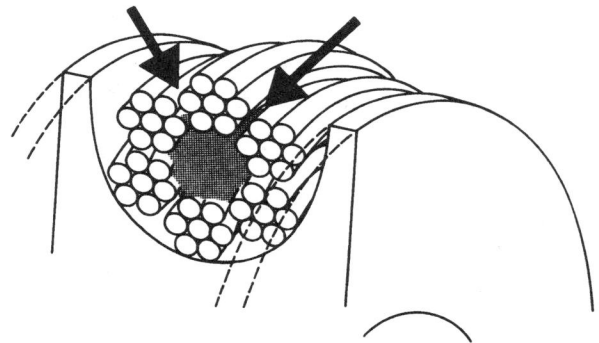

Figure 10.5 Opening of rope section during passage over sheave or drum. *Arrows indicate the access points for lubricant*

Manual – By can or by aerosol

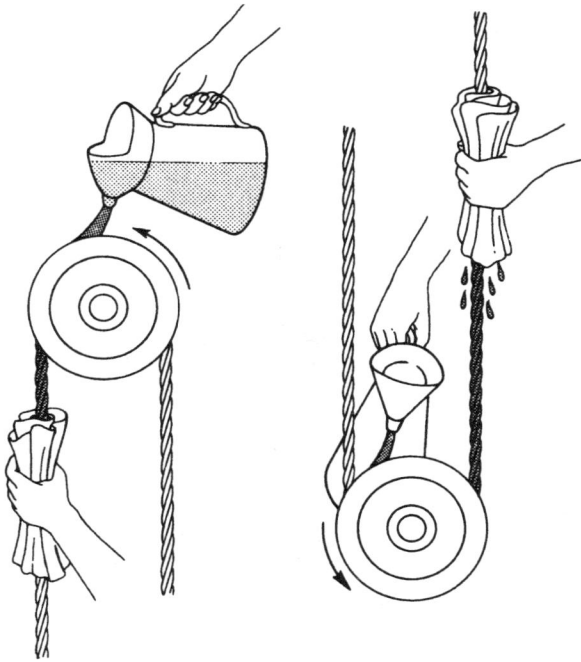

Figure 10.6 Manual application by can

Mechanical – By bath or trough. By drip feed. By mechanical spray

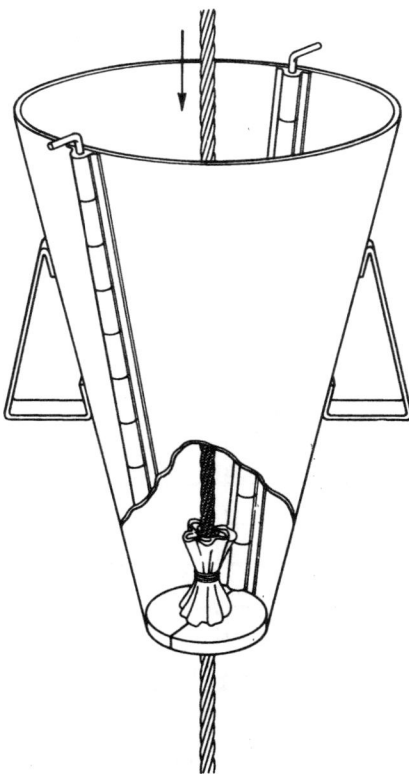

Figure 10.7 Mechanical application by trough

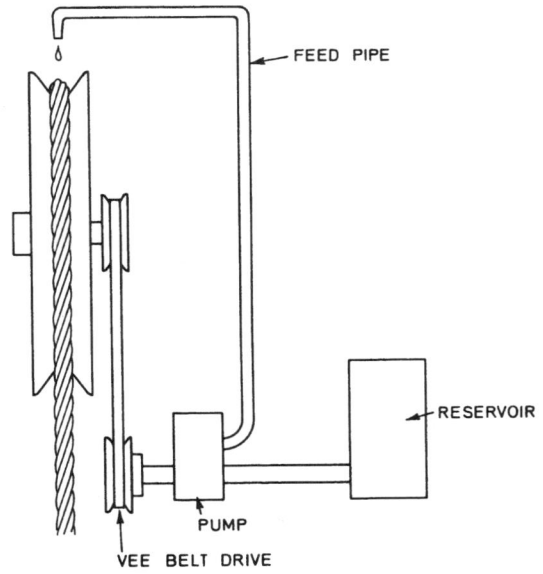

FEED PIPE

RESERVOIR

PUMP

VEE BELT DRIVE

Figure 10.8 Drip lubrication

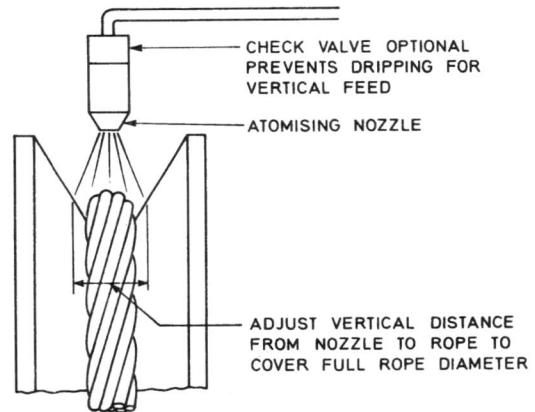

CHECK VALVE OPTIONAL PREVENTS DRIPPING FOR VERTICAL FEED

ATOMISING NOZZLE

ADJUST VERTICAL DISTANCE FROM NOZZLE TO ROPE TO COVER FULL ROPE DIAMETER

Figure 10.9 Sheave application by spray using fixed nozzle

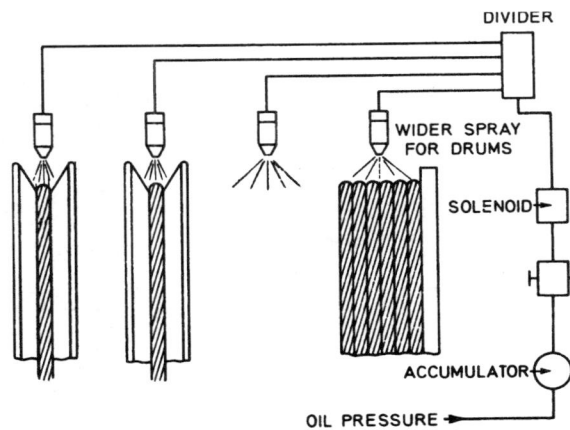

DIVIDER

WIDER SPRAY FOR DRUMS

SOLENOID

ACCUMULATOR

OIL PRESSURE

Figure 10.10 Multisheave or drum application by spray

FILLED COUPLINGS (GEAR, SPRING-TYPE, CHAIN)

Table 11.1 Recommendations for the lubrication of filled couplings

| Lubricant type | Limiting criteria | | | Lubricant change period | Remarks |
| | Centrifugal Effects | | Heat Dissipation | | |
	Pitch-line acceleration $d\omega^2/2$ (m/sec^2)	Range in practical units Dn^2 (ft/sec^2)	Pn		
No 1 Grease (mineral oil base)	0.15×10^3 $0.5\ \ \times 10^3$	25 max 25–80	– –	2 years 12 months	Soft grease preferred to ensure penetration of lubricant to gear teeth
No 3 Grease (mineral oil base)	$1.5\ \ \times 10^3$ $5.0\ \ \times 10^3$ 12.5×10^3	80–250 250–850 850–2000	– – –	9 months 6 months 3 months	Limitation is loss of oil causing hardening of grease; No 3 grease is more mechanically stable than No 1.
Semi-fluid polyglycol grease or mineral oil	$45.0\ \times 10^3$	3000–5000	230×10^3 max	2 years	Sealing of lubricant in coupling is main problem

d = pcd, m; D = pcd, ft; ω = rads/sec; n = rev/sec; P = hp transmitted

Limits

Grease lubrication, set by soap separation under centrifuging action. Semi-fluid grease lubrication, set by heat dissipation.

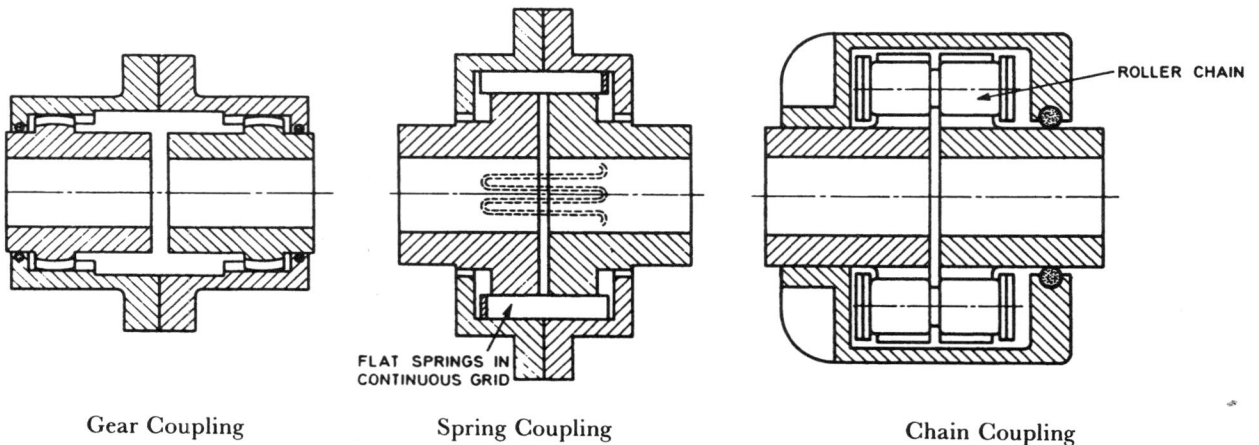

Gear Coupling   Spring Coupling   Chain Coupling

FLAT SPRINGS IN CONTINUOUS GRID

ROLLER CHAIN

Figure 11.1 Types of filled couplings

CONTINUOUSLY-LUBRICATED GEAR COUPLINGS

Lubrication depends on coupling type

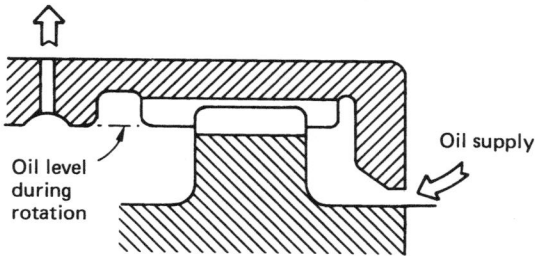

Figure 11.2 Dam-type coupling

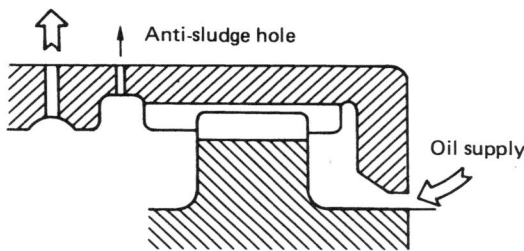

Figure 11.3 Dam-type coupling with anti-sludge holes

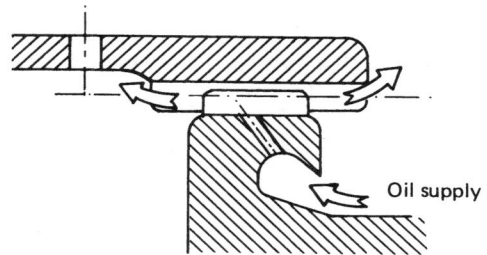

Figure 11.4 Damless-type coupling

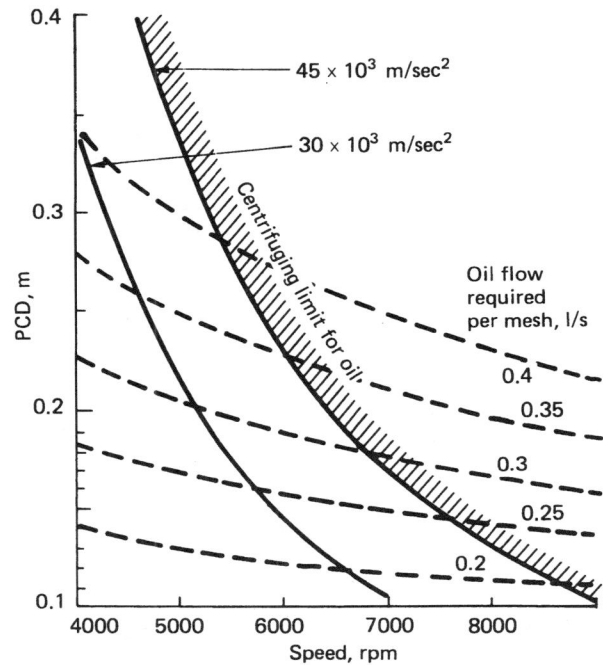

Figure 11.5 Lubrication requirements of gear couplings

Limits:

set by centrifuging of solids or sludge in oil causing coupling lock:

damless-type couplings	$45 \times 10^3 \, \text{m/sec}^2$
dam-type couplings	$30 \times 10^3 \, \text{m/sec}^2$

Lubricant feed rate:

damless-type couplings	Rate given on Figure 11.5
dam-type coupling with sludge holes	50% of rate on Figure 11.5
dam-type coupling without sludge holes	25% of rate on Figure 11.5

Lubricant:

Use oil from machine lubrication system (VG32, VG46 or VG68)

Slides are used where a linear motion is required between two components. An inherent feature of this linear motion is that parts of the working surfaces must be exposed during operation. The selection of methods of slide lubrication must therefore consider not only the supply and retention of lubricant, but also the protection of the working surfaces from dirt contamination.

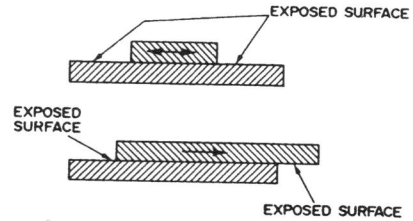

Figure 12.1 Slide movements expose the working surfaces to contamination

Table 12.1 The lubrication of slides in various applications

Application	Function of lubricant	Lubricant commonly used	Remarks
Slides and linear bearings on machine tools	To minimise the wear of precision surfaces. To avoid any tendency to stick slip motion	Depends on the type of slide or linear bearing. See next Table	Swarf must be prevented from getting between the sliding surfaces
Slides and linear bearings on packaging machines, textile machines, mechanical handling devices	To reduce friction and wear at the moving surfaces without contaminating the material being handled	Greases and solid lubricants are commonly used but air lubrication may also be possible	The sliding contact area should be protected from dirt by fitting scraper seals at each end if possible
Crossheads on reciprocating engines and compressors	To give low friction and wear by maintaining an adequate film thickness to carry the impact loads	The same oil as that used for the bearings	Oil grooves on the stationary surface are desirable to help to provide a full oil film to carry the peak load

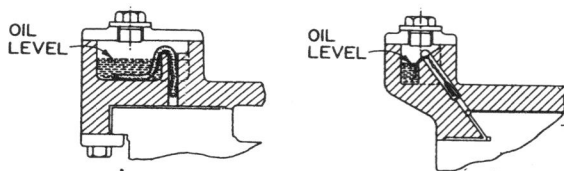

Figure 12.2 Typical wick lubricator arrangement on a machine tool

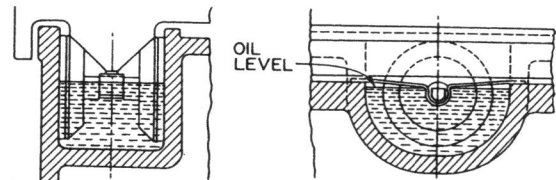

Figure 12.3 Typical roller lubricator arrangement on a machine tool

Table 12.2 The lubrication of various types of linear bearings on machine tools

Type of linear bearing	Suitable lubricant	Method of applying the lubricant	Remarks
Plain slide ways	Cutting oil	By splash from cutting area	Only suitable in machine tool applications using oil as a cutting fluid
	Mineral oil containing polar additives to reduce boundary friction	By wick feed to grooves in the shorter component	Requires scraper seals at the ends of the moving component to exclude swarf
		By rollers in contact with the bottom face of the upper slide member, and contained in oil filled pockets in the lower member	Only suitable for horizontal slides Requires scraper seals at the ends of the moving component to exclude swarf
		By oil mist	Air exhaust keeps the working surfaces clear of swarf
	Grease	By grease gun or cup, to grooves in the surface of the shorter component	Particularly suitable for vertical slides, with occasional slow movement
Hydro-static plain slideways	Air or any other conveniently available fluid	Under high pressure via control valves to shallow pockets in the surface of the shorter member	Gives very low friction and no stick slip combined with high location stiffness. Keeps working surfaces clear of swarf
Linear roller bearings	Oil	Lower race surface should be just covered in an oil bath	Not possible in all configurations. Must be protected from contamination
	Grease	Packed on assembly but with grease nipples for replenishment	Must be protected from contamination

To achieve efficient planning and scheduling of lubrication a great deal of time and effort can be saved by following a constructive routine. Three basic steps are required:

(a) A detailed and accurate survey of the plant to be lubricated including a consistent description of the various items, with the lubricant grade currently used or recommended, and the method of application and frequency

(b) A study of the information collected to attempt to rationalise the lubricant grades and methods of application

(c) Planning of a methodical system to apply lubrication

THE PLANT SURVEY

Plant identification

A clearly identifiable plant reference number should be fixed to the machinery. The number can incorporate a code of age, value and other facts which can later facilitate information retrieval.

A procedure to deal with newly commissioned or existing plant and a typical reference document is illustrated below, Figure 13.1.

Table 13.1 A convenient standardised code to describe the method of lubricant application

Main method	Code	Application
Representing:		
Oil can	← OC–C	Cup
	–H	Hole
	–B	Bottle (gravity) (wick feed) (drip feed) (syphon)
	–S	Surface: slideways hand oiled
	–A	Air mist
	–W	Well
Oil gun	← OG–N	Nipple
	–M	Multivalve
Oil filled	← OF–B	Bath. Sump. Ring oiler
	–L	Mechanical lubricator
	–S	Circulating system
	–H	Hydraulic
Grease gun	← GG–N	Nipple
	–M	Multivalve
	–S	Spring feed—grease cups
Grease hand	← GH–H	Hand applied
	–C	Cups—stauffer
Grease filled	← GF–S	System

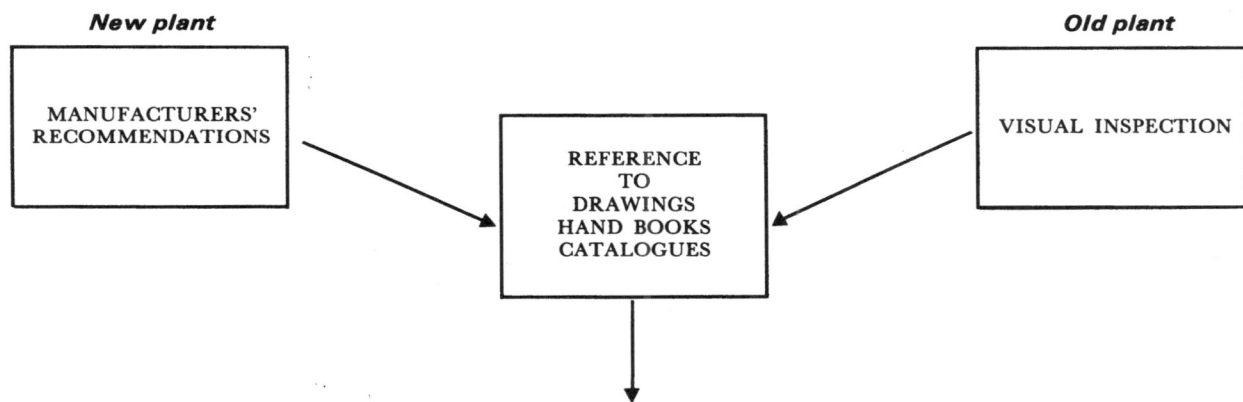

New plant — MANUFACTURERS' RECOMMENDATIONS

REFERENCE TO DRAWINGS HAND BOOKS CATALOGUES

Old plant — VISUAL INSPECTION

TITLE	*Machine description, type or model number and makers' name*		ITEM NO.	*Machine inventory number*
LOCATION	*Area 7*		DEPT. NO.	*XYZ*

PLANNED LUBRICATION MAINTENANCE INVENTORY SHEET

NO.	ITEM OF PLANT	METHOD TABLE 1	NO. OF POINTS	GRADE	QUANTITY	FREQUENCY
	(Details of part to be lubricated)					
1234	*Electric Motor Bearing*	GGN	2	Grease No. 3	2 shots	A
3256	*Spindle Bearings*	OGN	2	Oil No. 927	2 shots	D
3221	*Gear Box*	OFB	1	Oil No. 41	10 gallon sump	W

Figure 13.1

Method of application

In the case of new plant the proposed methods of lubrication should be subjected to careful scrutiny bearing in mind subsequent maintenance requirements. Manufacturers are sometimes preoccupied with capital costs when selling their equipment and so designed-out maintenance should be negotiated early on when the tribological conditions are studied. In this context it is possible to economise on the application costs of lubrication and problems of contamination and fire hazards can be forestalled. A standardised code for describing the method of application is given in Table 13.1. Confusion can arise unless a discipline is maintained both on surveying and scheduling.

Number of application points

The number of application points must be carefully noted.

(a) By adequate description – group together numbers of identical points wherever possible when individual point description serves no purpose. This simplifies the subsequent planning of daily work schedules.
(b) Highlighting of critical points by symbol or code identification as necessary.

Factors for lubricant selection

For the purpose of assessing the grade of lubricant, the following table suggests the engineering details required to determine the most suitable lubricant.

Table 13.2 Some factors affecting lubricant selection

Element	Type	Size	Material	Operating temperature	Operating conditions	Velocity	Remarks
Bearings	Plain, needle roller, ball	Shaft diameter				rev/min	
Chain drives	Links. Number and pitch	PCD of all wheels and distance between centres				Chain speed ft/min	
Cocks and valves	Plug, ball etc.		Fluid being controlled				Depends on properties of the fluid
Compressors	BHP, manufacturer's name			Gas temperature	Max gas pressure	rev/min	
Couplings	Universal or constant velocity					rev/min	
Cylinders		Bore, Stroke	Cylinder, piston, rings	Combustion and exhaust gas temperature	Combustion and exhaust gas pressure	Crank speed rev/min	
Gears	Spur, worm, helical, hyperbolic	BHP, distance between centres			Radiated heat and heat generated	rev/min	Method of lubricant application
Glands and seals	Stuffing box		Fluid being sealed				Depends on design
Hydraulic systems	BHP Pump type (gear, piston vane)		Hydraulic fluid materials 'O' rings and cups etc.				Lubricant type adjusted to loss rate
Linkages				Environmental heat conditions		Relative link speeds ft/s, angular vel. rad/s	
Ropes	Steel hawser	Diameter			Frequency of use and pollution etc.		
Slideways and guides						Surface relative speed ft/min	

LUBRICANT RATIONALISATION

Recommended grade of lubricant

Manufacturers recommended grades may have to be acceded to during the guarantee period for critical applications. However, a compromise must be reached in order to ensure the maintenance of an optimum list of grades which is essential to the economic sorting, handling and application of lubricants. In arriving at this rationalised list of grades, speeds, tolerances, wear of moving parts and seals create conditions where the viscosity and quality of lubricant required may vary. For a balanced and economic rationalisation, all tribological factors have therefore to be assessed. Where a special lubricant has to be retained, if economically viable it may form a compromise solution that will satisfy future development projects, particularly where demands are likely to be more critical than for existing equipment. Generally speaking, in most industries 98% of the bulk of lubrication can be met by six grades of oil and three greases.

A considerable range of lubricant grades exists largely blended to meet specific demands of manufacturers. Table 13.3 illustrates a typical selection. There are viscosity ranges, indices, inherent characteristics and additive improvers to be considered. Generally speaking, the more complex the grade, the more expensive, but often the more comprehensive its application. Advice is readily available from oil companies.

Quantity and frequency

In the main the quantity of lubricant applied is subjected to so many variable conditions that any general scale of recommendations would be misleading. 'Little and often' has an in-built safeguard for most applications (particularly new plant), but as this can be uneconomic in manpower, and certain items can be over-lubricated planning should be flexible to optimise on frequency and work loads. Utilisation of the machinery must also be allowed for.

Knowledge of the capacity and quantity required will naturally help when assessing the optimum frequency of application and a rough guide is given in Table 13.4.

Table 13.3 Range of lubricant grades commonly available showing factors to be taken into account for economic rationalisation

Table 13.4 Some factors affecting lubrication frequency (This is a general guide only – affected by local conditions and environment)

Frequency / Element	Shift	Daily	Weekly	Fortnightly	Monthly	Bi-annually	Biennially	Remarks
High-speed bearings	Oil top-up	Oil top-up	Oil top-up					High frequency where dia.* × rpm is greater than 6000. Where dia. × rpm is less than 6000 lower frequencies may be used unless extreme dirt conditions or temperatures above 60°C prevail
Low speed bearings			Apply grease with gun until slight resistance is felt	Grease if very slow speed or limited movement			Clean and repack, or at major overhaul	Weekly and fortnightly lubrication for relatively high speeds (dia.* × rpm) greater than 3000 but less than 6000
Chain drives		Clean off and renew lubricant						Use high frequency with very dirty conditions
Cocks and valves			When used frequently		When used less than ten times per day			Where usage less than once per day lubricate just prior to use
Compressors cylinders and sumps		Top-up if required	Oil change after 250h if sealing poor		Oil change after 500 h if sealing good			Change more frequently if very adverse conditions prevail
Couplings				Grease or top-up depending on sealing				Grease nipples 2–3 shots until resistance felt
Gears—open		Clean off lubricant and apply new		Check and top-up if necessary				Depending upon environment
Gears—enclosed				Check and top-up if necessary	Change oil			Depending upon operation conditions, temperature etc.
Glands and seals		For soft packed glands						Especially those handling fluids which re-act with the lubricant give one or two shots of grease
Hydraulic systems		Top-up if required				Change oil depending on operating conditions, temperatures etc.		Hydraulic fluid may be changed more frequently if the colour indicates contamination from dissolving seals etc.
Ropes			Clean and apply new lubricants					Previous experience will determine variations depending upon dirt and usage
Slideways, guides and linkages	Apply lubricants				Guides, lifts and hoists only			More or less frequently depending on conditions of dirt, swarf and usage

* Dia. in inches

For brevity and convenience the vast array of lubrication systems have been grouped under nine headings. These are each more fully discussed in other Sections of the Handbook.

Type	Title
A	Total loss grease
B	Hand greasing
C	Centralised greasing
D	Total loss oil
E	Wick/pad oil
F	Ring and disc oil
G	General mist and splash
H	Pressure mist
J	Circulating oil

TYPES OF LUBRICATION SYSTEM

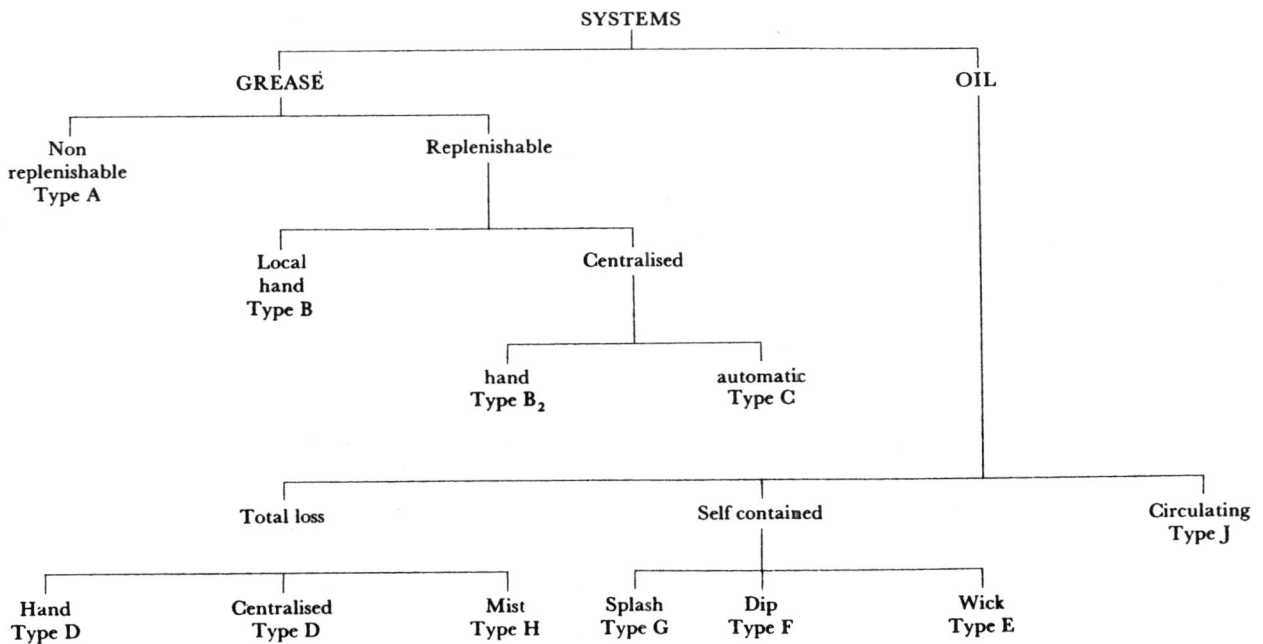

Type letters refer to subsequent tables

METHODS OF SELECTION

Table 14.1 Oil systems

	Type	Characteristics	Application
T O T A L L O S S	Hand Type D	Oil can	Simple bearings, Small numbers, low duty, easy access
	Centralised Type D	Intermittent feed	Small heat removal, difficult access, large numbers
	Mist Type H	Aerosol spray.	Rolling bearings, mechanisms
S E L F C O N T A I N E D	Splash Type G	Lubrication by mist	Enclosed mechanisms, gearboxes
	Dip Type F	Ring or disc systems. Oil circulates	Plain bearings— slow or moderate duty
	Wick Type E	Pad or wick feed from reservoir— total loss or limited circulation	Plain bearings, low duty. Light loaded mechanisms
	Circulating Type J	Oil from tank or sump fed by pressure pump to bearings or sprays	Almost all applications where cost is justified

Table 14.2 Grease systems

Type	Characteristics	Applications
Non replenishable Type A	Bearings packed on assembly. Refilling impossible without stripping	Rolling bearings, some plastic bushings
Local—hand Type B₁	Grease nipple to each bearing	Small numbers, easy, access, cheap
Centralised— hand Type B₂	Feed pipes brought to manifold or pump	Reasonable numbers. Inaccessible bearings
Centralised— automatic Type C	Grease pump feed to bearing and sets of bearings from automatic pump	Large numbers, important bearings, great distances. Where frequent relubrication is required

Table 14.3 Relative merits of grease and oil systems

Component or performance requirement	Grease	Oil
Removing heat	No	Yes. Amount depends on system
Keep out dirt	Yes. Forms own seal	No. But total loss systems can flush
Operate in any attitude	Yes. Check design of grease valves	No. Unless specially designed
Plain bearings	Limited (slow speed or light load)	Extensive
Rolling bearings	Extensive	Extensive
Sealed rolling bearings	Almost universal	No
Very low temperatures	No. Except for some special greases	Yes (check pour point of oil)
Very high temperatures	No. Except for some special greases	Yes. With correct system
Very high speeds	No. Except small rolling bearings	Yes
Very low or intermittent speeds	Yes	Hydrodynamic bearings— limited. Hydrostatic bearings—yes
Rubbing plain bearings	Yes	Yes. With limited feed

Table 14.4 Selection by heat removal

System	Effect
Total loss grease or oil	Will not remove heat
Dip or splash	Will remove heat from hot spots. Will remove some heat from the body of the components
Mist	Will remove considerable heat
Recirculating. Oil	May be designed to remove almost any amount of heat

Table 14.5 Selection by type of component to be lubricated

System		Rolling bearings	Fluid film plain bearings	Rubbing plain bearings Gears
Non-replenishable grease Type A	For general use	Very light duty	General	Light duty or slow speeds
Hand grease feed Type B	Heavier loads. Higher temperatures	Slow speed	General for high loads and temperatures	Rarely
Centralised grease feed Type C	Heaviest loads. Higher temperatures. Large sizes	Slow speed. Higher temperatures	General for high loads and temperatures	Slow, heavy duty
Total loss oil Type D	Most applications	All where small heat removal	Yes	Open gears
Wick/Pad, oil Type E	Limited (light duty)	Slow speed	Some	Small gears
Ring or disc, oil Type F	NA	Medium duty	NA	NA
General mist and splash Type G	Most	Light duty	Light duty	Wide
Pressure mist Type H	Almost all. Excellent	Rare	NA	Rare
Circulating oil Type J	All	All	NA	All

Table 14.6 Selection by economic considerations

Lubrication system	Initial purchase	Maintenance of system	First fill	Subsequent lubricant costs	Subsequent labour costs	Notes
Non-replenishable grease Type A	Very cheap	Nil	Cheap	Nil	Nil	Life of bearing is life of lubricant. Expensive if relubricated
Grease feed hand Type B	Cheap	Cheap	Cheap moderate if long liner	Moderate but can be expensive	Expensive	Regular attendance is vital. Neglect can be very expensive
Grease feed automatic Type C	Moderate to expensive	Moderate	Moderate to expensive	Moderate	Moderate	Needs comparatively skilled labour. Costs increase with complications
Total loss oil Type D	Cheap	Cheap	Cheap	Moderate but can be expensive	Moderate	Periodic refilling required. Neglect can be very expensive
Wick or pad Type E	Cheap	Cheap	Cheap	Cheap	Cheap	Also need topping-up but not so often and wick gives some insurance
Ring or disc Type F	Cheap	Nil to cheap	Cheap	Cheap	Cheap	Needs very little attention
General splash Type G	Cheap	Nil	Moderate	Cheap to moderate	Moderate	Oil level must be watched
Pressure mist Type H	Moderate to expensive	Expensive	Cheap	Cheap	Moderate to expensive	Needs comparatively skilled labour. Requires compressed air supply
Circulating oil Type J	Expensive	Expensive	Moderate to expensive	Cheap	Moderate	Simple system requires little attention. Costs increase with complications

Table 14.7 General selection by component. Operating conditions and environment

Type of component	High temp. (over 150°C)	Normal temp. (−10 to 120°C)	Low temp. (below −20°C)	High speed	Medium speed	Low speed	Dust and dirt	Wet and humid	Vacuum (special lubricants)
Rolling bearings	A (special grease), B, C, H*, J*	A, B, C, D, G, H*, J*	A*, D, G, H*, J	A (small), D, H*, J	A, B, C, D, E, H*, J*	A, B, C*, D, E, G, H, J	A, B, C*, D, J	A, B, C*, J	A*, E — Also dry
Fluid film, plain bearings	D, H, J*	A (slow), B, C, D, E, F, G, J	A (slow), D, E, F, G, J	D, G (small sizes), H (small), J	D, E, F, G, H, J*	A (light), B, C, D, E, F, G, J	A, B, C*, D, J	A, B, C, J*	A*, E, J (possible)
Rubbing plain bearings	B, C*, J*	A, B, C*, D, E, F	A, D, G	—	—	A, B, C*, D, E, G — Also dry	A, B, C*, D — Also dry*	A, B, C* — Also dry	A* — Also dry
Slides	C*, G, J*	A, B*, C, D, E, J	A, D, E, G, J	A (light), B, C, D, G, J	A, B, C, D, E (small), F, G, J*	A, B, C*, D, E, J	A, B, C*, D	A, B, C*, D, J	A*, E
Screws	B, G, J*, C	A, B*, C, D, E, G, J	A, D, G, J	A (light), D, G, H, J	A (light), B, C, D, G, H, J	A, B, C*, D, G, J	A, B, C*	A, B, C*, D	E, A*, J, E
Gears	B, C, G, J*	A, B, C, E, H, J*	A, D, G, H	D, G, H, J*, J	A−V (small), C, D, G, H, J	A, B, C, D, H, J	A (special greases), E, G, J, J	A, B, C, D	A*, J

* Preferred systems.

Selection of gear lubrication systems

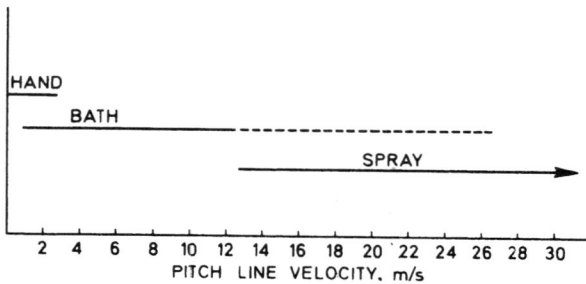

TYPES OF TOTAL LOSS GREASE SYSTEMS AVAILABLE

Description	Diagram	Operation	Drive	Suitable grease NLGI No.	Typical line pressures	Adjustability and typical limiting pipe lengths*
DIRECT FEED Individual piston pumps		Rotating cam or swash plate operates each piston pump in turn	Motor Machine Manual	0–2	700–2000 kN/m² (100–300 lbf/in²)	By adjustment of stroke at each outlet 9–15 m (30–50 ft)
Distributing piston valve system		Valve feeds output of a single piston pump to each line in turn	Motor Machine Manual	0–3 0–2	700–2000 kN/m² (100–300 lbf/in²)	None. Output governed by speed of pump 25–60 m (80–200 ft)
Branched system		Outputs of individual pumps split by distributing valves	Motor Machine	0–3	700–2800 kN/m² (100–400 lbf/in²)	Adjustment at each outlet/meter block 18–54 m (60–180 ft) ⎫ pump to ⎰ divider 6–9 m (20–30 ft) ⎫ divider ⎰ to bearing
INDIRECT FEED PROGRESSIVE Single line reversing		Valves work in turn after each operation of reversing valve R				
Single line		First valve block discharges in the order 1, 2, 3, . . ., etc. One unit of this block is used as a master to set the second block in operation. Second and subsequent blocks operate sequentially	Motor Machine Manual	0–2	14 000–20 000 kN/m² (2000–3000 lbf/in²)	Normally none. Different capacity meter valves available—otherwise adjustment by time cycle Main lines up to 150 m (500 ft) depending on grease and pipe size. Feeder lines to bearings 6–9 m (20–30 ft)
Double line		Grease passes through one line and operates half the total number of outlets in sequence. Valve R then operates, relieves line pressure and directs grease to the other line, operating remaining outlets				

Description	Diagram	Operation	Drive	Suitable grease NLGI No.	Typical line pressures	Adjustability and typical limiting pipe lengths*
INDIRECT FEED PARALLEL Single line		The line is alternately pressurised and relieved by a device on the pump. Two variations of this system exist in which 1 injection is made by the line pressure acting on a piston within the valves 2 injection is made by spring pressure acting on a piston within the valves	Motor Manual	0–1	Up to 17 000 kN/m² (2500 lbf/in²) Up to 8000 kN/m² (1200 lbf/in²)	Operation frequency adjustable (some makes). Output depends on nature of grease 120 m (400 ft)
Single line oil or air actuated	OIL OR AIR SUPPLY	Pump charges line and valves. Oil or air pressure operates valves	Motor with cycle timer	0–3	Up to 40 000 kN/m² (6000 lbf/in²)	Full adjustment at meter valves and by time cycle 600 m (2000 ft)
Double line		Grease pressure in one line operates half the total number of outlets simultaneously. Valve R then relieves line pressure and directs grease to the other line, operating the remaining outlets	Motor Manual	0–2	Up to 40 000 kN/m² (6000 lbf/in²)	As above. 60 m (200 ft) } Manual 120 m (400 ft) } Automatic

* Lengths shown may be exceeded in certain specific circumstances.

Considerations in selecting type of system

	Required performance of system	
Type of bearing	Grease supply requirements	Likely system requirements
Plain bearings, high loads	Near-continuous supply	Direct feed system
Rolling bearings	Intermittent supply	Indirect feed

PIPE-FLOW CALCULATIONS

To attempt these it is necessary that the user should know:

(a) The relationship between the apparent viscosity (or shear stress) and the rate of shear, at the working temperature;

(b) The density of the grease at the working temperature.

This information can usually be obtained, for potentially suitable greases, from the lubricant supplier in graphical form as below (logarithmic scales are generally used).

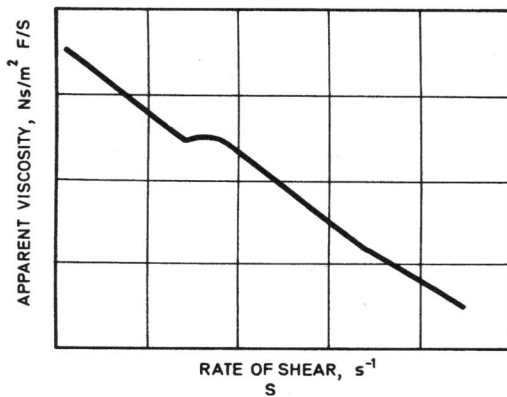

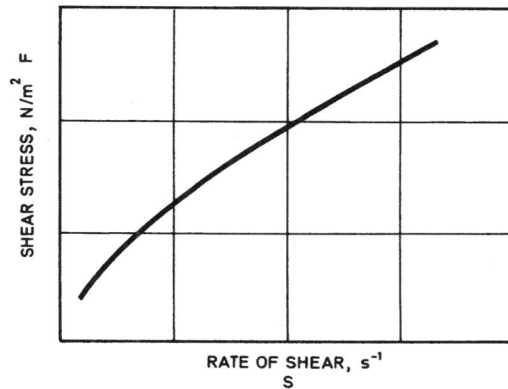

Given	To estimate	Procedure
Permissible pressure loss P/L, over a known length of pipe	Pipe inside diameter D, to give a certain rate of flow Q	(a) Calculate the value of $2.167 \sqrt[3]{Q\,P/L}$ (b) Plot graph of $(\sqrt[3]{S})\,F$ against F from supplier's data (if convenient on log–log paper) (c) Find point on $(\sqrt[3]{S})\,F$ scale equivalent to the value found in (a) above (d) Note value of F (e) Pipe diameter $D = 4LF/P$
Available pumping pressure P, length L and inside diameter D of pipe	Rate of grease flow Q	(f) Plot curve F against S from manufacturers' data (g) Calculate $F = PD/4L$ from the given P, D and L conditions (h) Find corresponding value of S from curve (i) Then $Q = \pi D^3 S/32$
Rate of flow Q and inside diameter D of pipe to be used	Pressure gradient P/L to sustain flow rate Q	(j) Calculate $S = \dfrac{32Q}{\pi D^3}$ for given conditions (k) Find corresponding value of S from the graph of F against S (l) Then $P/L = 4F/D$

Notes:

1 These formulae are correct for SI units, P in N/m^2, L in m, Q in m^3/s. F in N/m^2, S in S^{-1}.

2 D is inside diameter.

Typical pipe sizes used in grease systems

Material	Standard No.	Bore mm	Bore in	Wall thickness mm	Wall thickness in	Outside diameter mm	Outside diameter in
STEEL		**15***	$\frac{1}{2}$	3.25	0.128	21.5	0.756
		20	$\frac{3}{4}$	3.25	0.128	26.5	1.006
Heavyweight grade	ISO 65	**25**	**1**	4.05	0.160	33.1	1.320
		40	**1**	4.05	0.160	48.1	1.820
		50	**2**	4.50	0.176	59.0	2.350
Cold drawn, seamless, fully annealed	ISO 3304	3.35	0.132	0.70	0.028	4.80	$\frac{3}{16}$
		4.50	0.178	0.90	0.036	6.40	$\frac{1}{4}$
		6.10	0.240	0.90	0.036	7.90	$\frac{5}{16}$
		7.10	0.279	1.22	0.048	9.50	$\frac{3}{8}$
		10.30	0.404	1.22	0.048	12.70	$\frac{1}{2}$
COPPER	ISO 196	3.35	0.132	0.70	0.028	4.80	$\frac{3}{16}$
		4.50	0.178	0.70	0.028	6.40	$\frac{1}{4}$
As drawn (M) quality		6.10	0.240	0.90	0.036	7.90	$\frac{5}{16}$
Annealed (O) quality		7.10	0.303	0.90	0.036	9.50	$\frac{3}{8}$
	ISO 274	2.80	0.110	0.60	0.024	**4**	0.158
		3.40	0.130	0.80	0.032	**5**	0.197
		4.00	0.160	1.00	0.039	**6**	0.236
		6.00	0.240	1.00	0.039	**8**	0.315
		7.60	0.300	1.20	0.047	**10**	0.394
BRASS	ISO 196	3.35	0.132	0.70	0.028	4.8	$\frac{3}{16}$
		4.50	0.178	0.90	0.036	6.4	$\frac{1}{4}$
Half-hard or normalised		6.10	0.240	0.90	0.036	7.9	$\frac{5}{16}$
		7.10	0.279	1.22	0.048	9.5	$\frac{3}{8}$
NYLON	ISO 7628	3.10	0.122	0.80	0.033	4.8	$\frac{3}{16}$
		4.10	0.165	1.10	0.043	6.4	$\frac{1}{4}$
		5.40	0.213	1.30	0.050	7.9	$\frac{5}{16}$
		6.20	0.245	1.80	0.070	9.5	$\frac{3}{8}$
		2.40	0.090	0.80	0.032	**4**	0.158
		3.30	0.130	0.85	0.034	**5**	0.197
		4.00	0.160	1.00	0.039	**6**	0.236
		5.40	0.210	1.30	0.051	**8**	0.315
		6.50	0.260	1.75	0.069	**10**	0.394

* Figures in bold type indicate nominal bore or outside diameters used in ordering.

Typical data for flexible hoses used in grease systems

Type		Sizes mm	Sizes in	Typical working pressures at 20°C kN/m²	Typical working pressures at 20°C lbf/in²	Sizes mm	Sizes in	Typical working pressures at 20°C kN/m²	Typical working pressures at 20°C lbf/in²
SYNTHETIC RUBBER Reinforced with cotton, terylene or wire braid or spiral wrapped	Single-wire braid			23 000	3300			3500	500
	Multi-spiral wrapped	6	$\frac{1}{4}$ Nominal bore	70 000	10 000	50	2 Nominal bore	27 000	4000
PLASTICS Simple extruded for low pressure to braided high pressure	High pressure single braid double braid			24 500 / 28 000	3500 / 4000	12.5	$\frac{1}{2}$ Nominal bore	17 500 / 35 000	2500 / 5000
	Unreinforced	4	$\frac{1}{8}$ o.d.	3500	500	28	$1\frac{1}{8}$ o.d.	1750	250
POLYMER AND METAL Inner core of polymeric material. Stainless steel outer braid		3	$\frac{1}{8}$ Nominal bore	7000	1000				

CONSIDERATIONS IN STORING, PUMPING AND TRANSMITTING GREASE AND GENERAL DESIGN OF SYSTEMS

	Considerations
Storing grease	TEMPERATURE at which grease can be pumped
	CLEANLINESS of grease
	PACKAGE sizes available. Ease of handling
	RESERVOIRS: Exclusion of dirt; internal coating; positive pump prime; provision of strainer, follower plate and scraper; pressure relief; level indicator; emptying and cleaning access; connections for remote feed
Pumping grease	ACCESSIBILITY of pump for maintenance
	CORRECT ALIGNMENT of pump and motor
	COUPLING between pump and motor to be adequate size for horse power transmitted
	MOUNTING to be rigid and secure
	ACTUATING ROD (if any) to be free from deflection
	NAME PLATE on pump to be clearly visible and kept clean
	DIRECTION of rotation to be marked
Transmitting grease	PIPE SIZING: Pressure in lines; length of runs; volume of lubricants; number of bends; temperature fluctuation
Work conditions	INGRESS of water, scale, dust
	POSSIBLE DAMAGE to piping
	ACCESS to piping in onerous locations
Human factors	AVAILABILITY of labour
	RELIABILITY of labour
	Trade union practices
General design points	CORRECT positioning of lubrication points
	PRODUCTION requirements
	QUALITY OF PIPING and materials— use only the best
	AVOID splitting unmetered grease lines
	ENSURE that grease holes and grooves communicate directly with faces to be lubricated

GENERAL

Most total loss systems available from manufacturers are now designed to deliver lubricants ranging from light oils to fluid greases of NLGI 000 consistency.

Fluid grease contains approximately 95% oil and has the advantage of being retained in the bearing longer than oil, thus reducing the quantity required whilst continuing to operate satisfactorily in most types of system.

The main applications for total loss systems are for chassis bearings on commercial vehicles, machine tools, textile machinery and packaging plant.

Because of the small quantity of lubricant delivered by these systems, they are not suitable for use where cooling in addition to lubrication is required, e.g. large gear drives.

Fluid grease is rapidly growing in popularity except in the machine tool industry where oil is preferred.

All automatic systems are controlled by electronic or electric adjustable timers, with the more sophisticated products having the facility to operate from cumulated impulses from the parent machine.

Individual lubricant supply to each bearing is fixed and adjustment is effected by changing the injector unit. However, overall lubrication from the system is adjusted by varying the interval time between pump cycles.

Multi-outlet – electric or pneumatic

Operation: An electric or pneumatic motor drives cam-operated pumping units positioned radially on the base of the pump. The pump is cycled by an adjustable electronic timer or by electrical impulses from the parent machine, e.g. brake light operations on a commercial vehicle.

Individual 4 mm OD nylon tubes deliver lubricant to each bearing.

Applications: Commercial vehicles, packaging machines and conveyors.

Specification:
Outlets: 1–60 (0.01–1.00 ml).
Pressure: To $10 \, \text{MN/m}^2$.
Lubricants: 60 cSt oil to NLGI 000 grease (NLGI 2 pneumatic).
Failure warning: Pump operation by light or visual movement.
Cost factor: Low (electric), Medium (pneumatic).

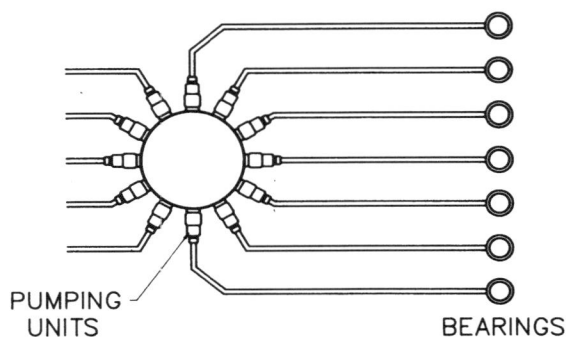

PUMPING UNITS   BEARINGS

Figure 16.1 Schematic

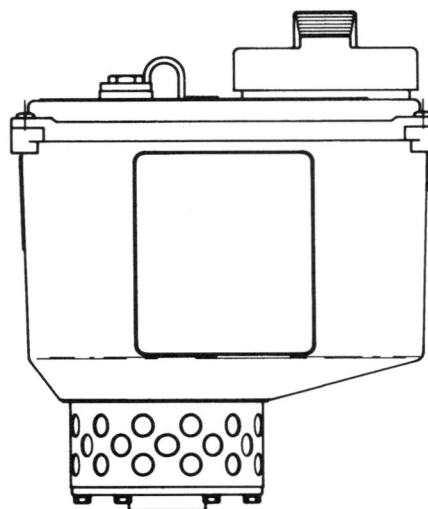

Figure 16.2 Pump

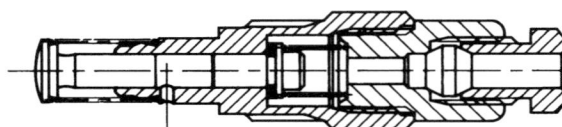

Figure 16.3 Pumping unit

Single line – volumetric injection

Operation: The pump delivers lubricant under pressure to a single line main at timed intervals. When the pressure reaches a predetermined level, each injector or positive displacement unit delivers a fixed volume of lubricant to its bearing through a tailpipe.

When full line pressure has been reached the pump stops and line pressure is reduced to a level at which the injectors recharge with lubricant ready for the next cycle.

Pumps are generally electric gear pumps or pneumatic piston type. All automatic systems are controlled by adjustable electronic timers but hand operated pumps are available.

Main lubricant pipework is normally in 6 or 12 mm sizes and tailpipes in 4 or 6 mm depending on the size of system.

Applications: All types of light to medium sized manufacturing plant and commercial vehicles.

Specification:
Outlets: 1–500 (0.005–1.5 ml).
Pressure: $2–5\,MN/m^2$.
Lubricants: 20 cSt oil to NLGI 000 grease.
Failure warning: Main line pressure monitoring.
Cost factor: Medium.

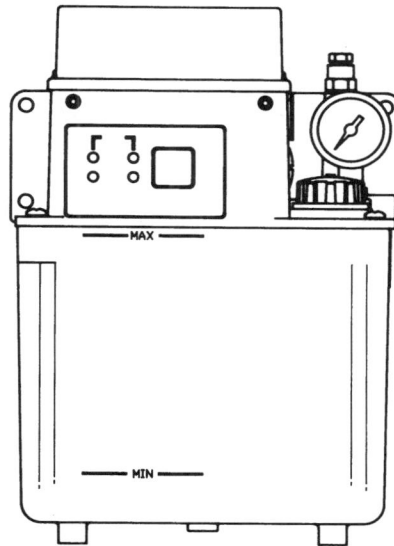

Figure 16.5

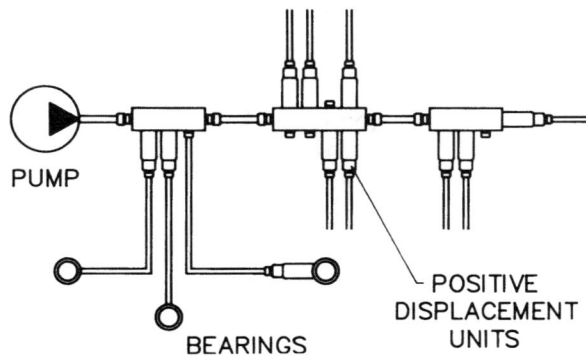

Figure 16.4

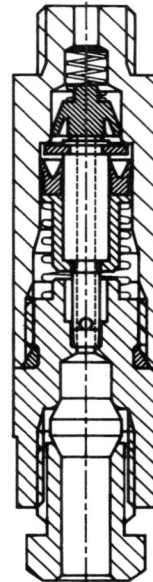

Figure 16.6 Positive displacement unit

Single line-resistance (oil only)

Operation: A motor driven piston pump discharges a predetermined volume of oil at controlled intervals to a single main line. Flow units in the system proportion the total pump discharge according to the relative resistance of the units.

Care needs to be taken in the selection of components to ensure each bearing receives the required volume of lubricant; as a result, this type of system is used predominantly for original equipment application.

Applications: Machine tools and textile machinery.

Specification:
Outlets: 1–100 (0.01–1 ml).
Pressure: 300–500 kN/m^2.
Lubricants: 10–1800 cSt (Oil only).
Failure warning: line pressure monitoring.
Cost factor: Low.

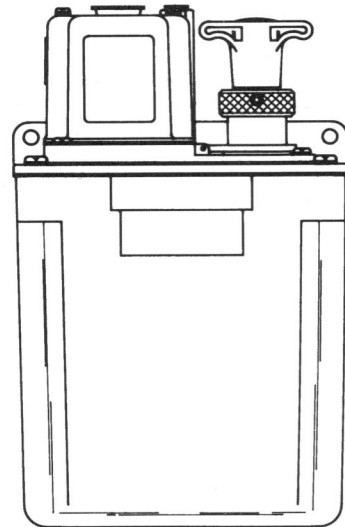

Figure 16.8 Pump

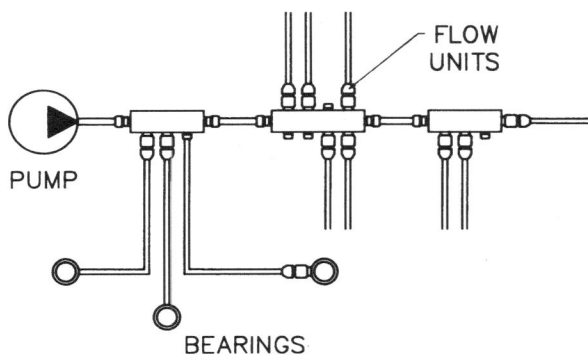

Figure 16.7 Schematic

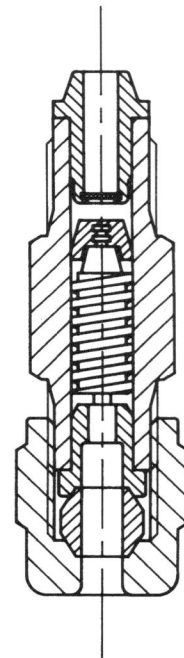

Figure 16.9 Flow unit

A16.3

Single line progressive

Operation: An electric or pneumatic piston pump delivers lubricant on a timed or continuous basis to a series of divider manifolds. The system is designed in such a way that if a divider outlet fails to operate, the cycle will not be completed and a warning device will be activated. Due to this feature, the system is widely used on large transfer machines in the automobile industry.

Applications: Machine tools and commercial vehicles.

Specification:
Outlets: 200 (0.01–2 ml).
Pressure: $10\,\text{MN/m}^2$.
Lubricants: 20 cSt oil to NLGI 2 grease.
Failure warning: Failure of individual injector can activate system alarm.
Cost factor: Medium to high.

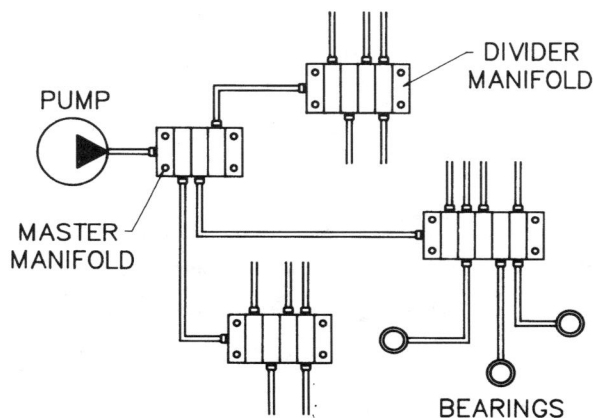

Figure 16.10 Schematic

CHECK LIST FOR SYSTEM SELECTION AND APPLICATION

Most economic method of operation – electric, pneumatic, manual, etc.

Cost installed – Max 2% of parent machine.

Degree of failure warning required – warning devices range from low lubricant level to individual bearing monitoring. Cost of warning devices must be balanced against cost of machine breakdown.

Check pressure drop in main lines of single line systems. Large systems may require larger pipe sizes.

Ensure adequate filtration in pump – Some types of systems are more sensitive to dirt in the lubricant.

Is the system protected against the operating environment? e.g.: High/low temperature, humidity, pressure steam cleaning, vibration/physical damage, electrical interference etc.

If using flexible tubing, is it ultra violet stabilised?

If air accidentally enters the system, is it automatically purged without affecting the performance?

Check for overlubrication – particularly in printing and packaging applications.

Is the lubricant suitable for use in the system? (Additives or separation.)

Reservoir capacity adequate? Minimum one month for machinery, three months for commercial vehicles.

Accessibility of indicators, pressure gauges, oil filler caps, etc.

Should system be programmed for a prelube cycle on machine start up?

Mist systems, generically known as aerosol systems, employ a generator supplied with filtered compressed air from the normal shop air main, to produce a mist of finely divided oil particles having little tendency to wet a surface. The actual air pressure applied to the inlet of the generator is controlled and adjusted to provide the desired oil output.

The mist must be transmitted at a low velocity below 6 m/s and a low pressure usually between 25 and 50 mbar gauge through steel, copper or plastic tubes. The tubes must be smooth and scrupulously clean internally.

At the lubrication point the mist is throttled to atmospheric pressure through a special nozzle whose orifice size controls the total amount of lubricant applied and raises the mist velocity to a figure in excess of 40 m/s. This causes the lubricant to wet the rubbing surfaces and the air is permitted to escape to atmosphere. Empirical formulae using an arbitrary unit – the 'Lubrication Unit', are used to assess the lubricant requirements of the machine, the total compressed air supply required and the size of tubing needed.

DESIGN

The essential parameters of components are indicated in Table 17.2 and the load factors for bearings are given in Table 17.1.

Table 17.1 Load factors

Type of bearing	No preload	Preloaded	W/bd values					
			< 0.7	> 0.7 < 1.5	> 1.5 < 3.0	> 3.0 < 3.5	> 3.5	
			Load factor F					
Ball	1	2						Consult aerosol equipment suppliers
Hollow roller	3	3						
Needle roller	1	3						
Straight roller	1	3						
Spherical roller	2	2						
Taper roller	1	3						
Plain journal (see Notes)			1	2	4	8		

Notes: for explanation $\dfrac{W}{bd}$ see Table 17.2. For plain journal bearings with high end losses or made of white metal, double these load factors.

Table 17.2 Information required for the calculation of lubricant flow rates

Essential parameters	Units	Formulae symbols	Ball and roller bearing	Ball nuts	Plain (journal) bearing	Gear pair	Worm and gear	Gear train	Rack and pinion	Cam and follower	Moving slide	Roller chain drive	Inverted tooth chain	Conveyor chain
Contact length of bearing surface	mm	b			X						X			
Shaft diameter	mm	d	X		X									
Load factor (see Table 17.1)		F	X		X									
Load on bearing	N	W			X									
Length of chain	m	l												X
Maximum diameter	mm	D								X				
Number of rows of balls or rollers or chain strands		N	X	X								X		
Pitch circle diameters —Gears—Worm— Sprocket or nut	mm	P		X		X	X	X	X			X	X	X
Chain link pitch	mm	C										X		
Speed of rotation	rev/s	n										X	X	
Contact width of slide or gear or chain width	mm	w				X	X	X	X	X	X		X	X

Note: 'X' denotes information that is required for each type of component

The Lubrication Unit (LU) rating of each component should be calculated from the formulae in Table 17.3, using the values in Tables 17.1 and 17.2.

Table 17.3 Lubrication unit rating

Type of component	Formulae for lubrication unit rating
Rolling bearings	$4(d.F.N).10^{-2}$
Ball nuts	$4P[(N-1)+10].10^{-3}$
Journal bearings	$2(b.d.F).10^{-4}$
Gears (see Note)	
Gear pair	$4w(P_1+P_2).10^{-4}$
Gear train	$4w(P_1+P_2+P_3\dots+P_n).10^{-4}$
Worm and gear	$4[(P_M.w_M)+(P_w.w_w)].10^{-4}$
Rack and pinion	$12(P.w).10^{-4}$
Cams	$2(D.w).10^{-4}$
Slides and ways	$8(b.w).10^{-5}$
Chains (see Note)	
Roller chain drive	$(N.P.C)(n)^{\frac{1}{2}}.10^{-4}$
Inverted tooth chain	$5(P.w)(n)^{\frac{1}{2}}.10^{-5}$
Conveyor chains	$5w(25l+P).10^{-4}$

Notes: if gears are reversing double the calculated LU rating. If the pitch diameter of any gear exceeds twice that of a mating gear, i.e. $P_1 > 2P_2$, consider $P_1 = 2P_2$. For chain drives if $n < 3$ use $n = 3$.

Total the LU ratings of all the components to obtain the total Lubrication Unit Loading (LUL). This is used later for estimating the oil consumption and as a guide for setting the aerosol generators.

Distribution piping

When actual nozzle sizes have been decided, the actual nozzle loadings (measured in Lubrication Units) can be totalled for each section of the pipework, and this determines the size of pipe required for that section. The actual relationship is given in Table 17.4 and Figures 17.2. Where calculated size falls between two standard sizes use the larger size. Machined channels of appropriate cross-sectional area may also be used as distribution manifolds.

Nozzle sizes

Select standard nozzle fitting or suitable drilled orifice size from Figure 17.1 for each component using its calculated LU rating. Where calculated LU rating falls between two standard fitting or drill sizes, use the larger size fitting or drill. Multiple drillings may be used to produce nozzles with ratings above 20 LU.

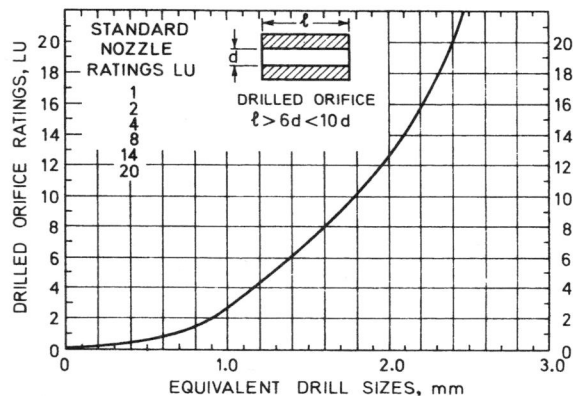

Figure 17.1 Drill size and orifice ratings

Maximum component dimensions (Table 17.1) for a single nozzle.

$b = 150\,mm$

$w = 150\,mm$ for slides, 12 mm for chains, 50 mm for other components.

Where these dimensions are exceeded and for gear trains or reversing gears use nozzles of lower LU rating appropriately sized and spaced to provide correct total LU rating for the component.

Table 17.4 Pipe sizes

	Distribution pipe size	
Nozzle loading, LU	Copper tubing to ISO 274 OD (mm)	Medium series steel pipe to ISO 65, nominal bore (mm)
10	6	—
15	8	6
30	10	8
50	12	10
75	16	—
100	20	15
200	25	20
300	32	25
500	40	32
650	50	40
1000	63	50

ISO 274 corresponds to BS 2871
ISO 65 corresponds to BS 1387

Figure 17.2 Sizing of manifolds and piping

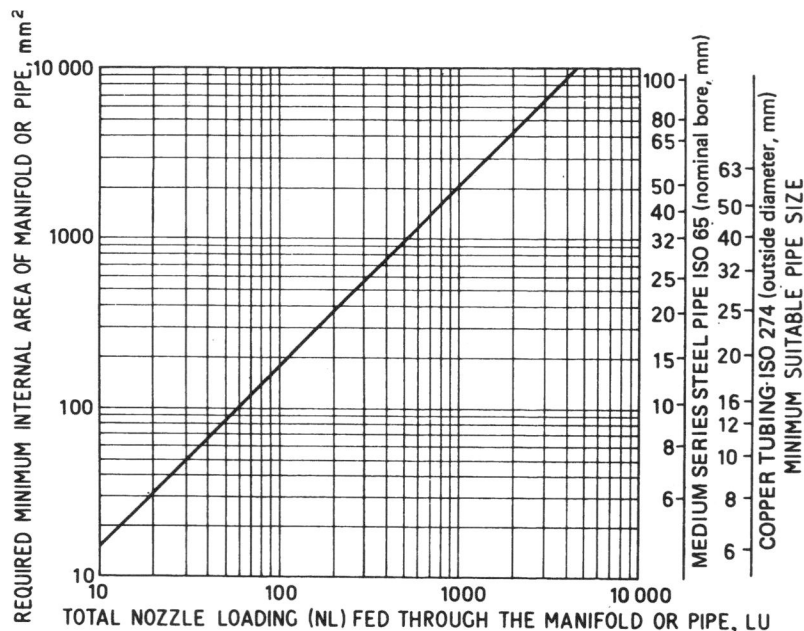

The chart axes read:
- Vertical left axis: REQUIRED MINIMUM INTERNAL AREA OF MANIFOLD OR PIPE, mm² (from 10 to 10 000)
- Horizontal axis: TOTAL NOZZLE LOADING (NL) FED THROUGH THE MANIFOLD OR PIPE, LU (from 10 to 10 000)
- Right axes: MEDIUM SERIES STEEL PIPE ISO 65 (nominal bore, mm); COPPER TUBING ISO 274 (outside diameter, mm); MINIMUM SUITABLE PIPE SIZE

Generator selection

Total the nozzle ratings of all the fittings and orifices to give the total Nozzle Loading (NL) and select generator with appropriate LU rating based on Nozzle Loading. Make certain that the minimum rating of the generator is less than NL.

Air and oil consumption

Air consumption is a function of the total nozzle loading (NL) of the system. Oil consumption depends on the concentration of oil in the air and can be adjusted at the generator to suit the total Lubrication Unit Loading (LUL).

Air consumption
Using the total Nozzle Loading (NL), the approximate air consumption can be calculated in terms of the volume of free air at atmospheric pressure, from:

Air consumption = 0.015 (NL) dm³/s

Oil consumption
Using the total Lubrication Unit Loading (LUL), the approximate oil consumption can be calculated from:

Oil consumption = 0.25 (LUL) ml/h

INSTALLATION

Locate nozzle ends between 3 mm and 25 mm from surface being lubricated. Follow normal practice in grooving slides and journal bearings. The positioning of the nozzles in relationship to the surface being lubricated should be similar to that used in circulation systems. See Table 17.5.

Appropriate vents with hole diameters at least 1.5 times the diameter of the associated supply nozzle must be provided for each lubrication point. If a single vent serves several nozzles, the vent area must be greater than twice the total area of the associated nozzles.

Follow instructions of aerosol generator manufacturer in mounting unit and connecting electrical wiring. Avoid sharp bends and downloops in all pipework. Consult BS 4807: 1991 'Recommendations for Centralised Lubrication as Applied to Plant and Mechinery' for general information on installation.

Select appropriate grade of lubricant in consultation with lubricant supplier and generator manufacturer.

Table 17.5 Nozzle positioning

ROLLING BEARINGS EXCEPT TAPERED ROLLER		
TAPERED ROLLER BEARINGS		
JOURNAL BEARINGS		
GEARS AND WORM DRIVES		
SLIDES		
CHAIN DRIVES		

SPUR, BEVEL AND HELICAL GEARS

All gears, except very slow running ones, require complete enclosure. In general, gears dip into oil for twice tooth depth, to provide sufficient splash for pinions, bearings, etc. and to reduce churning loss to a minimum.

Typical triple reduction helical gear unit

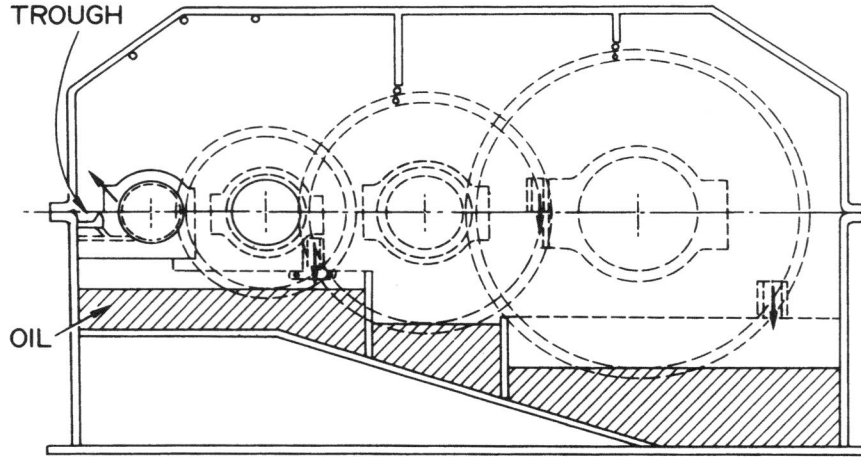

Figure 18.1

This has guards and tanks for individual gears. Bearings are fed by splash and from a trough round the walls of the case. Suitable for up to 12.5 m/s (2500 ft/min) peripheral speed.

Typical single reduction bevel gear unit

Suitable for up to 12.5 m/s (2500 ft/min).

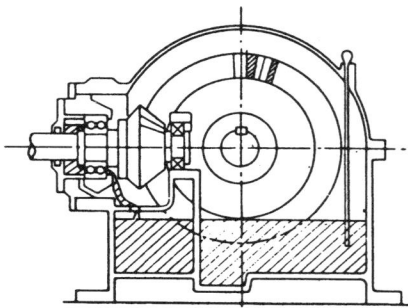

Figure 18.2 Bevel unit with double row bearings on pinion shaft

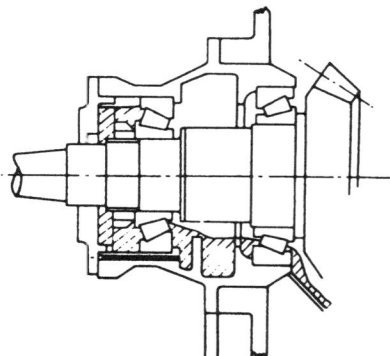

Figure 18.3 Typical bearing lubrication arrangement with taper roller bearings

Typical high-speed gear unit guard

Normally satisfactory up to 25 m/s (5000 ft/min). With special care can be used up to 100 m/s (20 000 ft/min).

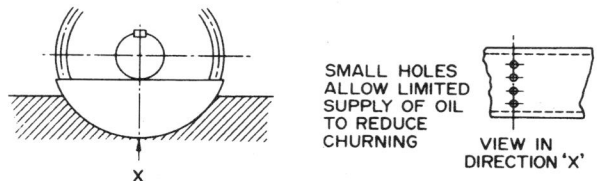

Figure 18.4

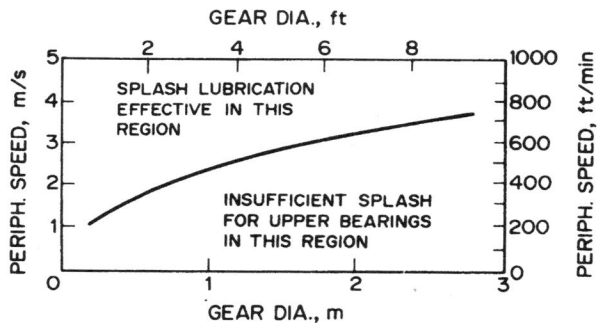

Figure 18.5 Peripheral speed against gear diameter for successful splash lubrication to upper bearings in dip-lubricated gear units

WORM GEARS

Typical under-driven worm gear unit

Oil is churned by the worm and thrown up to the top and sides of the case. From here it drips down via the wheel bearings to the sump.

A simple lip seal on a hard, ground shaft surface, prevents leakage.

Oil level generally just below worm centre-line.

An oil scraper scrapes oil into a trough to feed the wheel bearings.

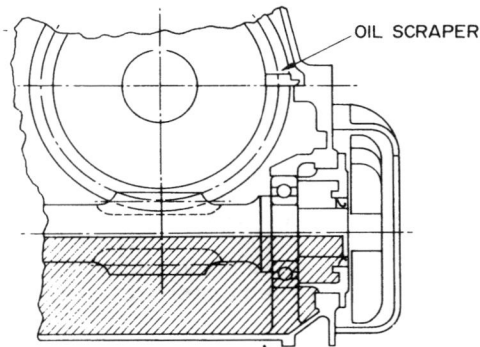

Figure 18.6

Typical over-driven worm gear unit

Similar to under-driven worm gear unit except that the worm is over the wheel at the top of the unit, and the oil level varies in depth from just above wheel tooth depth to almost up to the centre line of the wheel, depending upon speed. The greater the speed, the higher the churning loss, therefore the lower should be the oil level. At low speeds, the churning loss is small and a large depth of oil ensures good heat-transfer characteristics.

GENERAL DESIGN NOTES

Gears

In dip-splash systems, a large oil quantity is beneficial in removing heat from the mesh to the unit walls and thence to the atmosphere.

However, a large quantity may mean special care has to be paid to sealing, and churning losses in gears and bearings may be excessive. It is necessary to achieve a balance between these factors.

Other applications

The cylinders and small-end bearings of reciprocating compressors and automotive internal combustion engines are frequently splash lubricated by oil flung from the rotating components. In these applications the source of the oil is usually the spill from the pressure-fed crankshaft bearings. In some small single-cylinder compressors and four-stroke engines, the cap of the connecting rod may be fitted with a dipper which penetrates up to 10 mm into the oil in the sump and generates splash lubrication as a result. In lightly loaded applications the big-end bearings may also be splash lubricated in this way, and in some cases the dipper may be in the form of a tube which scoops the oil directly into the big-end bearing. In small domestic refrigeration compressors, a similar system may also be used to scoop oil into the end of the crankshaft, in order to lubricate all the crankshaft bearings.

A circulation system is defined as an oil system in which the oil is returned to the reservoir for re-use. There are two groups of systems: *group 1*, lubrication with negligible heat removal; and *group 2*, lubrication and cooling.

GROUP 1 SYSTEMS

Virtually any form of mechanically or electrically driven pump may be used, including piston, plunger, multiplunger, gear, vane, peristaltic, etc. The systems are comparatively simple in design and with low outputs. Various metering devices may be used.

Multiplunger pump systems

These systems utilise the plunger-type oil lubricators of the rising or falling drop type, employing a separate pumping element for each feed, giving individual adjustment and a positive feed to each lubrication point. Generally, the lubricators have up to some 32 outlets with the discharges being adjustable from zero to maximum.

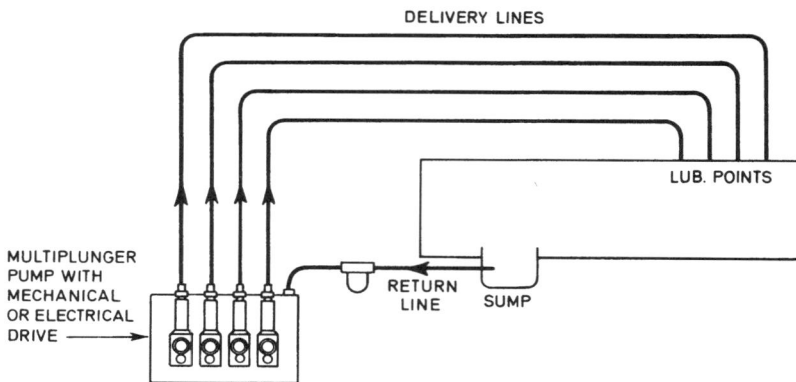

Figure 19.1 Simple multipoint lubricator – system contains barest elements of pump reservoir and interconnecting pipework

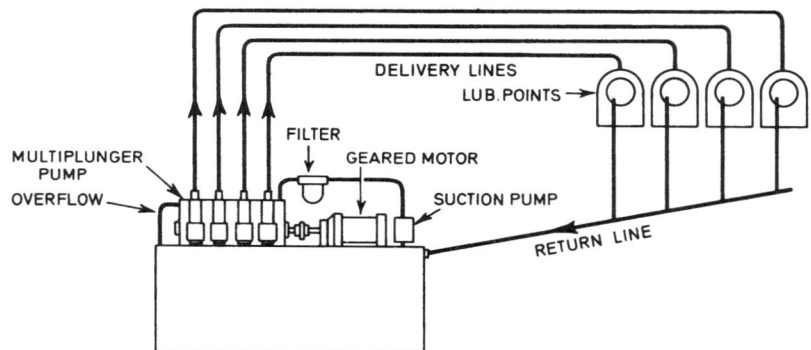

Figure 19.2 Multi-point lubricators mounted on receiving tank – system has the advantage of providing setting time and better filtration of returned oil

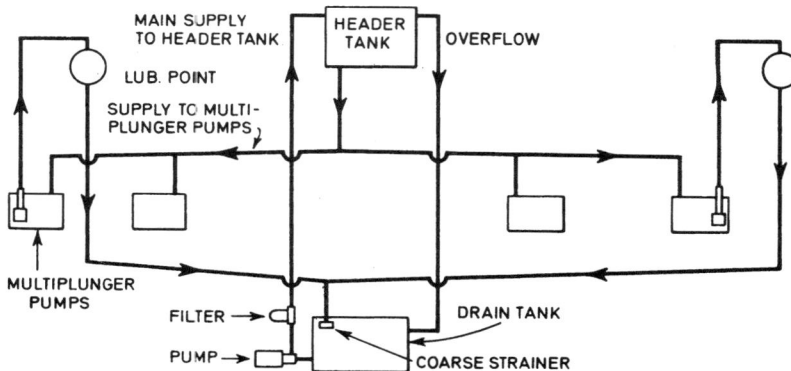

Figure 19.3 Extensive system for large numbers of bearings, using multi-point lubricator system can be extended within the limitation of the gravity feed from the header tanks

Typical applications

Paper machines, large kilns, calenders, and general machinery with a large number of bearings requiring a positive feed with feed adjustment.

Positive-split systems

In general these systems deliver a larger quantity of lubricant than multiplunger systems. They comprise a small high-pressure pump with or without stand-by, fitted with integral relief valve supplying lubricant to the bearings via positive dividers. These dividers then deliver the oil to the bearings in a predetermined ratio of quantities. By use of a microswitch operated by the indicator pin on the master divider (or single divider if the number of points is small), either timed automatic or continuous operation is available. The microswitch can also be used to give a warning of failure of the system.

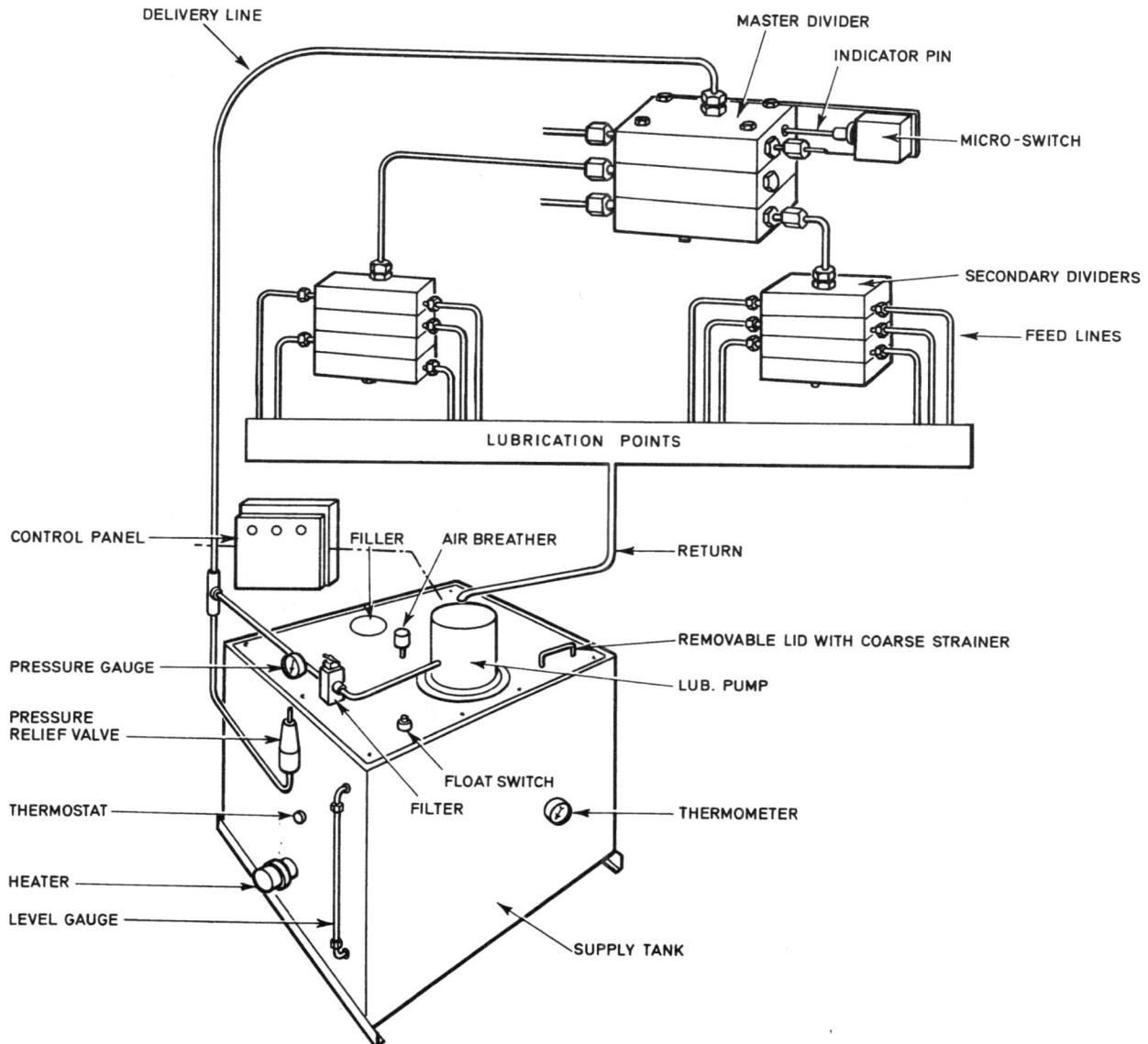

Figure 19.4

Typical applications

Machine tools, sugar industry, gearboxes, printing machines, and special-purpose machinery.

Double-line systems

Double-line elements can be used in conjunction with a reversing valve and piston or gear pump to lubricate larger numbers of points spread over longer distances geographically. These elements and their operation are similar to those previously described under grease systems.

Typical applications

Machine tools, textile plant, and special-purpose machinery.

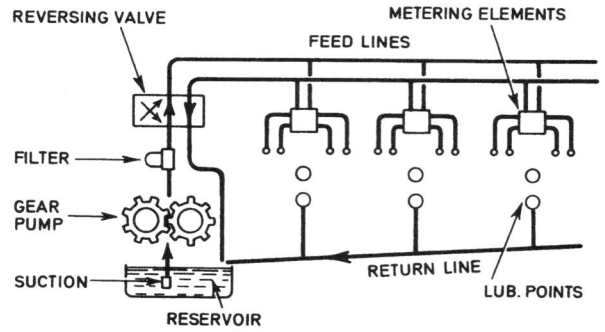

Figure 19.5

Simple low-pressure systems

The simple form illustrated uses a gear pump feeding the points from connections from a main feed line through needle valves with or without sight glasses.

Typical applications

Special-purpose machinery and machine tools.

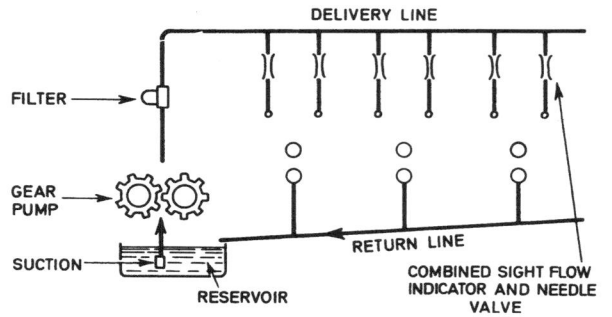

Figure 19.6

Gravity-feed systems

Gravity-feed systems consist of a header tank, piped through to one or more lubrication points. The level in the header tank is maintained by a gear or other pump with relief valve and filter mounted at the collection tank.

   This may be used as a back-up for a forced-feed system where important bearings have a long run-down period after removal of the power source, e.g. large air fans.

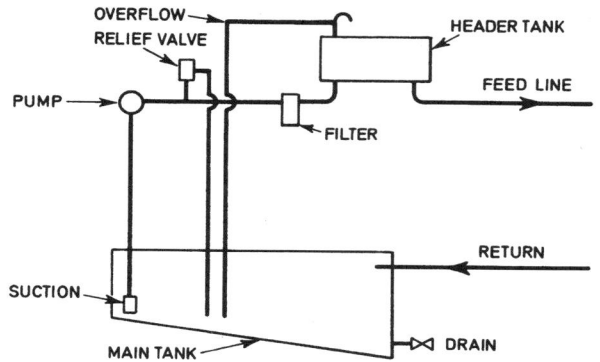

Figure 19.7

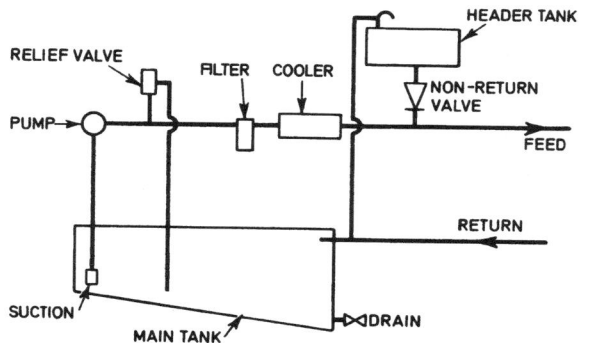

Figure 19.8

GROUP 2 SYSTEMS

The larger and usually more complex type of oil-circulatory system, used for both lubrication and cooling, falls into two distinct classes. The first type, known as the self-contained system, is usually limited in size by the weights of the components. For this reason the storage capacity of this type does not usually exceed 1000 gal. The second type covering the larger systems has the main components laid out at floor level, e.g. in the oil cellar. The detailed design considerations of the main components are discussed elsewhere, but in laying out the system the possible need for the equipment in Table 19.1 should be considered.

Self-contained systems

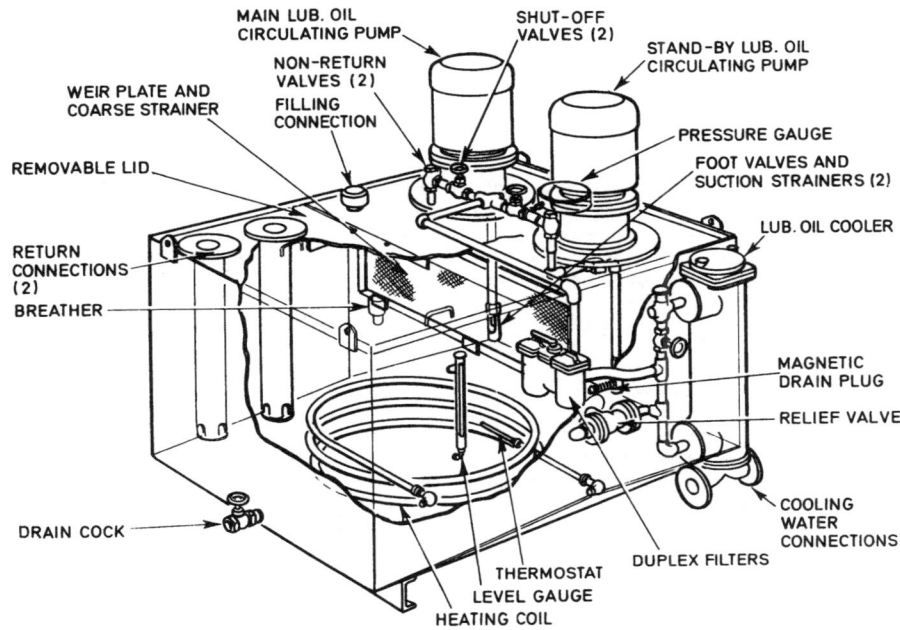

Figure 19.9 A typical self-contained oil-circulatory system, incorporating a 200 gal tank. These types of system may be used, if required, with a pressure vessel which would be mounted as a separate unit

Large oil-circulatory systems

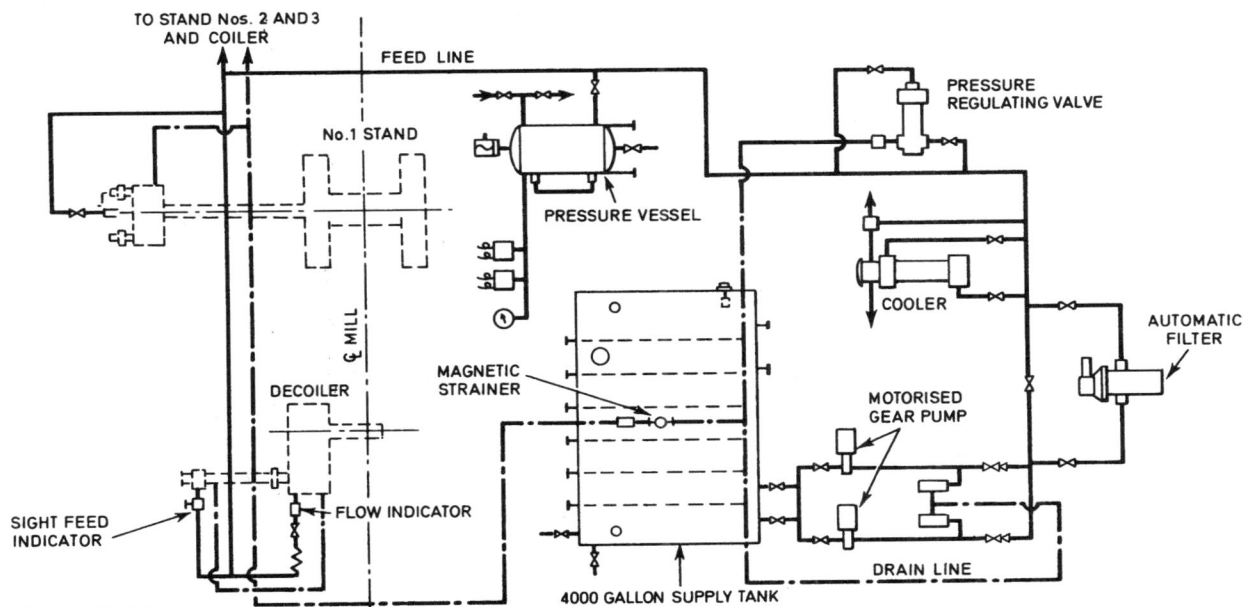

Figure 19.10

The large oil-circulatory systems typical of those in use in steelworks, marine applications and power stations are illustrated diagrammatically above.

Table 19.1 Main components of group 2 systems

Storage tank	One or more storage tanks, dependent on water contamination, will be required with a capacity of between 20 and 40 times the throughput per minute of the system
Tank heating	Electric, steam or hot-water heating are used for raising the temperature of the lubricant in the storage tank
Pumps	One or more main pumps and a stand-by pump are required. The main pumps should have a capacity of at least 25% in excess of basic system requirement
Pressure control valve	System pressure will be maintained by the use of pressure control valves
Non-return valves	A non-return valve is required after each pump unit
Self-cleaning strainer	Either manual or automatic, single or duplex self-cleaning strainers will be used for cleaning the oil
Magnetic strainer	Particularly in the case of gear lubrication, magnetic strainers may be fitted whether in the supply line and/or on the return oil connection to the tank
Pressure vessel	Pressure vessels may be required in order to maintain a flow of oil in the event of a power failure, to allow the run-down of machinery or the completion of a machine operation
Cooler	A cooler may be used to extract the heat taken away from the equipment to the lubricant and maintain the viscosity in the system prior to the supply to the lubrication points
Pressure-reducing stations	On extensive systems the lubrication points can be split into groups for controlling the flow rate by means of a pressure-reducing station, followed by either orifice plates or simple pipe sizing flow control
Water/sludge trap	Where water contamination is likely a water/sludge trap should be fitted in the return line immediately before the tank
Valving	All major equipment should be capable of being shut off by the use of gate valves for maintenance purposes. In the case of filters and coolers a bypass arrangement is necessary
Instrumentation	Normal instrumentation will cater for the specific system requirements, including pressure gauges, thermometers, thermostats, pressure switches, recording devices, etc.

CONTROL OF LUBRICANT QUANTITIES

The quantity fed to the lubrication point can be controlled in a number of ways; typical examples are shown below:

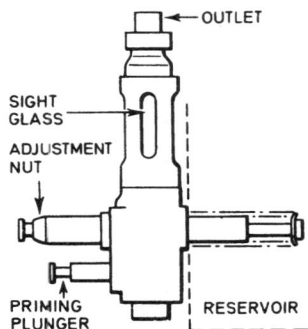

Figure 19.11

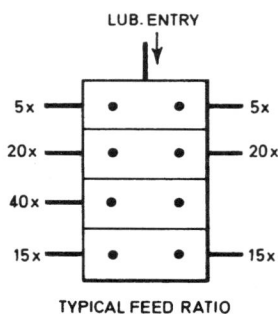

Figure 19.12

The output of a metering pump is itself adjustable by some form of manual adjustment on each pump unit. Sight glasses, of the rising or falling drop type, or of the plug and taper tube type, are normally fitted.

The positive dividers may have sections which have different outputs, and may be cross-drilled to connect one or more outlets together to increase the quantity available for each cycle.

Figure 19.13 *Typical flow ratios*

Orifice plates may be used at the entry to the bearing or gear system. The actual flow rates will vary with viscosity unless knife-edge orifices are used, in which case the viscosity variation is negligible.

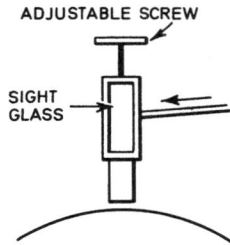

Figure 19.14

Combined needle and sight flow indicators used for adjusting small quantities of lubricant giving only a visual indication of the flow of lubricant into the top of a bearing.

With larger flow rates it may be adequate, with a controlled pressure and oil temperature, simply to alter the bore of the pipe through which the supply is taken. The actual flow rates will vary with viscosity, and pipework configuration, i.e. increased number of fittings and directional changes.

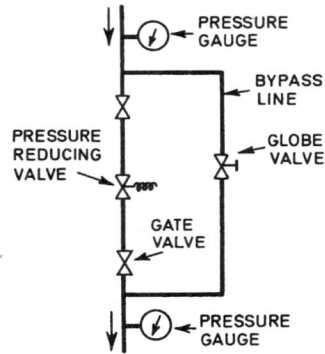

Figure 19.15

The layout of a typical pressure control station is shown above.

TANK VOLUME AND PROPORTIONS

The diagram shows a tank with dimensions: height 0.7w to 1.3w, depth w, width w to 3w.

FREE SPACE	10 to 20% of total volume for ventilation, foam containment and thermal expansion — Stationary level
SYSTEM CAPACITY	Volume of oil draining back on shutdown — Running level
RUNNING CAPACITY = total oil flow rate × dwell time	TYPICAL DWELL TIMES (min)
FITMENT ALLOWANCE	= Volume of tank's internal fitments

TYPICAL DWELL TIMES (min)

Piston compressors 1 - 8   Steam turbines 5 - 10
Hydraulic systems 2 - 4   Large electric motors 5 - 10
Gas turbines (land and marine) 5   Steel mill machinery 20 - 60
Gas turbines (aircraft) ½   Paper mill machinery 40 - 60

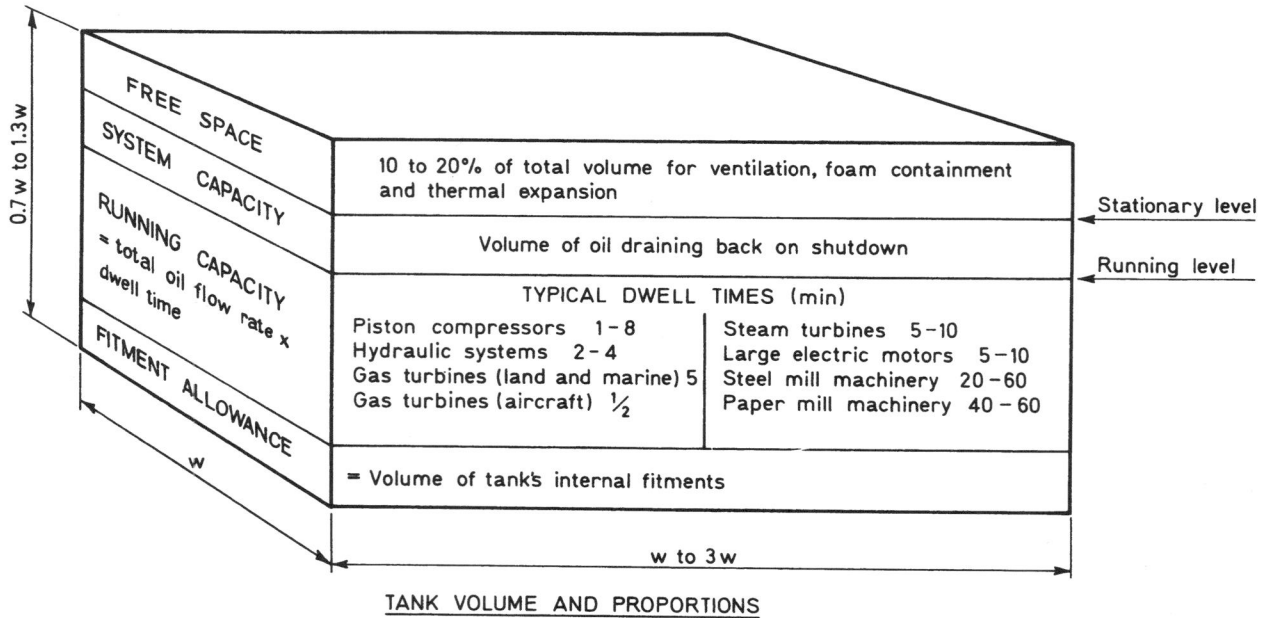

Table 20.1 Tank materials

Stainless steel or anodised aluminium alloy	Mild steel plate
Material cost relatively high, but expensive preparation and surface protective treatment is not needed. Maintenance costs relatively low. Thinner gauge stainless steel may be used	Most widely used material for tanks; low material cost, but surface requires cleaning and treatment against corrosion, e.g. by shot or sand blasting and (a) lanolin-based rust preventative (b) oil resistant paint (c) coating with plastic (epoxy resin) (d) aluminium spraying

Table 20.2 Tank components

Component	Design features	Diagram
RETURN LINE	*Location:* at or just above running level *Size and slope:* chosen to run less than half full to let foam drain, and with least velocity to avoid turbulence *Alternative:* allow full bore return below running level *Refinement:* perforated tray prevents aeration due to plunging	PERFORATED TRAY / RUNNING OIL LEVEL
SUCTION LINE	*Location:* remote from return line *Inlet depth:* shallow inlet draws clean oil; deep inlet prolongs delivery in emergency. Compromise usually two-thirds down from running level *Refinement:* filter or strainer; floating suction to avoid depth compromise (illustrated); anti-vortex baffle to avoid drawing in air	PIVOT STOP FLOAT
BAFFLES AND WEIRS	*Purpose:* see diagram. Also provide structural stiffening, inhibit sloshing in mobile systems, prevent direct flow between return and suction *Design:* Settling needs long, slow, uniform flow; baffles increase flow velocity. Avoid causing constrictions or 'dead pockets'. Arrange baffles to separate off cleanest layer. Consider drainage and venting needs	① WEIR TO LOCALISE TURBULENCE ② BAFFLE TO TRAP FOAM AND FLOATING CONTAMINANTS ③ BAFFLE TO TRAP SINKING CONTAMINANTS

Table 20.2 Tank components (continued)

VENTILATORS	*Purpose:* to allow volume changes in oil; to remove volatile acidic breakdown products and water vapour *Number:* normally one per 5 m² (50 ft²) of tank top *Filtration:* treated paper or felt filters, to be inspected regularly, desirable. Forced ventilation, by blower or exhauster, helps remove excess water vapour provided air humidity is low
DRAINAGE POINTS AND ACCESS	*Drainage:* Located at lowest point of tank. Is helped by bottom slope of 1:10 to 1:30. Gravity or syphon or suction pump depending on space available below tank. Baffle over drain helps drainage of bottom layer first *Access:* Size and position to allow removal of fittings for repair and to let all parts be easily cleaned. Manholes with ladders needed in large tanks
LEVEL INDICATORS	*Dipstick:* suitable only for small tanks *Sight-glass:* simple and reliable unless heavy or dirty oil obscures glass. Needs protection or 'unbreakable' glass. Ball-check stops draining if glass breaks *Float and pressure gauges:* various proprietary gauges can give local or remote reading or audible warning of high or low levels
DE-AERATION SCREEN	*Purpose:* removal of entrained air in relatively clean systems, to prevent recirculation *Position:* typically as shown. Should be completely immersed as surface foam can penetrate finest screen *Material:* wire gauze of finest mesh that will not clog too quickly with solid contaminants. Typically 100-mesh
HEATERS	*Purpose:* aid cold-start circulation, promote settling by reducing viscosity, assist water removal *Design features:* Prevent debris covering elements. Provide cut-out in case of oil loss. Take care that convection currents do not hinder settling. Consider economics of tank lagging
COOLERS	*Purpose:* reducing temperature under high ambient temperature conditions. Usually better fitted as separate unit
STIFFENERS	*Purpose:* prevention of excessive bulging of sides and reduction of stresses at edges in large tanks *Location:* external stiffening preferred for easy internal cleaning and avoidance of dirt traps, but baffles may eliminate need for separate stiffening
INSTRUMENTS, ETC.	*Thermometer or temperature gauge:* desirable in tanks fitted with heaters *Magnetic drain-plug or strainers:* often fitted both for removing ferrous particles and to collect them for fault analysis *Sampling points:* May be required at different levels for analysis

Table 21.1 System factors affecting choice of pump type

Factor	*Way in which pump choice is affected*	*Remarks*
Rate of flow	Total pump capacity = maximum equipment requirement+known future increase in demand (if applicable)+10 to 25% excess capacity to cater for unexpected changes in system demand after long service through wear in bearings, seals and pump Determines pump size and contributes to determining the driving power	Actual selection may exceed this because of standard pumps available. Capacity of over 200% may result for small systems
Viscosity	Lowest viscosity (highest expected operating temperature) is contributing factor in determining pump size Highest viscosity (lowest expected operating temperature) is contributing factor in determining driving power	May influence decision whether reservoir heating is necessary
Suction conditions	May govern selection of pump type and/or its positioning in system Losses in inlet pipe and fittings with highest expected operating oil viscosity+static suction lift (if applicable) not to exceed pump suction capability	See Figure 21.1 below for determining positive suction head, or total suction lift
Delivery pressure	Total pressure at pump = pressure at point of application+static head+losses in delivery pipe, fittings, filter, cooler, etc. with maximum equipment oil requirement at normal viscosity Determines physical robustness of pump and contributes in deciding driving power	See Figure 21.1 below for determining delivery head
Relief valve pressure rating (positive displacement pumps)	Relief valve sized to pass total flow at pressure 25% above 'set-pressure'. Set pressure = pumping pressure + 70 kN/m^2 (10 p.s.i.) for operating range 0–700 kN/m^2 (0–100 lbf/in^2) or, +10% for operating pressure above 700 kN/m^2 (100 lbf/in^2) Determine actual pressure at which full flow passes through selected valve	

Driving power	Maximum absorbed power is determined when considering:
	   (1) total flow
	   (2) pressure with total flow through relief valve
	   (3) highest expected operating oil viscosity
	Driver size can then be selected	

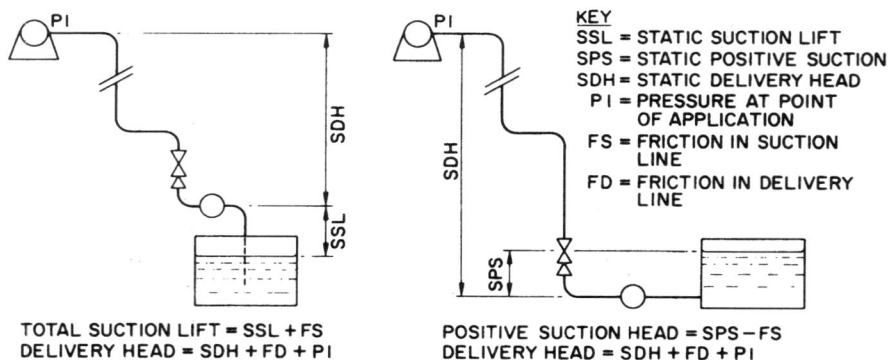

KEY
SSL = STATIC SUCTION LIFT
SPS = STATIC POSITIVE SUCTION
SDH = STATIC DELIVERY HEAD
P I = PRESSURE AT POINT OF APPLICATION
FS = FRICTION IN SUCTION LINE
FD = FRICTION IN DELIVERY LINE

TOTAL SUCTION LIFT = SSL + FS
DELIVERY HEAD = SDH + FD + PI

POSITIVE SUCTION HEAD = SPS − FS
DELIVERY HEAD = SDH + FD + PI

Figure 21.1 Definition of pump heads

Table 21.2 Comparison of the various types of pump

Gear pump
Spur gear relatively cheap, compact, simple in design. Where quieter operation is necessary helical or double helical pattern may be used. Both types capable of handling dirty oil. Available to deliver up to about $0.02 \, \text{m}^2/\text{s}$ (300 g.p.m.).

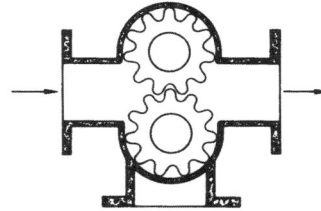

Lobe pump
Can handle oils of very viscous nature at reduced speeds.

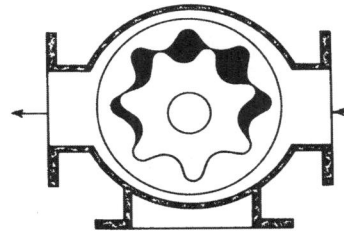

Screw pump
Quiet running, pulseless flow, capable of high suction life, ideal for pumping low viscosity oils, can operate continuously at high speeds over very long periods, low power consumption. Adaptable to turbine drive. Available to deliver up to and above $0.075 \, \text{m}^3/\text{s}$ (1000 g.p.m.).

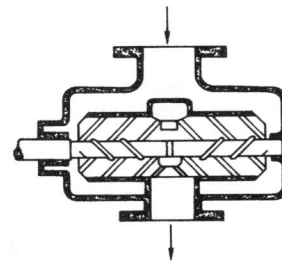

Vane pump
Compact, simple in design, high delivery pressure capability, usually limited to systems which also perform high pressure hydraulic duties.

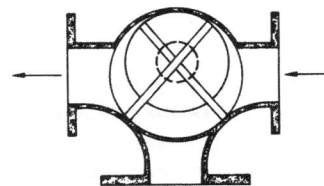

Centrifugal pump
High rate of delivery at moderate pressure, can operate with greatly restricted output, but protection against overheating necessary with no-flow condition. Will handle dirty oil.

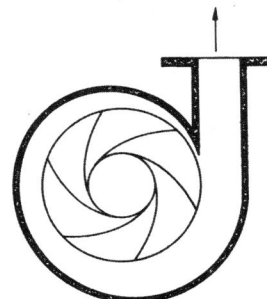

Table 21.3 Pump performance factors affecting choice of pump type

Positive displacement	Centrifugal
Rate of delivery at given speed substantially unaffected by changes in delivery pressure	Rate of delivery affected by change in delivery pressure. Therefore flow demand and temperature which influence pressure must be accurately controlled
Rate of delivery varies nearly directly with speed	
Delivery pressure may be increased within material strength limitation of the pump, by increasing drive power	Because wide variation in output results from change of pressure, pump is well suited for installations requiring large flows and subject to occasional transient surge conditions, e.g. turbine hydraulic controls
Very high delivery pressures can be produced by pumps designed to reduce internal leakages	

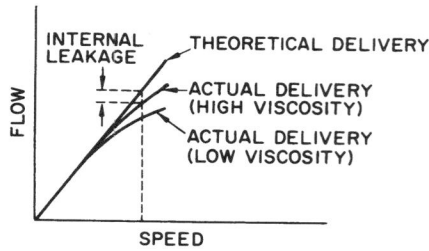

Figure 21.2 Delivery against speed and viscosity for a positive displacement pump

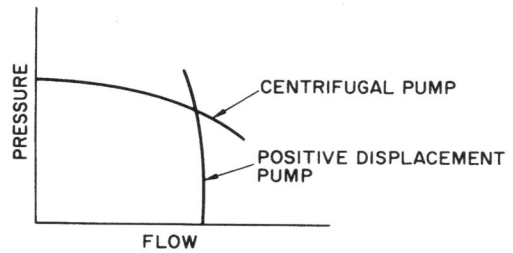

Figure 21.3 Pressure against delivery for positive displacement and centrifugal pumps

Table 21.4 Selection by suction characteristics

Pump type	Maximum suction lift, m					Self-priming
	2	4	6	8	10	
Gear						Yes if wet
Lobe rotor						Depends on speed
Screw						and viscosity
Vane						Yes
Radial piston						Depends on viscosity
Centrifugal						No

Table 21.5 Selection by head or pressure

Pump type	1000 (150)	3000 (430)	5000 (710)	7000 (1000)	20 000 (2850)	70 000 (10 000)	kN/m² (lbf/in²)
Gear							
Lobe rotor							
Screw							
Vane							
Radial piston							
Centrifugal							

Table 21.6 Selection by capacity

Pump type	0.01 (130)	0.03 (400)	0.05 (660)	0.07 (930)	0.08 (1050)	0.15 (2000)	m³/s (gpm)
Gear							
Lobe rotor							
Screw							
Vane							
Radial piston							
Centrifugal							

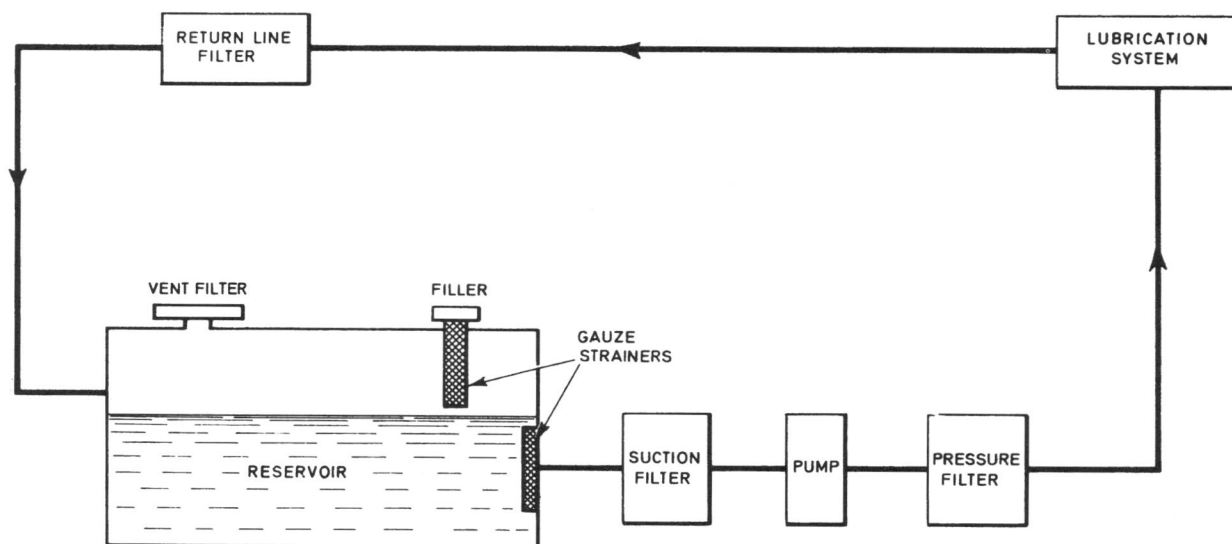

Figure 22.1 Typical circuit showing positions of various filters

Table 22.1 Location and purpose of filter in circuit

Location	Degree of filtration	Type	Purpose
Oil reservoir vent	Coarse	Wire wool Paper Oil bath	Removal of airborne contaminant
Oil reservoir filler	Coarse	Gauze	Prevention of ingress of coarse solids
Suction side of pump	Medium	Paper Gauze	Protection of pump
Delivery side of pump	Fine	Sintered metal Felt Paper	Protection of bearings/system
Return line to reservoir	Medium	Gauze Paper	Prevention of ingress of wear products to reservoir
Separate from system	Very fine	Centrifuge	Bulk cleaning of whole volume of lubricant

Table 22.2 Range of particle sizes which can be removed by various filtration methods

Filtration method	Examples	Range of minimum particle size trapped micrometres (μm)
Solid fabrications	Scalloped washers, wire-wound tubes	5–200
Rigid porous media	Ceramics and stoneware Sintered metal	1–100 3–100
Metal sheets	Perforated Woven wire	100–1000 5–200
Porous plastics	Plastic pads, sheets, etc. Membranes	3–100 0.005–5
Woven fabrics	Cloths of natural and synthetic fibres	10–200
Cartridges	Yarn-wound spools, graded fibres	2–100
Non-woven sheets	Felts, lap, etc. Paper—cellulose —glass Sheets and mats	10–200 5–200 2–100 0.5–5
Forces	Gravity settling, cyclones, centrifuges	Sub-micrometre

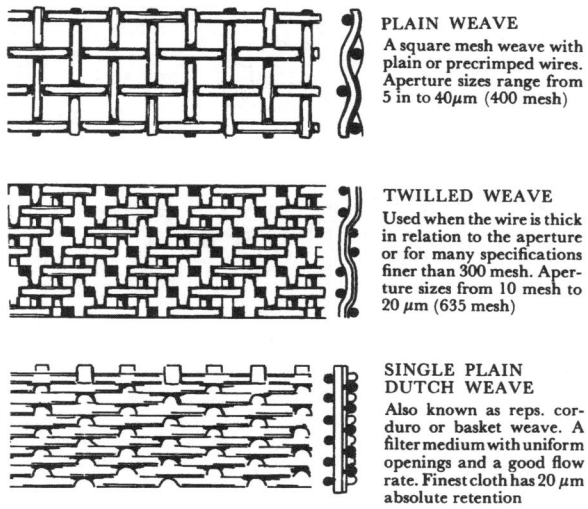

PLAIN WEAVE
A square mesh weave with plain or precrimped wires. Aperture sizes range from 5 in to 40μm (400 mesh)

TWILLED WEAVE
Used when the wire is thick in relation to the aperture or for many specifications finer than 300 mesh. Aperture sizes from 10 mesh to 20 μm (635 mesh)

SINGLE PLAIN DUTCH WEAVE
Also known as reps. corduro or basket weave. A filter medium with uniform openings and a good flow rate. Finest cloth has 20 μm absolute retention

Figure 22.2 Various forms of woven wire mesh

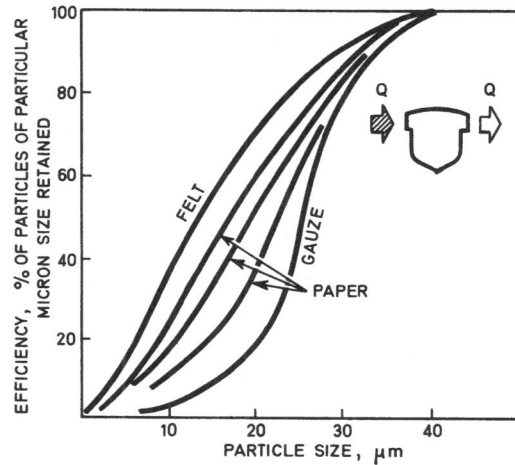

Figure 22.3 Typical filter efficiency curves

PRESSURE FILTERS

Pressure filter specification and use

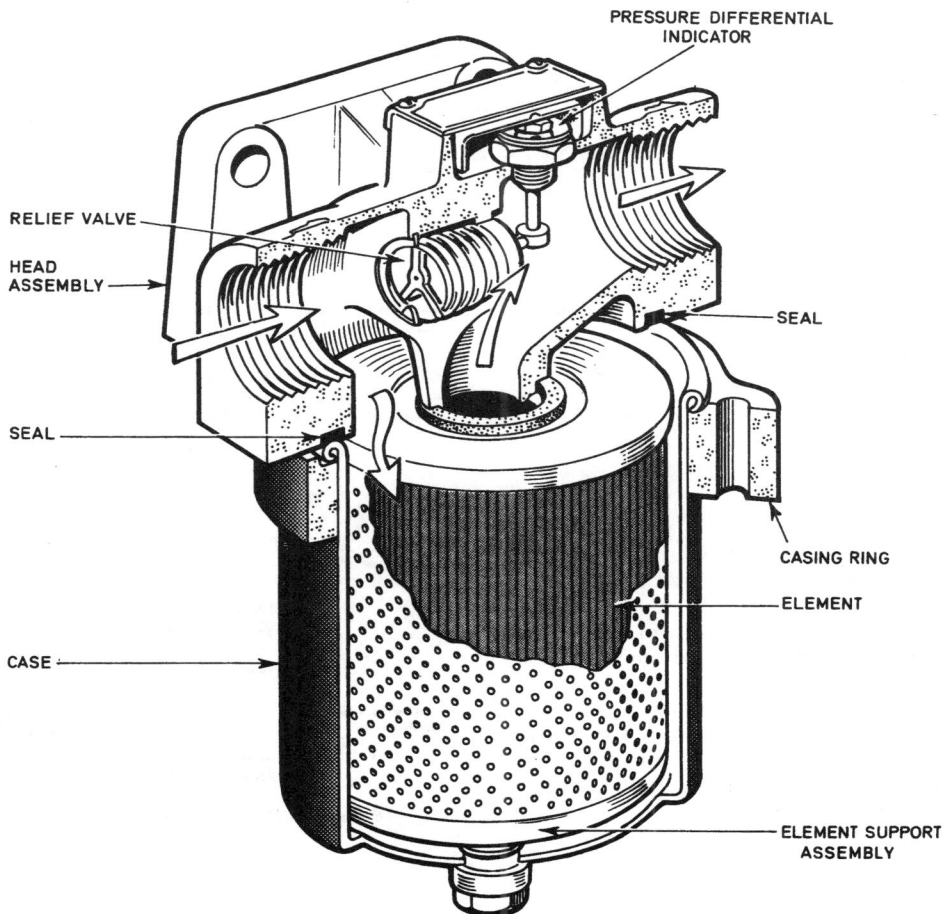

Figure 22.4 Typical full-flow pressure filter with integral bypass and pressure differential indicator

In specifying the requirements of a filter in a particular application the following points must be taken into account:

1. Maximum acceptable particle size downstream of the filter.
2. Allowable pressure drop across the filter.
3. Range of flow rates.
4. Range of operating temperatures.
5. Viscosity range of the fluid to be filtered.
6. Maximum working pressure.
7. Compatibility of the fluid, element and filter materials.

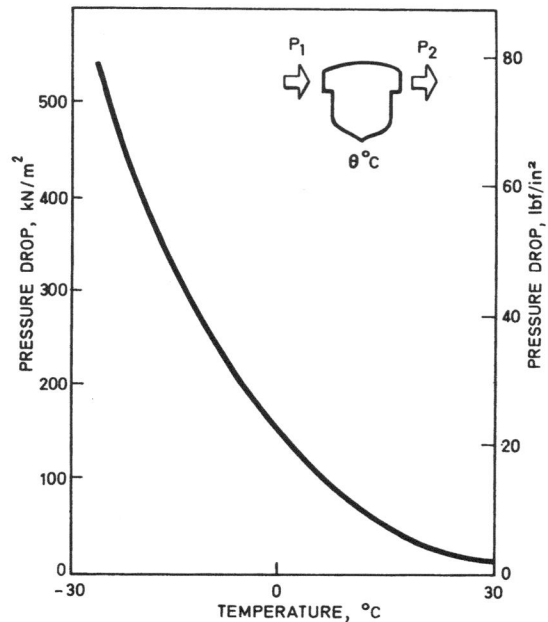

Figure 22.5 Curve showing effect of temperature on pressure drop when filtering lubricating oil

In-line filtration

In many systems, the lubricating oil flows under pressure around a closed circuit, being drawn from and returned to a reservoir. The same oil will then pass through the system continuously for long periods and effective filtration by one of two approaches is possible, i.e. full-flow filtration and bypass filtration.

Full-flow filtration

A full-flow filter will handle the total flow in the circuit and is situated downstream of the pump. All of the lubricant is filtered during each circuit.

ADVANTAGE OF FULL FLOW
All particles down to specified level are removed.

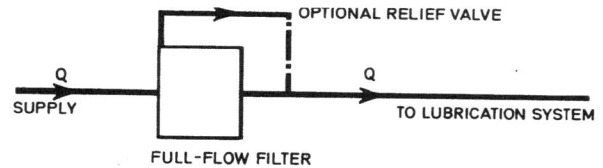

Figure 22.6 Simplified circuit of full-flow filter

Bypass filtration

In bypass filtration only a proportion of the oil passes through the filter, the rest being bypassed unfiltered. In theory, all of the oil will eventually be filtered but the prevention of the passage of particles from reservoir to bearings, via the bypass, cannot be guaranteed.

ADVANTAGES OF BYPASS
Small filter may be used. System not starved of oil under cold (high viscosity) conditions. Lower pressure drop for given level of particle retention. Filter cannot cut off lubricant supply when completely choked.

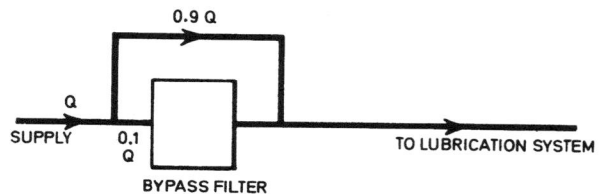

Figure 22.7 Simplified circuit of bypass filter

A22.3

CENTRIFUGAL SEPARATION

Throughput specification

Selection of a centrifugal separator of appropriate throughput will depend on the type of oil and the system employed. A typical unit of nominal 3000 l/h (660 gal/h) should be used at the following throughput levels:

	l/h	gal/h
Lubricating oil, straight:		
Bypass system, maximum	3000	660
Bypass system, optimum	1200	265
Batch system recommended	1900	420
Lubricating oil, detergent:		
Bypass system, maximum	1800	395
Bypass system, maximum	750	165
Batch system, recommended	1150	255

Operating throughputs of other units may be scaled in proportion.

Recommended separating temperatures

Straight mineral oils, 75°C (165°F).
Detergent-type oils, 80°C (175°F).

Fresh-water washing

Water washing of oil in a centrifuge is sometimes advantageous, the following criteria to be used to determine the hot fresh-water requirement:

	Quantity	Requirement
Straight mineral oil	3–5% of oil flow	About 5°C (9°F) above oil temperature
Detergent-type oil*	Max. 1% of oil flow	About 5°C (9°F) above oil temperature

* Only on oil company recommendation.

Lubricating oil heaters and coolers are available in many different forms. The most common type uses steam or water for heating or cooling the oil, and consists of a stack of tubes fitted inside a tubular shell. This section gives guidance on the selection of units of this type.

LUBRICATING OIL HEATERS

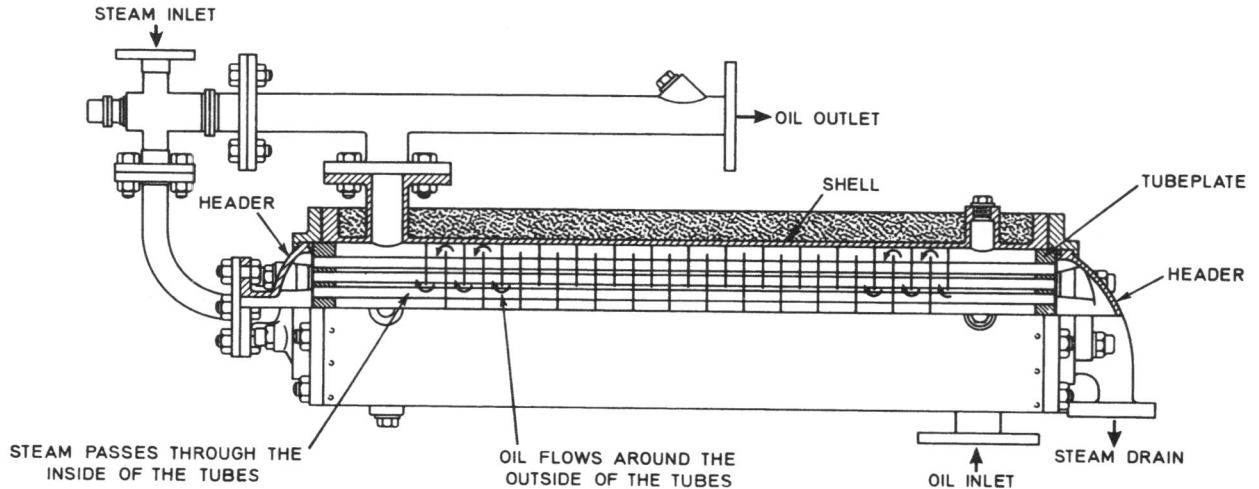

Figure 23.1 Cross-section through a typical oil heater

The required size of the heater and the materials of construction are influenced by factors such as:

Lubricating oil circulation rate.
Lubricating oil pressure and grade or viscosity.
Maximum allowable pressure drop across the heater.
Inlet lubricating oil temperature to heater.
Outlet lubricating oil temperature from heater.
Heating medium, steam or hot water.
Inlet pressure of the steam or hot water to the heater.
Inlet steam or hot water temperature.

Table 23.1 Guidance on materials of construction

Component	Suitable material	Remarks
Shell	Cast iron Mild steel fabrication	In contact with oil
Tubes	Mild steel	In contact with steam and hot water, but corrosion is not usually a problem with treated boiler water
Tube plates	Mild steel	
Headers	Gunmetal Cast iron Mild steel fabrication	

Guidance on size of heat transfer surface required

The graph shows how the required heat transfer surface area varies with the heat flow rate and the oil velocity, for a typical industrial steam heated lubricating oil heater, and is based on:

Heating medium	Dry saturated steam at $700 \, kN/m^2$ (100 p.s.i.)
Oil velocity	Not exceeding 1 m/s
Oil viscosity	SAE 30
Oil inlet temperature	20°C
Oil outlet temperature	70°C

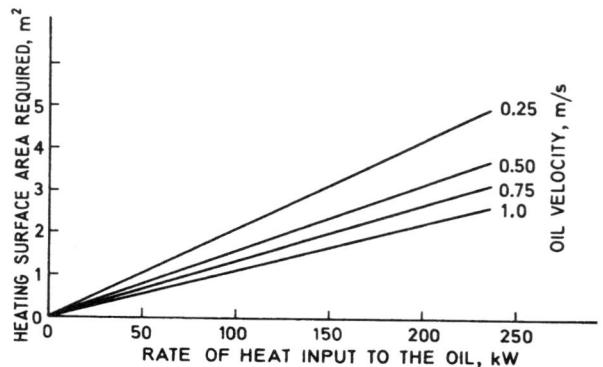

Figure 23.2 Guide to the heating surface area for a desired rate of heat input to oil flowing at various velocities

LUBRICATING OIL COOLERS

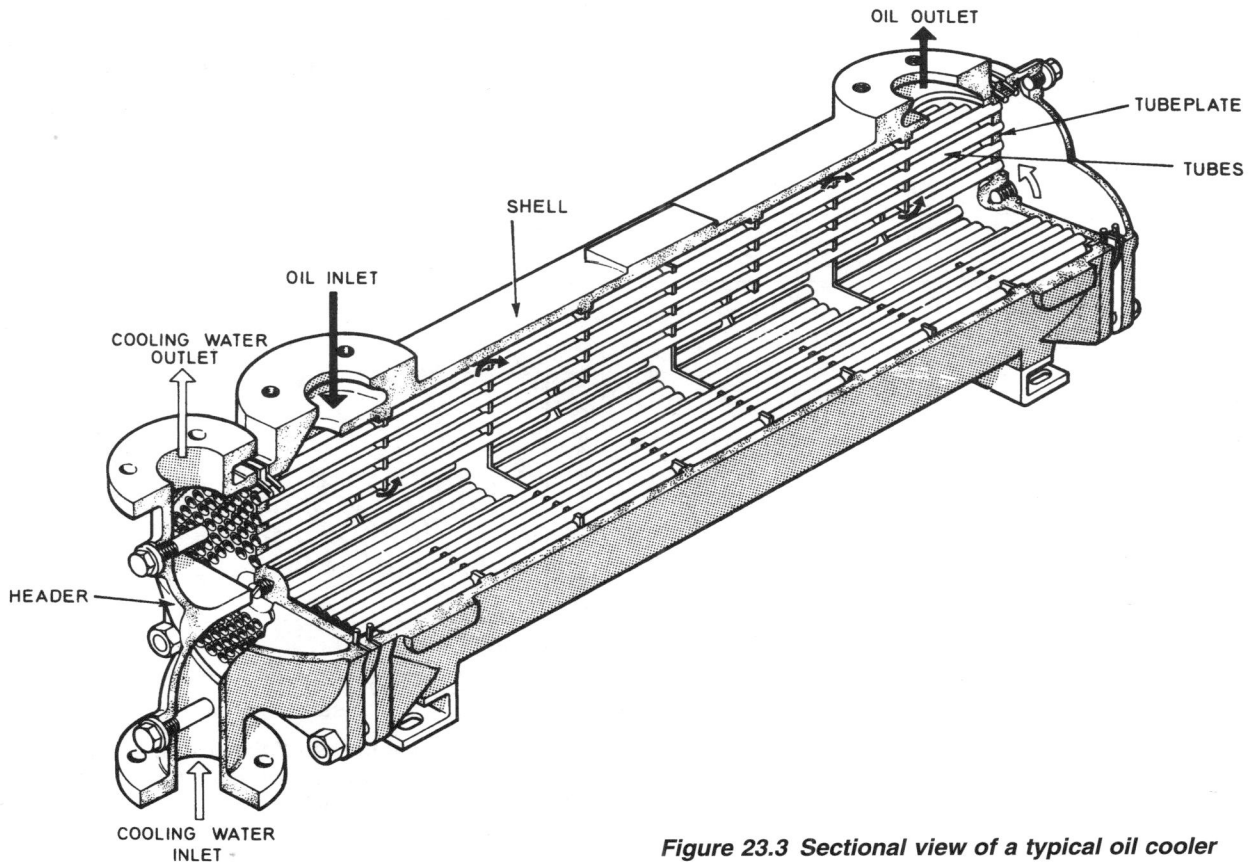

Figure 23.3 Sectional view of a typical oil cooler

The required size of cooler and the materials of construction are influenced by factors such as:

Lubricating oil circulation rate.
Lubricating oil pressure and grade or viscosity.
Maximum allowable pressure drop across the cooler.
Inlet lubricating oil temperature to cooler.
Outlet lubricating oil temperature from cooler.
Cooling medium (sea water, river water, town water, etc.)
Cooling medium pressure.
Cooling medium inlet temperature to cooler.
Cooling medium circulation rate available.

Table 23.2 Guidance on materials of construction

Component	Suitable material	Remarks
Shell	Cast iron Aluminium Gunmetal Mild steel fabrication	In contact with oil
Headers	Cast iron Gunmetal Bronze	In contact with cooling water but are of thick section so corrosion is less important
Tubes	Copper based alloys Titanium	The risk of corrosion by the cooling water is a major factor in material selection. Guidance is given in the next table
Tube plates	Copper based alloys Titanium	

Table 23.3 Choice of tube materials for use with various types of cooling water

Material	Cooling water		
	Aerated non saline waters	Aerated saline waters	Polluted waters
Copper	S	NR	NR
70/30 Brass	S	NR	NR
Admiralty brass	R	NR	NR
Aluminium brass	R	R	NR
90/10 Copper nickel	X	X	S
68/30/1/1 Copper nickel	X	X	R
Titanium	X	X	R
	River, canal, town's main water, deionised and distilled water	Estuarine waters seawater	Polluted river canal, harbour and estuarine waters, often non aerated

NR = Not recommended, S = Satisfactory, R = Recommended, X = Satisfactory, but more expensive

Guidance on the size of cooling surface area required

The graph shows how the cooling area required varies with the heat dissipation required, and the cooling water temperature for typical lubricating oil system conditions of:

Oil velocity	0.7 m/s
Oil viscosity	SAE 30
Water velocity	1 m/s
Oil inlet temperature	70°C
Oil outlet temperature	60°C

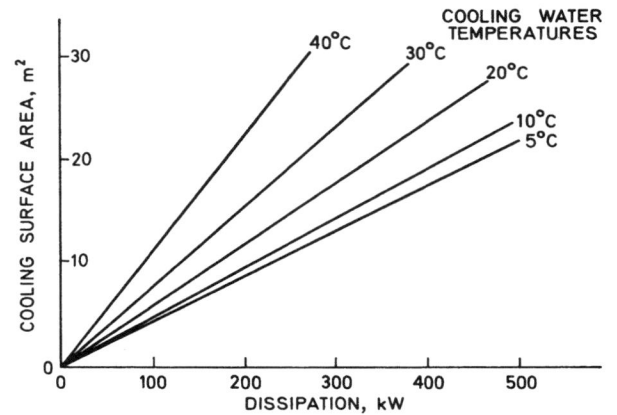

Figure 23.4 Guide to the cooling surface area required for a desired dissipation rate at various cooling water temperatures

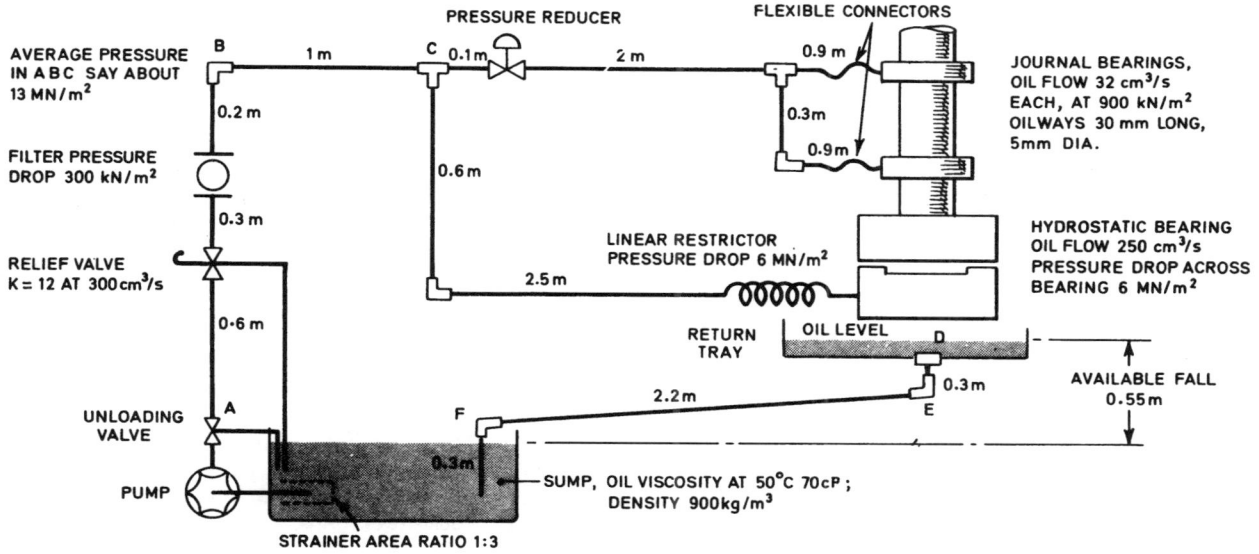

AVERAGE PRESSURE IN A B C SAY ABOUT 13 MN/m²

PRESSURE REDUCER

FLEXIBLE CONNECTORS

B   1 m   C 0.1m   2 m   0.9 m

JOURNAL BEARINGS, OIL FLOW 32 cm³/s EACH, AT 900 kN/m² OILWAYS 30 mm LONG, 5mm DIA.

0.2 m

0.3m

FILTER PRESSURE DROP 300 kN/m²

0.9m

0.3 m

0.6 m

LINEAR RESTRICTOR PRESSURE DROP 6 MN/m²

RELIEF VALVE K = 12 AT 300 cm³/s

2.5 m

HYDROSTATIC BEARING OIL FLOW 250 cm³/s PRESSURE DROP ACROSS BEARING 6 MN/m²

0·6 m

RETURN TRAY   OIL LEVEL   D

AVAILABLE FALL 0.55m

UNLOADING VALVE   A   F   2.2 m   E   0.3 m

PUMP

0.3m

SUMP, OIL VISCOSITY AT 50°C 70cP ; DENSITY 900kg/m³

STRAINER AREA RATIO 1:3

Figure 24.1 Typical lubrication system

Table 24.1 Selection of pipe materials

| Material | Range | | Approx. relative price | | Applications | Notes |
	Bore dia. in	Pressure * lbf/in²	Material only	Installed		
Mild steel	No limit	No limit	2	3	Not recommended. Occasionally used for mains in large high-pressure systems where cost important	Thoroughly clean and de-scale by acid pickling. Seal ends after pickling until installation
Stainless steel	No limit	No limit	12	6	Large permanent systems	Best material for high flow rates. Can be untidy on small systems
Half-hard copper and brass	0.07–2	5000–1000	4	4	Universal. Especially favoured for low flow rates	Brass resists corrosion better than copper. Neat runs and joints possible
Flexible reinforced high-pressure hose	0.125–0.75	5000–2500	9	4	As final coupling to bearing to assist maintenance. Connection to moving part or where subject to heavy vibration	Nylon or rubber base obtainable
Hard nylon	0.125–1.125	700–350	2	2	As cheap form of flexible coupling. Large centralised low-flow lubrication systems for cheapness	Deteriorates in acid atmosphere. Heavily pigmented variety should be used in strong light
PVC reinforced	0.125–0.75	150	2	1	Low-pressure systems. Gravity returns	Readily flexible. Pigmented variety best in sunlight
PVC unreinforced	0.125–2	20–10	1	1	Gravity returns	Pigmented variety best in sunlight. Can be untidy. Relatively easily damaged

* The pressures quoted are approximate working pressures corresponding to the minimum and maximum bores, respectively, and are intended as a guide to selection only. Use manufacturers' values for design purposes.

Table 24.2 Selection of control valves

Valve type	Operation	Common use	Notes	
Bypass (unloading)	Bleeds to drain when primary pressure exceeds given value	Control main circuit pressure	Designed to operate in open position without oil heating. Often included in pump unit, otherwise requires return to sump	
Relief (safety)	As above	Safety device to protect pipework fittings and pump	If used to control circuit pressure, oil heating may result. Do not undersize or forget return to sump	
Non- return (check)	Prevents reverse flow	Prevent back-flow through a bypass	Sometimes incorporated in bypass or relief valve	
Pressure-reducing (regulating)	Gives fixed pressure drop or fixed reduced pressure	Cater for different bearing needs from one supply pressure	Requires drain to sump	
Flow-regulating and -dividing	Controls or divides flow rate irrespective of supply pressure	Pass fraction of flow to open (non-pressure) system or to filter	Approximate control only. Orifices are cheaper alternative, as effective for many applications	
Sequence	Admits oil to secondary circuit only after primary reaches given pressure	Protection of a particular bearing circuit in complex system	Rather uncommon. May be used in hydraulic power/bearing circuit	
Directional	Switches flow on receipt of signal	Activate various sections of circuit as required	Remote or manual operation. For simple systems, stop valves may be better	
Volume metering	Meter quantity of lubricant to bearing on receipt of pressure pulse	Intermittent lubrication system only, placed at lubrication points		

Table 24.3 Pipe sizes and pressure-drop calculations

Item	Equation,* figure	Example	Notes
Determination of pipe bore			
(a) Draw flow diagram		See Figure 24.1	Essential for all but the simplest cases
(b) Size lines assuming velocities:			
(i) 3 m/s (10 ft/s) for delivery	$d = \sqrt{\dfrac{4Q}{\pi v}}$ or	For *ABC* in Figure 24.1:	Adjust sizes, if necessary, after calculation of pressure drops.
(ii) 1 m/s (3 ft/s) for pump suction	use Figure 24.2	$Q = 314$ cm^3/s From Figure 24.2, $d \simeq 12$ mm or 0.5 in	Use reducer, if necessary, in pump suction. If air entrainment likely in return line, size to run half full
(iii) 0.3 m/s (1 ft/s) for return			
(iv) 20 m/s (60 ft/s) for linear restrictors			
Pressure drop—delivery and pump suction lines			Keep suction lines short, free of fittings
(a) Determine correct viscosity for working conditions of temperature and pressure	$\eta = X \times \eta_0$ for pressure, Figure 24.3	From Figure 24.3 at 13 MN/m^2: $X = 1.4$, $\eta = 1.4 \times 70 = 98$ cP	Use supplier's value for viscosity at working temperature (usually 30–50°C above ambient)
(b) Evaluate pipe viscous losses	$P_p = 32\,\dfrac{\eta v l}{d^2}$ or use Figure 24.4	From Figure 24.4 for *ABC*: $P_p = 65$ kN/m^2/m $\times 2.1 = 137$ kN/m^2	Check to ensure flow is laminar (Reynolds number < 2000). Find Re from Fig. 24.4
(c) Determine losses in valves, fittings and strainers	$P_t = k \times \frac{1}{2}\rho v^2$, find K from Table 24.4 Manufacturer's data	For the 90° bend at *B*: $P_c = 2 \times \frac{1}{2} \times 900 \times 9 = 8.1$ kN/m^2 For the relief valve: $P_c = 12 \times \frac{1}{2} \times 900 \times 9 = 48.6$ kN/m^2 For the filter: $P_c = 300$ kN/m^2	Use manufacturers' figures for K or P_c, where available. Where the flow changes direction, e.g. in a bend, losses increase at low Reynolds number ($2\times$ at Re $= 200$, $4\times$ at Re $= 100$ and $8\times$ at Re $= 50$)
(d) Determine total loss	$P = \sum P_p + \sum P_c$	For *ABC*: $P = 137 + 48.6 + 300 + 8.1 \simeq 494$ kN/m^2	The total loss is simply the arithmetical sum of the individual losses
Pressure drop—return lines			
(a) Calculate pressure losses as for delivery lines	As for 2(a) to (d)	For *DEF*: $P_p = 2.8 \times 0.8 = 2.24$ kN/m^2 K (2 bends, entrance and exit) $= 2+2+1+1 = 6$ $P_c = 6 \times \frac{1}{2} \times 900 \times 0.3^2 = 0.25$ kN/m^2 $P = 2.49$ kN/m^2	Keep lines as direct as possible, with minimum fittings. Self-draining lines are best ($1\frac{1}{2}°$ minimum slope)
(b) Express *P* as hydrostatic head of oil	$h = \dfrac{P}{\rho g}$	$h = \dfrac{2490}{900 \times 9.81} = 0.28$ m	h must be less than vertical drop available, i.e. < 0.55 m in the example of Figure 24.1
(c) Check to ensure positive pressure after each fitting		For the entrance *D*: $K = 1$, $P_c = 40$ N/m^2, $h = 45$ mm Thus, return tray must be at least 45 mm deep	Increase bore size if this condition cannot be met. The entrance to the return pipe is often a source of difficulty
Pressure drop—coiled pipes			
Linear restrictors for hydrostatic bearing circuits are often coiled capilliary tubing. Coiling increases the pressure drop and must be allowed for when calculating the length of the restrictor	$P_p = \dfrac{32\eta v l}{d^2}$ or use Fig. 4 $P_t = P_p \times Z$, Z from Fig. 6	For *G* in Figure 24.1: $Q = 250$ cm^3/s; $d = 4$ mm, say, when $v = 18$ m/s (Figure 24.2) $P_p = 3.5$ MN/m^2/m (Figure 24.4), giving $l = 1.71$ m for 6 MN/m^2 Now Re $= 650$ (Figure 24.5) and if D (dia. of coil) $= 36$ mm, $Z = 2$ Thus required $l = 1.71/2 = 0.85$ m	Fine bore tubes make better restrictors than orifices for large pressure drops Base calculations on actual bore sizes not nominal bores If l impractically long or short, increase or reduce v, respectively, and repeat calculation

* When using the equations with Imperial units, these must be self-consistent. For example, Q in in^3/s; v in in/s; d, h, e in in; η in reyns (1 reyn $= 69\,000$ poise); ρ in slugs/in^3 (1 slug/in$^3 = 32.2$ lb/in^3); p in lbf/in^2; $g = 386$ in/s^2

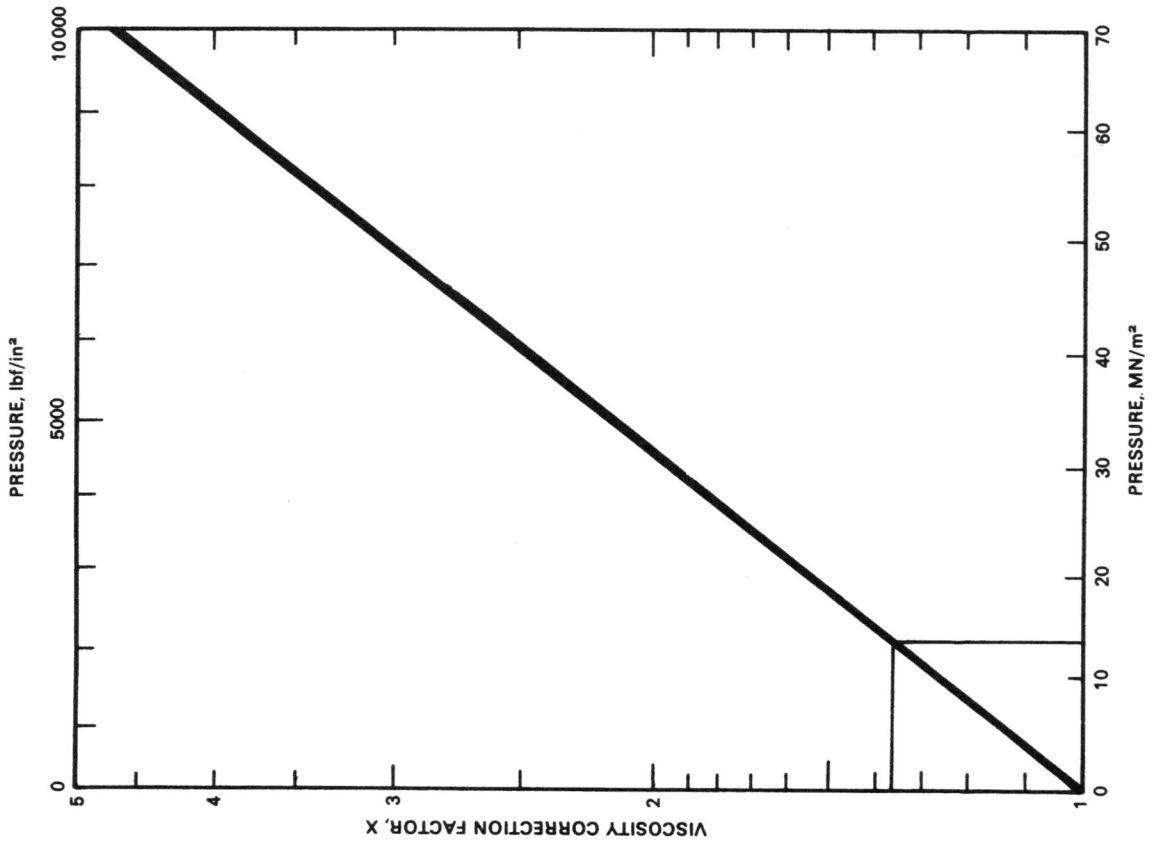

Figure 24.3 Viscosity correction factor, X, for mineral oils only

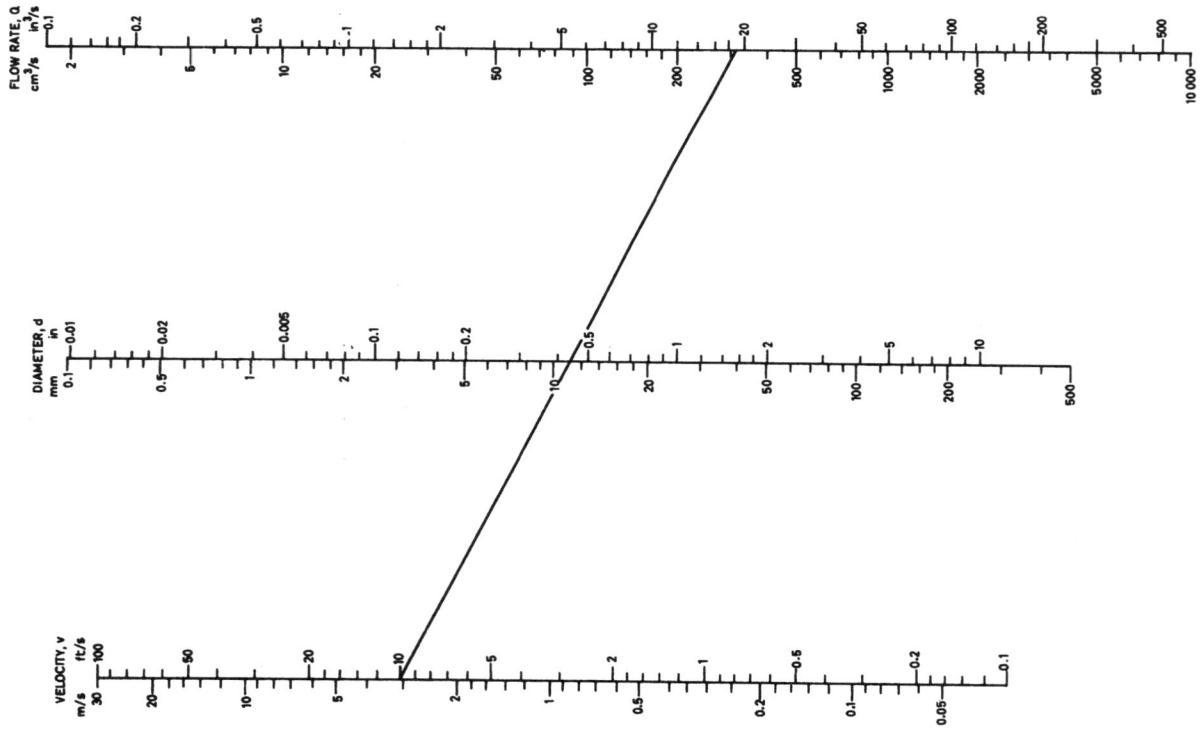

Figure 24.2 Nomogram for determination of pipe bore

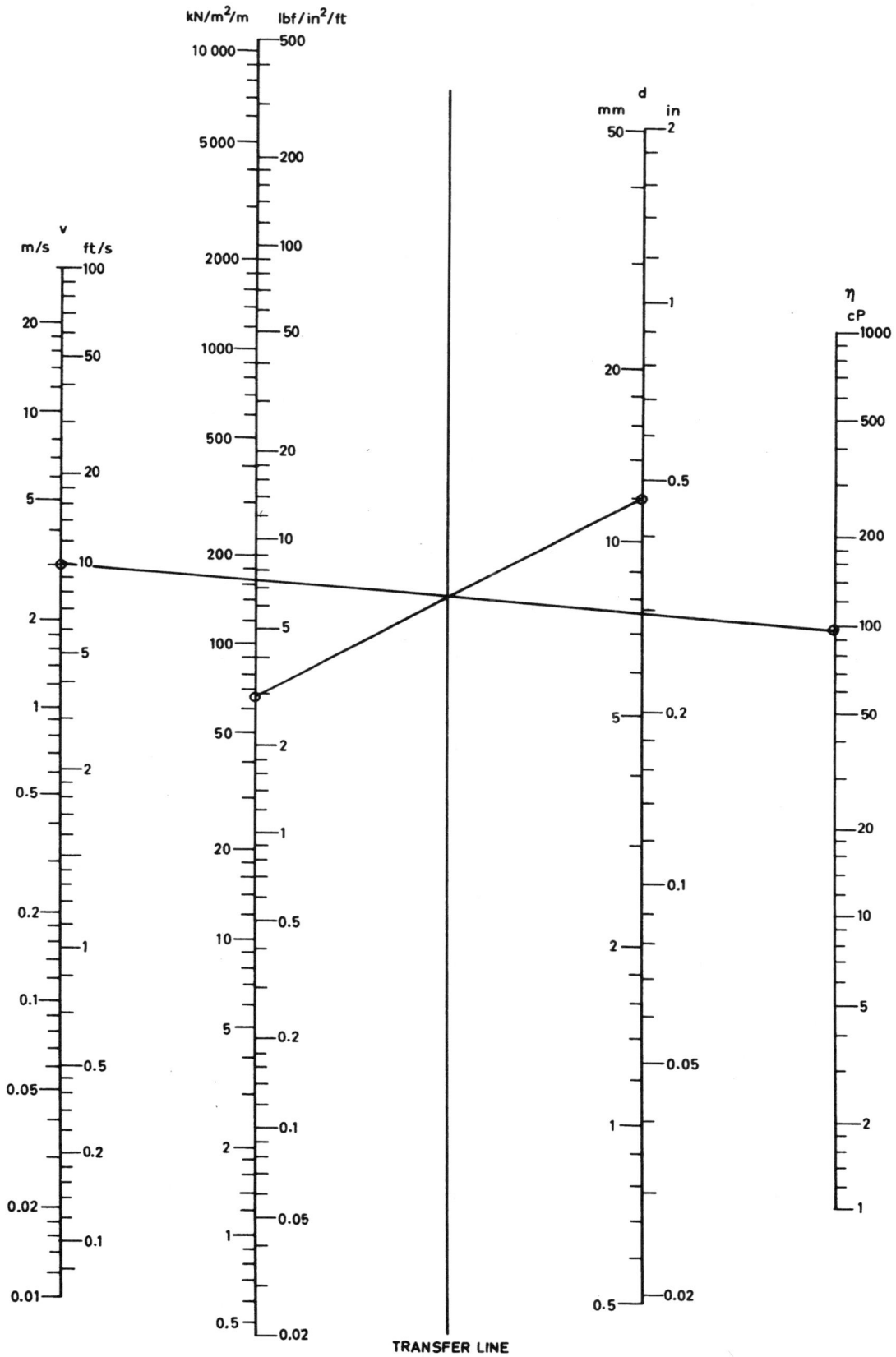

Figure 24.4 Pressure losses per unit length in pipes (Re < 2000)

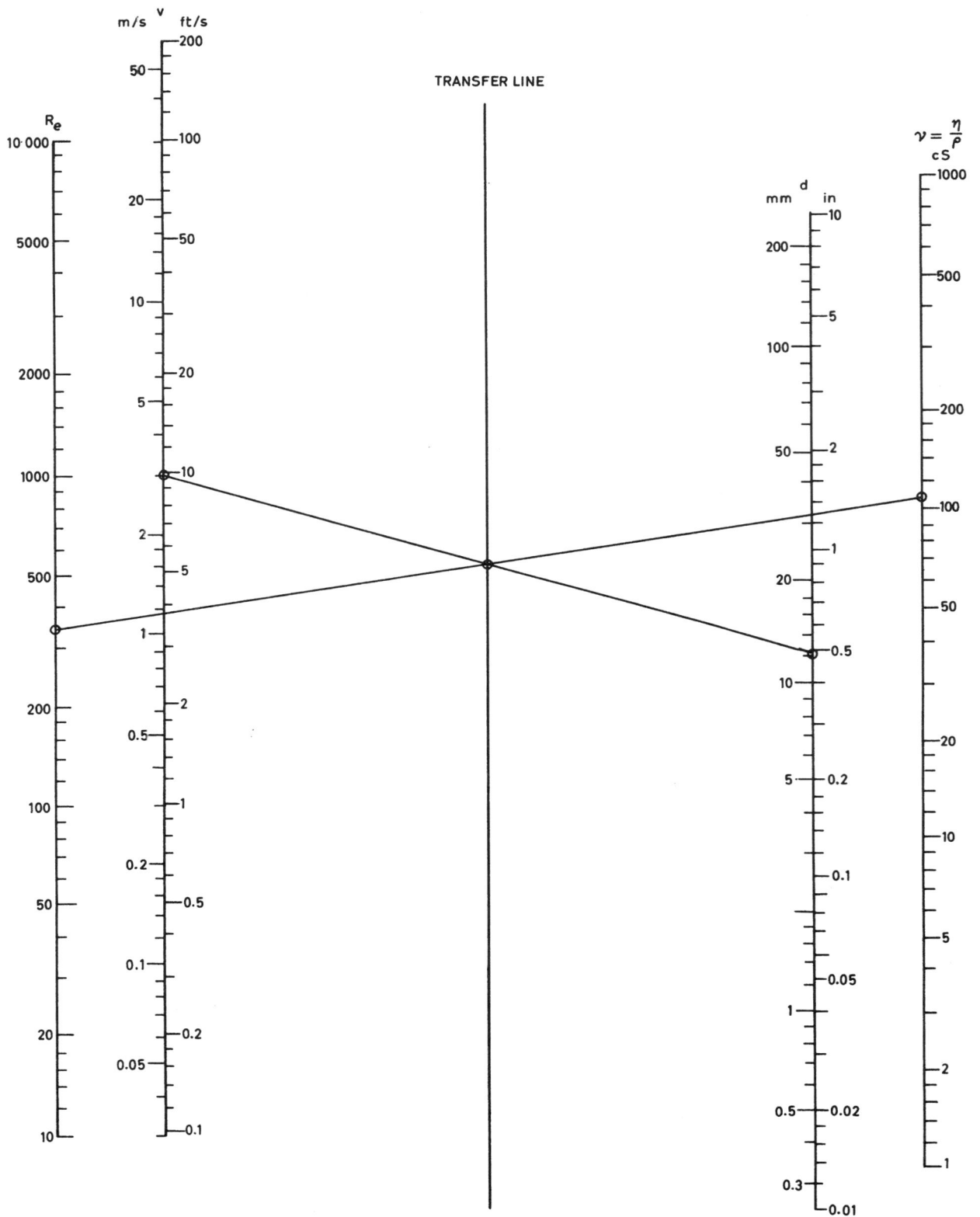

Figure 24.5 Nomogram for Reynolds No. $Re = \dfrac{\rho V d}{\eta} = \dfrac{V d}{\nu}$

Table 24.4 Loss coefficients

	Enlargement	Contraction		Restrictions		Bends

(a)

FITTINGS

	Enlargement	Contraction	Restrictions	Bends
$K =$	1	0·5	$K \simeq (A/a)^2$ where A is the area of the pipe and a the open area available	2

	Entrances			Exit	T-junction	
$K =$	0	0.5	1	1	3.0	0.5

(b)

VALVES

		Stop		Non-return		Spool
	Gate	Plug	Needle	Flap	Ball	
$K =$	2	20	60	10	50–100	5–100

Use manufacturers' figures, where available. Approximate loss coefficient in doubtful cases may be obtained from the formula

$$K = \left(\frac{\text{Cross-sectional area of approach pipe}}{\text{Minimum cross-sectional area of valve}} \right)^2$$

(c)

STRAINERS

Approximate loss coefficient $K = \left(\dfrac{\text{Cross-sectional area of pipe}}{\text{Open area of strainer}} \right)^2$

(d)

FILTERS

Not susceptible to calculation—use manufacturers' data. If a relief bypass valve is included the actual pressure drop may be two or three times the nominal setting.

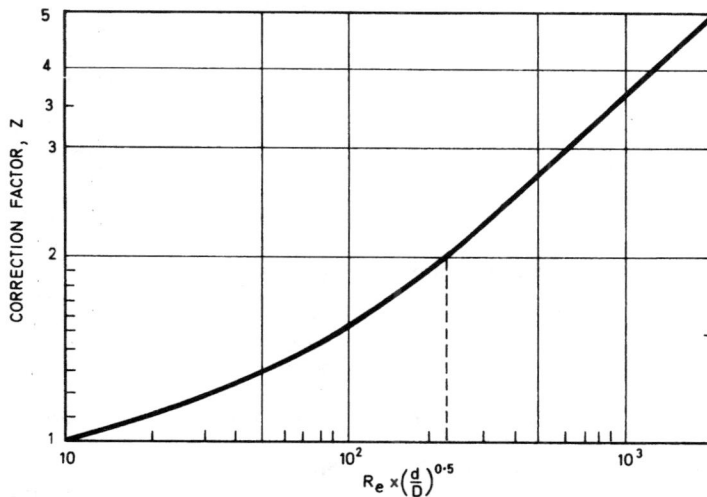

Figure 24.6 Correction factor Z for flow through curved capillary tubes of bore diameter d and coil diameter D

A25 Selection of warning and protection devices

Satisfactory operation of a centralised recirculatory lubrication system requires adequate control and instrumentation to ensure continuous delivery of the correct volume of clean oil at the design pressure and temperature.

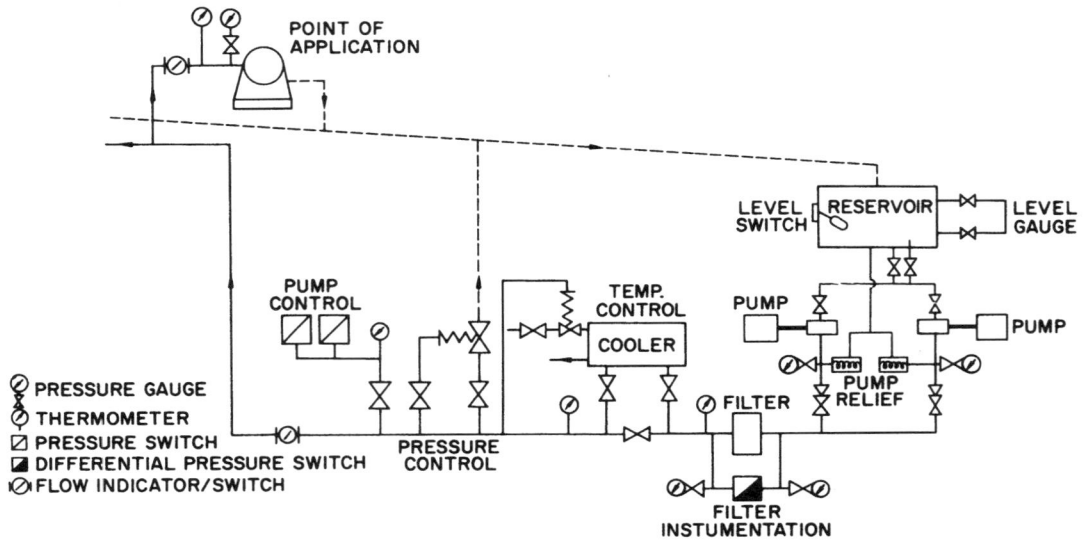

Figure 25.1 A basic lubrication system complete with warning and protection devices

Table 25.1 The function of each major system component and the device required to provide the information or control necessary to maintain that function

System component	Function of component	Information or control required to maintain function	Device required	Comment on application
Reservoir	Maintenance of required volume of oil	Level indication	Level gauge	Visual indication of contents
		Level control	Level switch	(1) High level warning of overfilling or ingress of water
				(2) Low level warning of system leak or normal consumption
Pump(s)	Delivery of the required quantity of oil at prescribed pressure	Pressure indication	Pressure gauge	(1) Situated at pump discharge, is necessary when adjusting pump relief valve
				(2) Situated at a point downstream of system equipment, is necessary when adjusting pressure control valve
		Pressure control	(1) Pressure switch(es)	To switch in standby pump on falling pressure. Give warning of falling pressure
			(2) Pressure control valve or other method	To spill-off surplus oil and regulate pressure variations due to temperature fluctuations and changes in system demand
		Flow indication and control	Flow switch	Indicates satisfactory flow is established
		Pump protection	Spring relief valve	Protection from overpressure due to system malfunction

Table 25.1 The function of each major system component and the device required to provide the information or control necessary to maintain that function (continued)

System component	Function of component	Information or control required to maintain function	Device required	Comment on application
Filter(s)	Maintenance of cleanliness of oil	Condition of filter	(1) Pressure gauges	Visual indication of pressure drop across filter
			(2) Differential pressure switch	To signal when pressure drop has increased to a pre-determined value and filter requires cleaning
Cooler(s)	Maintenance of prescribed temperature of oil supply	Temperature indication	Thermometer	Visual indication of temperature of oil into and out of cooler. Is necessary when adjusting temperature control valve
		Temperature control	Temperature control valve	Regulates temperature of oil leaving cooler
Point of application to lubricated parts	Dispense correct quantity of oil at prescribed pressure and temperature	Oil pressure, oil temperature, oil flow	Pressure gauge Thermometer Flow indicator	Local visual indication or fitted with contacts to give malfunction signal

Table 25.2 Some protective devices available with guidance on their selection and installation

Device	Some types	Comments on selection	Installation
Level gauge	(1) Dip-stick	Cheap, simple to make	Through reservoir top
	(2) Glass tube	Direct reading, simple in design, requires protecting with metal tube or cage	On reservoir side complete with shut-off cocks
	(3) Dial. Float actuated	Direct reading. For accuracy of calibration reservoir dimensions, shape and any internal obstructions, also specific gravity of oil must be considered	On reservoir side or top
	(4) Dial. Hydrostatic, pneumatic operated	Remote reading. Comments as for (3). Absolutely essential capillary joints are positively sealed against leak	Through reservoir top or side with dial remote panel mounted
Level switch	(1) Float actuated	Usually magnetic operation, glandless, therefore leakage from reservoir into switch housing not possible. Some types are level adjustable	Through reservoir top or side, or, on reservoir side in float chamber complete with isolating valves
	(2) Sensing probe	Techniques generally used; capacitance, conductivity or resistance. Complete lubricant characteristics must be provided to supplier for selection purposes. Not adjustable after fitting	Through reservoir top or side
Pressure gauge	(1) Bourdon tube	Available for front flange, back flange or stem mounting. Any normal pressure range	Local stem mounted or remote panel mounted, complete with shut-off cock
	(2) Diaphragm actuated	More robust construction, more positive indication than Bourdon tube type, mainly stem mounting. Withstand sudden pressure surges, overload pressures, overheating. Any normal pressure range	As for (1)
Pressure switch	(1) Bellows actuated (2) Piston actuated (3) Bourdon tube	Select switch to satisfy pressure and differential range requirements, also oil temperature. (1) and (3) suitable for oil and air. (2) limited to oil applications; will withstand higher pressures	Local stem mounted or remote panel mounted

Table 25.2 Some protective devices available with guidance on their selection and installation (continued)

Device	Some types	Comments on selection	Installation
Pressure control valve	(1) Standard high-lift spring-loaded relief	Simple design, cheap. Is viscosity sensitive but normally satisfactory for control of small simple systems	Piped-in branch from main delivery after filter. Discharge back to reservoir
	(2) Direct operated diaphragm	Used on larger systems, this valve will maintain control within acceptable limits provided viscosity remains reasonably stable	Fit as for (1). Pressure signal transmitted to diaphragm from tapping downstream of filter and cooler
	(3) Pneumatically controlled diaphragm	Employs same type of valve as in (2) but the diaphragm is actuated by low pressure air via a control instrument, providing more positive control. Used in place of a direct operated valve, where large diaphragm loads might occur	Fit as (1). Air control remote panel mounted. Pressure signal transmitted as (2)
Header tank		Control by use of static-head and provide emergency oil supply on failure of pumps	Erect at height equivalent to system pressure requirement
Flow switch	(1) Vane actuated	Simple design, robust construction. Can be obtained to give visual as well as electrical indication. Vane angle should be between 30° to 60° for maximum sensitivity	Pipe in line
	(2) Orifice with differential pressure switch	Used when very precise control is required. Orifice size and positioning of pressure tappings is critical	Pipe in line
Spring relief valve	(1) Standard safety valve	Simple design, either chamfered or flat seat	Integral with pump arranged for internal relief, or, fitted remote for external relief
	(2) High lift	Skirted flat seat, designed to give small difference between opening pressure and full flow pressure. Use where pressure accumulation must be minimal	
Differential pressure switch	(1) Opposed bellows actuated	Comments as for pressure switches	Local or remote panel mounted complete with shut-off cocks
	(2) Opposed piston actuated	Comments as for pressure switches	Pressure tappings close to filter inlet and outlet
Thermometer	(1) Bi-metallic	Dial calibration, almost equal spacing	Local vertical or co-axially mounted in separable pocket
	(2) Vapour-pressure	Dial calibration, logarithmic-scale. Changes of ambient have no effect on reading Allowance must be made if difference in height between bulb and dial exceeds 2 m. Maximum capillary length about 10 m	Bulb in separable pocket. Dial either local or remote panel mounted
	(3) Mercury-in-steel	Dial calibration, equally spaced. Ambient changes have minimal effect on reading No error of importance results from difference of height between bulb and dial. Maximum capillary length about 50 m	Install as for (2)
Temperature control valve	(1) Direct operating with integral bi-metallic element	Reverse acting. Control is not instantaneous, but is generally acceptable for many systems. Restricts flexibility of pipe routing. Care necessary in adjusting when control is operating	Valve in cooling water supply, operating element in the cooler oil outlet
	(2) Vapour pressure actuated, plunger operated	Reverse acting, reliable, protected against over-temperature. Can adjust when control is operating	Valve in cooling water supply. Remote sensing bulb in separable pocket in cooler oil outlet
	(3) Pneumatically controlled diaphragm	Employs valve similar to pressure control, reverse acting Diaphragm is actuated by low pressure air, providing more sensitive control	Valve as (2). Air controller remote panel mounted. Remote sensing bulb in separable pocket in cooler oil outlet

TOTAL-LOSS SYSTEMS

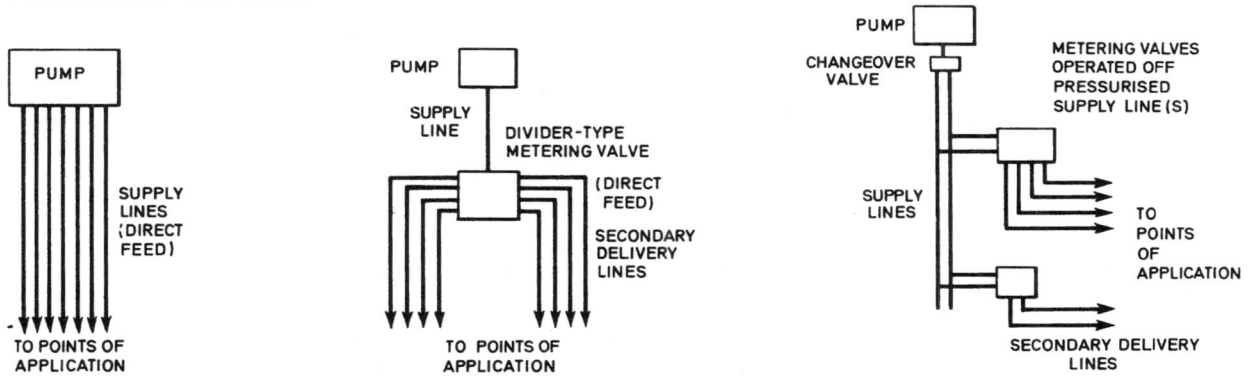

Figure 26.1 Schematic diagrams of typical total-loss systems – lubricant is discharged to points of application and not recovered

Commissioning procedure

1 Check pumping unit.
2 Fill and bleed system. *Note:* it is not normally considered practicable to flush a total-loss system.
3 Check and set operating pressures.
4 Test-run and adjust.

No special equipment is required to carry out the above procedure but spare pressure gauges should be available for checking system pressures.

Pumping unit

PRIME MOVER
For systems other than those manually operated, check for correct operation of prime mover, as follows.

(*a*) Mechanically operated pump – check mechanical linkage or cam.
(*b*) Air or hydraulic pump:
   (i)   check air or hydraulic circuit,
   (ii)   ascertain that correct operating pressure is available.
(c) Motor-operated pump:
   (i)   check for correct current characteristics,
   (ii)   check electrical connections,
   (iii)   check electrical circuits.

PUMP
(*a*) If pump is unidirectional, check for correct direction of rotation.
(*b*) If a gearbox is incorporated, check and fill with correct grade of lubricant.

CONTROLS
Check for correct operation of control circuits if incorporated in the system, i.e. timeclock.

RESERVOIR
(*a*) Check that the lubricant supplied for filling the reservoir is the correct type and grade specified for the application concerned.
(*b*) If the design of the reservoir permits, it should be filled by means of a transfer pump through a bottom fill connection via a sealed circuit.

(*c*) In the case of grease, it is often an advantage first to introduce a small quantity of oil to assist initial priming.

Filling of system

SUPPLY LINES
These are filled direct from the pumping unit or by the transfer pump, after first blowing the lines through with compressed air.

In the case of direct-feed systems, leave connections to the bearings open and pump lubricant through until clean air-free lubricant is expelled.

In the case of systems incorporating metering valves, leave end-plugs or connections to these valves and any other 'dead-end' points in the system open until lubricant is purged through.

With two-line systems, fill each line independently, one being completely filled before switching to the second line via the changeover valve incorporated in this type of system.

SECONDARY LINES (Systems incorporating metering or dividing valves)
Once the main line(s) is/are filled, secure all open ends and after prefilling the secondary lines connect the metering valves to the bearings.

System-operating pressures

PUMP PRESSURE
This is normally determined by the pressure losses in the system plus back pressure in the bearings.

Systems are designed on this basis within the limits of the pressure capability of the pump.

Check that the pump develops sufficient pressure to overcome bearing back pressure either directly or through the metering valves.

In the case of two-line type systems, with metering valves operating 'off' pressurised supply line(s), pressures should be checked and set to ensure positive operation of all the metering valves.

Running tests and adjustments

SYSTEM OPERATION
Operate system until lubricant is seen to be discharging at all bearings. If systems incorporate metering valves, each valve should be individually inspected for correct operation.

ADJUSTMENT
In the case of direct-feed systems, adjust as necessary the discharge(s) from the pump and, in the case of systems operating from a pressure line, adjust the discharge from the metering valves.

RELIEF OR BYPASS VALVE
Check that relief or bypass valve holds at normal system-operating pressure and that it will open at the specified relief pressure.

CONTROLS
Where adjustable electrical controls are incorporated, e.g. timeclock, these should be set as specified.

ALARM
Electrical or mechanical alarms should be tested by simulating system faults and checking that the appropriate alarm functions. Set alarms as specified.

Fault finding

Action recommended in the event of trouble is best determined by reference to a simple fault finding chart as illustrated in Table 26.2.

CIRCULATION SYSTEMS

Commissioning procedure

1 Flush system. *Note:* circulation systems must be thoroughly flushed through to remove foreign solids.
2 Check main items of equipment.
3 Test-run and adjust.

No special equipment is required to carry out the above but spare pressure gauges for checking system pressures, etc., and flexible hoses for bypassing items of equipment, should be available.

Flushing

1 Use the same type of oil as for the final fill or flushing oil as recommended by the lubricant supplier.
2 Before commencing flushing, bypass or isolate bearings or equipment which could be damaged by loosened abrasive matter.

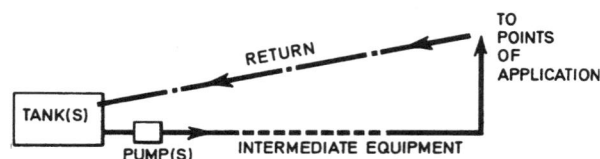

Figure 26.2 Schematic diagram of typical oil-circulation system. Oil is discharged to points of application, returned and re-circulated.

3 Heat oil to 60–70°C and continue to circulate until the minimum specified design pressure drop across the filter is achieved over an eight-hour period.
4 During flushing, tap pipes and flanges and alternate oil on an eight-hour heating and cooling cycle.
5 After flushing drain oil, clean reservoir, filters, etc.
6 Re-connect bearings and equipment previously isolated and refill system with running charge of oil.

Main items of equipment

RESERVOIR
(a) Check reservoir is at least two-thirds full.
(b) Check oil is the type and grade specified.
(c) Where heating is incorporated, set temperature-regulating instruments as specified and bring heating into operation at least four hours prior to commencement of commissioning.

ISOLATING AND CONTROL VALVES
(a) Where fitted, the following valves must initially be left open: main suction; pump(s) isolation; filter isolation; cooler isolation; pressure-regulator bypass.
(b) Where fitted, the following valves must initially be closed: low suction; filter bypass; cooler bypass; pressure-regulator isolation; pressure-vessel isolation.
(c) For initial test of items of equipment, isolate as required.

MOTOR-DRIVEN PUMP(S)
(a) Where fitted, check coupling alignment.
(b) Check for correct current characteristics.
(c) Check electrical circuits.
(d) Check for correct direction of rotation.

PUMP RELIEF VALVE
Note setting of pump relief valve, then release spring to its fullest extent, run pump motor in short bursts and check system for leaks.
    Reset relief valve to original position.

CENTRIFUGE
Where a centrifuge is incorporated in the system, this is normally commissioned by the manufacturer's engineer, but it should be checked that it is set for 'clarification' or 'purification' as specified.

FILTER
(a) Basket and cartridge type – check for cleanliness.
(b) Edge type (manually operated) – rotate several times to check operation.
(c) Edge type (motorised) – check rotation and verify correct operation.
(d) Where differential pressure gauges or switches are fitted, simulate blocked filter condition and set accordingly.

PRESSURE VESSEL
(a) Check to ensure safety relief valve functions correctly.
(b) Make sure there are no leaks in air piping.

PRESSURE-REGULATING VALVE
(a) Diaphragm-operated type – with pump motor switched on, set pressure-regulating valve by opening isolation valves and diaphragm control valve and slowly closing bypass valve.
Adjust initially to system-pressure requirements as specified.
(b) Spring-pattern type – set valve initially to system-pressure requirements as specified.

COOLER
Check water supply is available as specified.

Running tests and adjustments
(1) Run pump(s) check output at points of application, and finally adjust pressure-regulating valve to suit operating requirements.
(2) Where fitted, set pressure and flow switches as specified in conjunction with operating requirements.
(3) Items incorporating an alarm failure warning should be tested separately by simulating the appropriate alarm condition.

Fault finding
Action in the event of trouble is best determined by reference to a simple fault finding chart illustrated in Table 26.1.

FAULT FINDING

Table 26.1 Fault finding – circulation systems

Condition	Cause	Action
(1) System will not start	Incorrect electrical supply to pump motor	Check electrics
(2) System will not build up required operating pressure	(a) Leaking pipework	Find break or leak and repair
	(b) Regulating bypass valve open	Check and close
	(c) Pressure-regulating valve wrongly adjusted	Check and re-set to increase pressure
	(d) Loss of pump prime	Check suction pipework for leaks
	(e) Blocked system filter	Inspect and clean or replace
	(f) Low level of oil in reservoir	Top up
	(g) Pump relief valve bypassing	Check and repair or replace as necessary
	(h) Faulty pump	Check and repair or replace as necessary
(3) System builds up abnormal operating pressure	(a) Pressure-regulating isolation and diaphragm-control valves closed	Check and open
	(b) Pressure-regulating valve wrongly adjusted	Check and re-set to decrease pressure
(4) Oil fails to reach points of application		
(i) via orifices or sprays	Blockage or restriction	Clean out
(ii) via flow control valves	(a) Control valve wrongly adjusted	Check and re-set
	(b) Blockage or restriction in delivery line	Clean out or replace as necessary
(iii) via metering valves	(a) Metering valve contaminated or faulty	Clean out or replace
	(b) Blockage or restriction in delivery line	Clean out or replace
(iv) via pressurised sight feed indicators	Air supply failure	Check and restore correct air supply

Table 26.2 Fault finding – total-loss systems

Condition	Cause	Action
1 System will not start (automatically operated systems)	(a) Incorrect supply to pump prime mover	Check electrical, pneumatic or hydraulic supply
	(b) System controls not functioning or functioning incorrectly	Check controls and test each function separately
2 System will not build up required operating pressure, or normal pumping time is prolonged	(a) Faulty pump	Isolate pump, check performance and repair or replace as necessary
	(b) Loss of pump prime	(1) Bleed air from pump (2) Check for blockage in reservoir line strainer
	(c) Relief valve bypassing	If with relief valve isolated pump builds pressure, repair or replace relief valve
	(d) Broken or leaking supply line(s)	Find break or leak and repair
	(e) Air in supply line(s)	Bleed air from line(s) as for filling
(systems incorporating metering valves)	(f) Bypass leakage in metering valve(s)	Trace by isolating metering valves systematically and repair or replace faulty valve(s)
3 System builds up operating pressure at pump but will not complete lubrication cycle	(a) Changeover control device faulty	Isolate from system, check and repair or replace
(systems incorporating electrical and/or pressure control devices in their operation)	(b) Incorrect piping to control device	Check and rectify as necessary
	(c) Incorrect wiring to control device	Check and rectify as necessary
4 System builds correct operating pressures but metering valve(s) fails/fail to operate	If with delivery line(s) disconnected, valve(s) then operates/operate:	
(systems incorporating· metering valves operating 'off' pressurised supply line/s)	(a) Delivery piping too small	Change to larger piping
	(b) Blockage or restriction in delivery line	Clean out or replace
	(c) Blockage in bearing	Clean out bearing entry and/or bearing
	If with delivery line(s) disconnected, valve(s) still fails/fail to operate	Clean out or replace
	(d) Metering valve contaminated or faulty	
5 System builds excessive pressure	(a) Blockage or restriction in main supply line	Trace and clean out or replace faulty section of pipework
	(b) Supply pipe too small	(1) Change to larger piping (2) Lubricant too heavy Seek agreement to use lighter lubricant providing it is suitable for application
(direct-feed systems)	(c) Delivery line piping too small	Change to larger piping
,,	(d) Blockage or restriction in delivery line	Clean out or replace
,,	(e) Blockage in bearing	Clean out bearing entry and/or bearing
(direct-feed systems incorporating divider-type metering valves)	(f) Metering valve(s) contaminated or faulty [see 4(d) above]	Clean out or replace

1 GENERAL REQUIREMENTS

Condition required	Requirements not met by new assemblies because of:	Condition produced by running-in
Macro-conformity	Latitude in fits, tolerances, assembly, alignment Mechanical deflections of shafts, gear teeth, etc. Thermal and mechanical distortion of bearing housings, cylinder liners, etc. Differential expansion of bearing materials	Increase in apparent area of contact and the relief of contact stresses Reduction in friction and temperature
Micro-conformity	Initial composite surface roughness exceeds oil film thickness obtainable with smooth surfaces	Decrease in roughness. Increase in effective bearing area (Figure 26.1) coherence of oil film and, ideally, full film lubrication
Cleanliness	Casting sand, machining swarf, wear products of running-in, etc.	Safe expulsion of contaminants to filters
Protective layers	Scuff- and wear-resistant low-friction surfaces are not naturally present on freshly machined surfaces	Safe formation of work-hardened layers, low-friction oxide films, oil-additive reaction films, substrate extrusion films, e.g. graphite

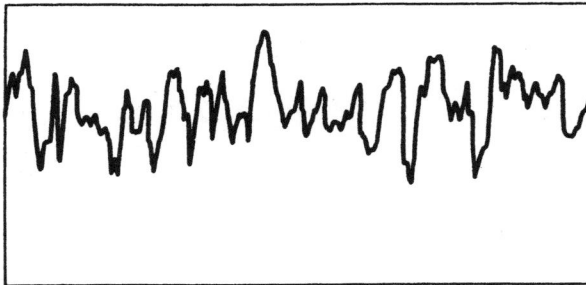

Un-run cylinder liner

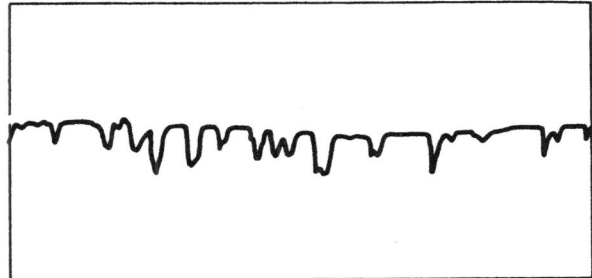

Run-in cylinder liner

Figure 27.1 Profilometer traces (vertical magnification 5 times the horizontal)

Running-in to achieve micro-conformity can be monitored by surface finish measurement and analysis before and after the running-in process. Surface finish criteria such as R_a (CLA) and bearing area curves are likely to be the best. The comparison of these parameters with subsequent reliability data can guide manufacturers on any improvements needed in surface finish and in running-in procedures. No generally applicable rule of thumb can be given.

2 RELATIVE REQUIREMENTS

The running-in requirement of assembled machinery is that of its most critical part. The list below rates the ease of running-in of common tribological contacts.

Most critical		
	Piston rings and liners	Highly critical especially in IC engines—see subsection 3
	Gears	Watch hypoids—see subsection 4
	Cams and tappets	Commonly catered for by needs of piston rings, otherwise see subsection 4
	Rubbing plain bearings Porous plain bearings Fluid film plain bearings	See subsection 5
Least critical	Hydrodynamic thrust bearings Rolling element bearings Externally pressurised bearings Gas bearings	Assuming correct design, finish and assembly no running-in is required. For certain high-precision rolling element (e.g. gyro) bearings, and for operation in reactive gas environment, consult makers

3 RUNNING-IN OF INTERNAL COMBUSTION ENGINES

The most effective running-in schedule for new and rebuilt engines depends to a large extent on the individual design of engine and materials used. It is therefore important to follow the maker's recommendations. In the absence of a specific schedule the following practice is recommended.

Running-in on dynamometer

Stage	Duration h (see Note 2)	% Max. b.m.e.p.	% Rated max. speed	Remarks
1	Until temperatures stabilise	0	30	
2	0.5	25	30	
3	0.5	50	50	
4	1.0	75	75	
5	0.1	100	100	Final power check
6	10	75	75	In service: limited max. duty period

Notes:
1 Where possible Stage 1 should be preceded by 0.5 h motoring at 30% rated max. speed.
2 Durations should be increased by (cumulative) factors according to engine design as follows:

Factor	Increase durations by:
Bores exceeding 150 mm (6 in) dia.	2 times
Bores exceeding 600mm (25 in) dia.	4 times
Max. load exceeding 10 bar (150 lb/in^2) b.m.e.p.	2 times
Max. load exceeding 15 bar (200 lb/in^2) b.m.e.p	4 times
Max. mean piston speed exceeding 20 m/s (3000 ft/min)	2 times

3 Change oil, tighten cylinder head, re-set valve clearances, etc., after Stage 6.
4 It may be possible to shorten these running-in times by progressive analysis of the amount and shape of wear particles in the lubricating oil. Ferrography, described later in this section, can be used for this.

Running-in a road vehicle

Stage	Duration h	Duty
1	Until temperatures stabilise	Vehicle stationary. 30% max. rated speed
2	2	Main road (avoiding hills, cities, motorways). No payload. Max. speed 50% rated max. Avoid 'slogging'. Set fast idle
3	10	Main road. With payload. Max. speed 75% rated max. Avoid 'slogging'. Set fast idle. Change oil, tighten cylinder head, reset valve clearances, etc., after this stage

Note: Durations should be modified according to engine design, as in dynamometer schedule—see previous Note 2.

Monitoring running-in

The following observations provide a guide as to the completeness of the running-in process:

Observe	Expect
Coolant temperature	Possible overheating
Oil temperature	Gradual decrease and stabilisation
Blow-by rate	Target: 20–40 l/kW(0.5–1 ft^3/h per b.h.p.)
Crankcase pressure	Gradual decrease and stabilisation
Exhaust smoke	Gradually clears
Specific fuel consumption Maximum power	Approach rated values
Oil consumption	Target: 0.5–1% of fuel consumption
Compression pressure	Gradual increase and stabilisation

In research and development the following additional observations provide valuable guidance:

Observe	By
Surface roughness	Profilometer traces on liner replicas (Figure 26.1)
Surface appearance	Photographs of liner replicas (Figure 26.2)
Wear debris generation	Radioactive tracers, iron and chromium analysis of oil samples
Engine friction	Willan's line,* Morse test,† hot motoring

* *Willan's line*—Applicable to diesel engines. At constant speed, plot fuel consumption *v.* b.m.e.p. Extrapolation to zero consumption gives friction m.e.p.

† *Morse test*—Applicable to multi-cylinder gasoline engines. At constant speed, observe reduction in torque when cutting each cylinder in turn. Assume reduction is the indicated m.e.p. of idle cylinder. Calculate friction m.e.p. as difference between average indicated m.e.p.'s and b.m.e.p.

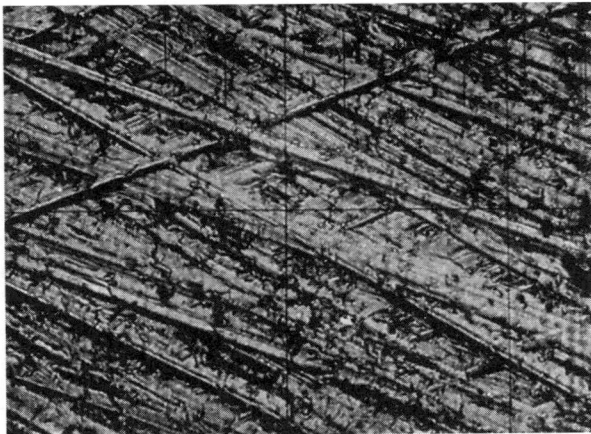

Figure 27.2(a) Un-run cylinder liner ×140

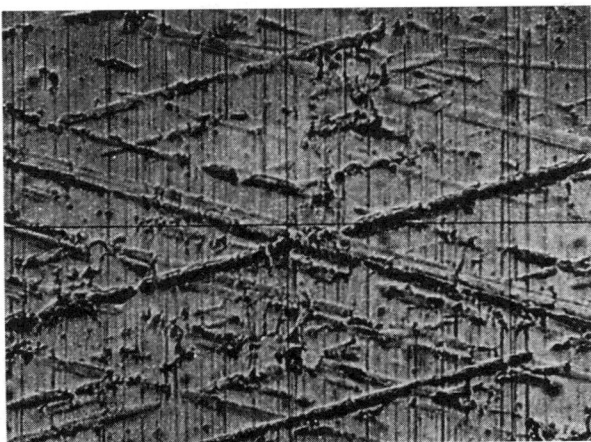

Figure 27.2(b) Run-in cylinder liner × 140

Running-in accelerators

Running-in accelerators should only be used in consultation with the engine maker. Improper use can cause serious damage.

Accelerator	Precautions
Abrasive added to intake air	Ensure even distribution and slow feed rate Danger of general abrasive wear Change oil and filters afterwards
Running-in additive in fuel	Consult oil supplier and engine maker to establish dosage and procedure Change oil and filters afterwards
Running-in oil	Use as recommended. Ensure change to service oil afterwards
Abrasive and other ring coatings	No special precautions
Taper-faced rings	Ensure that they are assembled with lower edge in contact with bore

Ferrography

Ferrography is a technique of passing a diluted sample of the lubricating oil over a magnet to extract ferrous particles. It has found useful application in running-in studies aimed at shortening running-in of production engines and so making possible large cost savings. The principle is to examine suitably diluted samples of engine oil to obtain, during the process of running-in, a measure of the content of large (L) and small (S) particles. Over a large number of dynamometer tests on new production engines a trend of 'Wear Severity Index' $(I_s = L^2 - S^2)$ with time may be discerned which allows comparison to be made between the effectiveness of running-in schedules.

4 RUNNING-IN OF GEARS

Procedures

It is not feasible to lay down any generally applicable running-in procedure. The following guiding principles should be applied in particular cases:

Avoid	Otherwise
High speed—high load combinations	High temperatures develop—scuffing
Low speed—high load combinations	Poor elastohydrodynamic conditions and possible plastic flow
High speed, shock loading	Scuffing—particularly met on over-run of hypoids

Observing progress of running-in

Observe	Remarks
Oil temperature	See Fig. 3
Wear debris analysis Depletion of oil additives	Ensure sample is homogeneous and representative of bulk oil
Surface roughness (of replicas) Additive deposition by autoradiography Gear efficiency increase	Research and development techniques

Materials and lubricants

See also Sections A23, 24, 25. Running-in has been found to be influenced by materials and lubricants broadly as follows:

Factor	Effect
Excessive hardness (>400 VPN) Carburising Nitriding Anti-wear additives in oil	Can delay running-in
EP additives in oil Thermo-chemical treatments, e.g. phosphating Increased oil viscosity	Can prevent scuffing during running-in

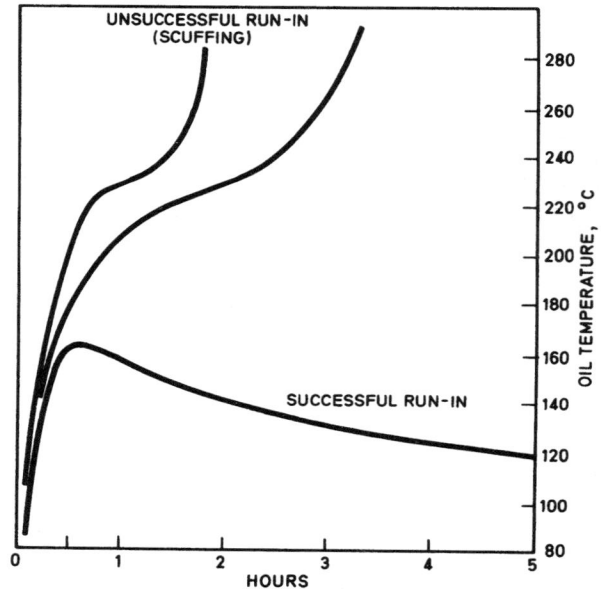

Figure 27.3 Examples of oil temperature variation during early life of hypoid axles

5 RUNNING-IN OF PLAIN BEARINGS
Special running-in requirements

Type of bearing	Requirement
Rubbing (non-metallic)	Running-in desirable for successful transfer of protective layer, e.g. PTFE to countersurface
Porous	Running-in desirable to achieve conformity without overheating impregnant oil
Fluid film	Usually none unless marginally lubricated—mainly a precaution against minor assembly errors. For piston rings, see subsection 3

Procedure

Principle	Restrict speed and load to limit temperature rise
Practice* (in order of preference)	1 Run at reduced speed and load 2 Run at reduced speed, normal load 3 Run at reduced load, normal speed 4 Run at normal load and speed for short bursts, cooling between
Observe	Bearing temperature (Figure 26.4) When curves repeat, run-in at those conditions is complete

* In all cases run with increased oil flow if possible

6 RUNNING-IN OF SEALS

Rubbing seals, both moulded and compression, undergo a bedding-in process. No general recommendations can be given but the following table summarises experience:

Typical bedding-in duration	2 h
Typical friction reduction	50%
Reduced speed operation during bed-in	Desirable but not essential
Pre-lubrication during fitting	Essential
Renewal of seals	If possible, never refit a bedded-in seal*
Compression seals	Allow excess leakage to start with, tighten as bedding-in proceeds

* This is because seals often harden in service, making re-accommodation difficult; flexing during removal may cause cracking

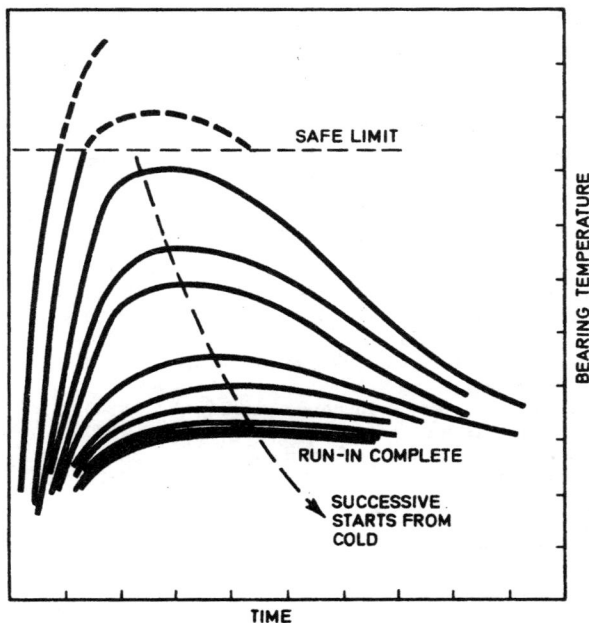

Figure 27.4 Typical effects of running-in on warm-up of plain journal bearings

Table 28.1 Effects of atmospheric conditions

Atmosphere	Effect	Precautionary measures
High temperature, ambient and radiation from hot sources	Lowers lubricant viscosity. Plastics and lubricants deteriorate. Expansion problems	Use cooling system. Choose high temperature materials. Allow adequate clearances in design
High temperature, radiation from hot sources	Unequal expansion in addition to above	Keep source/object distance large. Use heat reflectors or insulation. Surfaces preferably reflective, smooth and of light colour. Use large characteristic volume
High humidity	Etching, corrosion and electrical breakdown	Use protective coatings or materials that do not rust or corrode. Keep temperature changes to a minimum
Corrosive gases and vapours	Corrosion	Use inert materials or protective coatings
Dust	Accelerates wear. Movement seizure	Use adequate seals or filter air

TEMPERATURE

The main problems in industry arise with radiation from hot processes. Typical examples of heat sources are as follows:

Table 28.2 Temperatures of some industrial processes

Sources of heat	Temperature °C	Sources of heat	Temperature °C
Drying steel wire	150	Calorising (baking in aluminium powder)	930
Drying lacquer	150	Heating sheet bars	930
Japanning	80–230	Normalising sheet steel	930
Core baking for iron castings	150–230	Carburising	950
Blueing	150–230	Pack-heating sheet steel	950
Carbolic and creosote oil distillation	200–300	Coke oven	950–1050
Hot-dip tinning	260	Short cycle annealing, malleable iron	980
Tempering in oil	260	Cyaniding	980
Tempering high-speed steel	330	Glasing porcelain	1000
Hydrocarbon synthesis	300–400	Gas turbines	1000–1200
Tissue paper drying	400	Vitreous enamelling (castings)	1010
Annealing aluminium	400–500	Bar and pack heating, stainless steel	1040
Cracking petroleum	400	Normalising stainless steel	930–1090
Heating aluminium for rolling	450	Rolling stainless steel	950–1230
Petroleum processing	500–600	Forging titanium alloys	870–1060
Nitriding steel	510	Annealing manganese-steel castings	1040
Annealing brass	540	Heating tool steel for rolling	1040
Annealing glass	620	Heating sheet steel for pressing	1050
Annealing copper	620–700	Heating spring steel for rolling	1090
Annealing german silver	650	Glost-firing porcelain	1120
Enamelling wet process	650	Firebrick kiln	1150–1400
Strain relieving	650–700	Hardening high-speed steel	1200
Annealing cold-rolled strip	680–760	Bisque-firing porcelain	1230
Porcelain decorating	760	Heating steel blooms and billets for rolling, also rivet heating	1250
Heating brass for rolling	790		
Annealing nickel or monel wire or sheets	800	Heating steel for drop forging or die pressing	1300
Annealing high-carbon steel	820	Calcining limestone	1370
Heat-treating medium-carbon steel	840	Welding steel tubes from preformed skelp	1400
Patenting wire	840	Burning firebrick	1320–1480
Box-annealing sheet steel	840	Burning portland cement	1430
Vitreous-enamelling sheet steel	870	Glass melting	1300–1500
Annealing malleable iron, long cycle	870	Basic refractories	1550–1750
Heating copper for rolling	870	Steel melting	1680
Forging commercially pure titanium	870	Melting chromium steel	1790
Annealing steel castings	900	High alumina brick manufacture	1900–2000
Building brick kiln	900–1050	Spray steel making	2100
Normalising steel pipes	900		

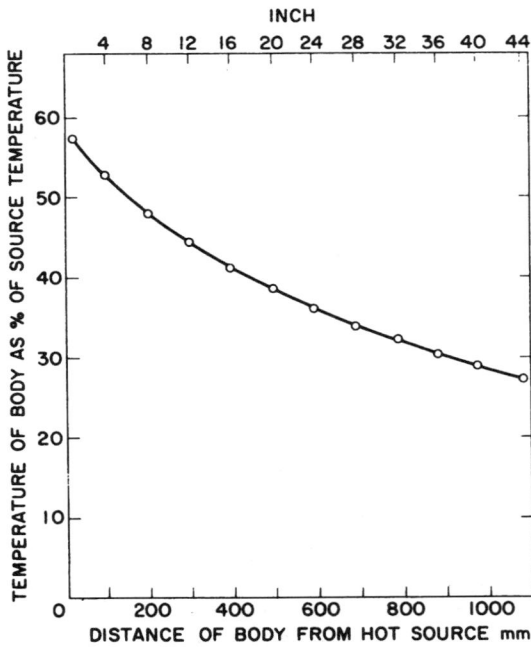

Figure 28.1 Applicable to furnace walls from 150 to 300°C

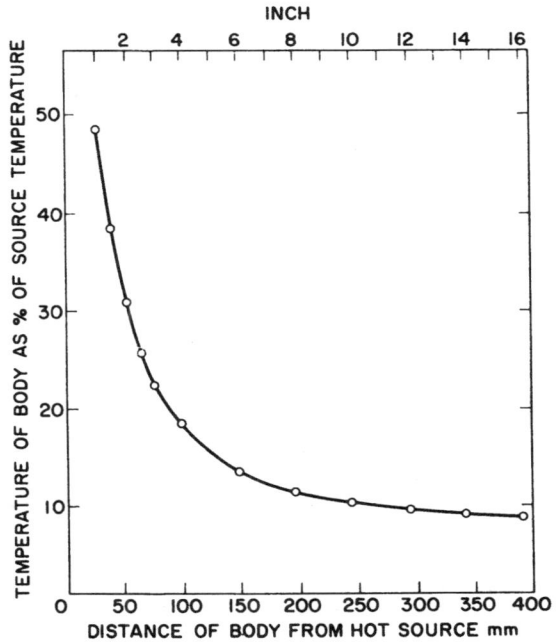

Figure 28.2 Applicable to sources from 300 to 1400°C

These graphs are based on laboratory and field measurements where a blackened metallic body was used with convective cooling. Figure 28.2 is for a source area of 20 in². Increasing the source area will reduce the slope of the graph towards that of Figure 28.1 which approximates to an infinite plane source. The multiplicity of variables associated with radiative heat transfer precludes a simple accurate calculation of the temperature any body will reach when placed near any source of heat. However, the graphs will indicate if temperature is likely to be a problem. The heat generated by the body itself must, of course, not be overlooked.

HUMIDITY

Relative humidities above 45% often lead to condensation problems.

Table 28.3 Typical values of relative humidity and dry bulb temperatures for working areas found in industry

Industry	Relative humidity %			Dry bulb temperature °C		
	Minimum	Maximum	Average	Minimum	Maximum	Average
Abrasive manufacture	—	—	50	—	—	24
Bakery	40	85	70	4	41	24
Brewery	55	75	70	0	10	7
Confectionery	13	63	45	4	66	24
Ceramics	35	90	60	16	66	27
Coke ovens	30	70	55	13	46	27
Distilling	35	60	55	0	24	21
Electrical products	15	70	50	20	24	23
Engineering	35	50	40	18	38	24
Ordnance	—	—	40	21	88	24
Pharmaceuticals	15	50	35	21	32	26
Plastics	25	65	45	4	27	24
Printing	45	55	50	21	27	24
Rubber goods	25	50	40	16	32	27
Textiles	50	85	60	21	29	27
Tobacco	60	85	70	24	32	27
Steel	35	90	55	7	35	21

CORROSIVE ATMOSPHERES

Table 28.4 Industries and processes with which corrosive atmospheres are often associated

Anodising and bright dipping aluminium	Pharmaceutical manufacture
Batteries and electrolytic cell production	Plastics manufacture
Chemical manufacture	Sheet steel pickling and phosphating
Electroplating	Smelting processes
Explosive manufacture	Textile and paper bleaching
Fertiliser manufacture	

DUST

Table 28.5 Industries in which dust problems may be excessive

Abrasive manufacture	Metal smelting
Asbestos manufacture and usage	Millers of grain
Brick manufacture including refractories	Potteries and earthenware manufacture
Cement manufacture	Quarries, mining and stone working
Foundries of all kinds	Roofing felt manufacturers
Gas and coke production	Steel manufacture
Glass fibre manufacture and usage	Users of mineral fillers in mastic and rubber industries

Table 28.6 Particle sizes of common materials as a guide to the specification of seals and air filters

Particles	Size μm	Particles	Size μm
Carbon black	0.01–2	Milk powder	1–10
Cement dust	5–120	Oil smoke	0.03–1
Coal dust	10–5000	Pigments	0.07–7
Flour*	5–110	Pulverised coal	10–400
Foundry dust	1–1000	Pulverised coal fly ash	1–50
Ground limestone	30–800	Sand blasting dust	0.5–2
Metallurgical dust	0.5–100	Stoker fly ash	10–800
Metallurgical fume	0.01–2		

Note: * Flour dust can be a particular problem when it drifts into small gaps, becomes damp and then sets into a hard solid mass.

PRESSURE

Level of pressure	Effect
Pressures in tens of MPa (thousands of psi)	Oil lubricants increase considerably in viscosity and density; they may even go solid
Pressures in MPa (hundreds of psi)	Gases in contact with lubricants dissolve in them causing property changes and foaming on decompression

Effect of pressure on lubricants

Type of lubricant	Factor	Effect	Tribological significance
Gas	Pressure	Increased density	Aerodynamic gas-lubricated bearings
Oil	Pressure	Increased density (volume change) (Figure 29.3)	Very high pressure hydraulic systems: elastohydrodynamic lubrication
		Increased viscosity (Figures 29.1 and 29.2) and raised pour point	
	Gas environment	Solubility of gas is increased with consequent fall in viscosity	Compressors where lubricant is in contact with gas (e.g. reciprocating piston-ring compressors, sliding vane rotary compressors)
Solid	Pressure	None	—

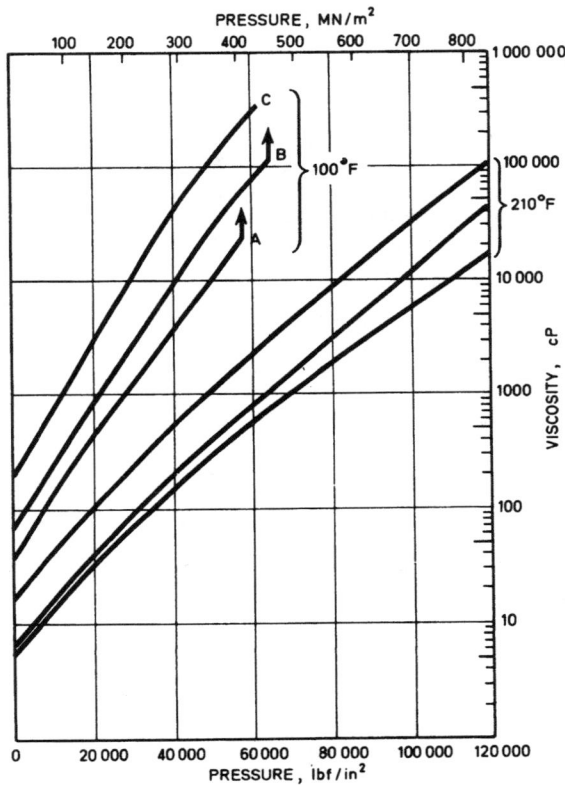

Figure 29.1 Effect of pressure on viscosity of HVI paraffinic oils

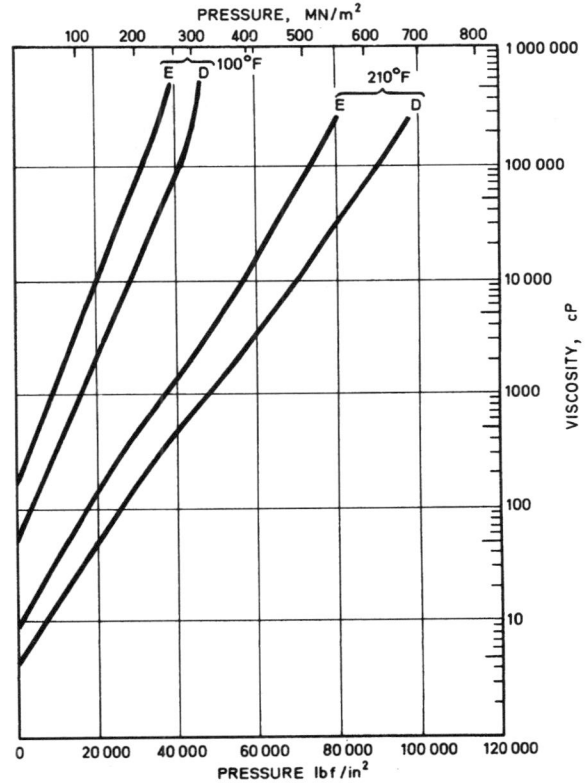

Figure 29.2 Effect of pressure on viscosity of LVI naphthenic oils

Oil	Viscosity, cS		
	100°F	210°F	VI
A	33.1	5.32	102
B	55.3	7.17	96
C	165.3	15.12	99

Oil	Viscosity, cS		
	100°F	210°F	VI
D	55.1	5.87	23
E	143.1	9.48	8

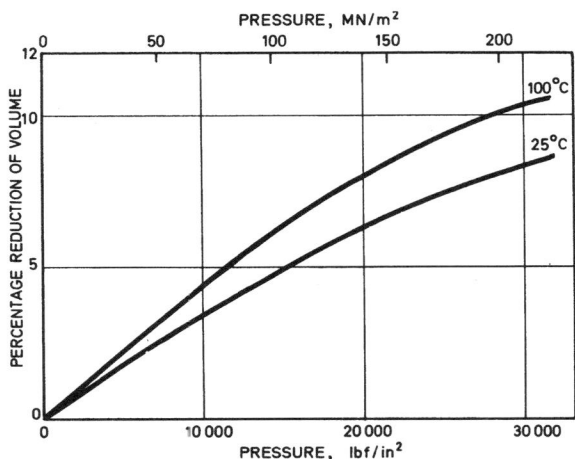

Figure 29.3 Compressibility of typical mineral oils

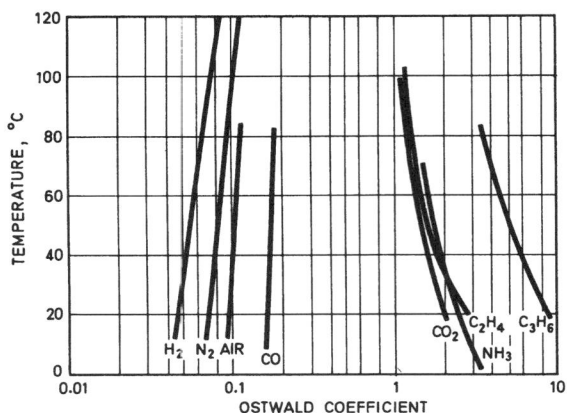

Figure 29.4 Ostwald coefficients for gases in mineral lubricating oils

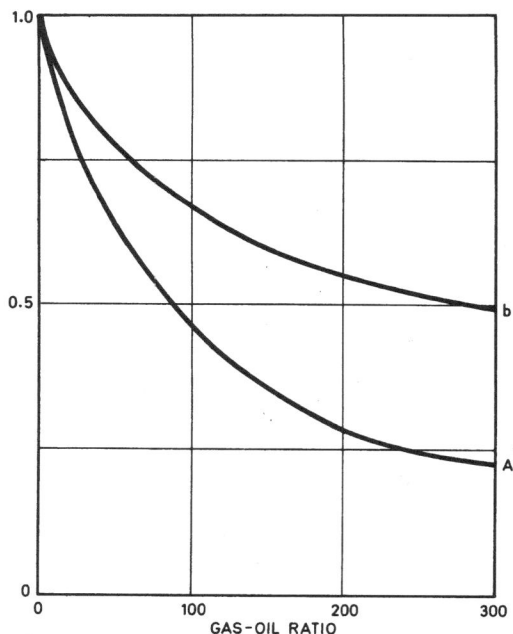

Figure 29.5 Constants for eqn (2)

Effect of dissolved gases on the viscosity of mineral oils

An estimate of the viscosity of oils saturated with gas can be obtained as follows:

(*i*) Determine Ostwald coefficient for gas in mineral oil from Figure 29.4.

(*ii*) Calculate gas: oil ratio from:

Gas: oil ratio

$$= \text{Ostwald coefficient} \times p \cdot \frac{293}{\theta + 273} \quad \dots \quad (1)$$

where p is the mean gas pressure (bar), and θ the mean temperature (°C).

(*iii*) Obtain viscosity of oil saturated with gas(es) from:

$$v_s = A v_o^b \quad \dots \quad (2)$$

where v_o is the viscosity of oil at normal atmospheric pressure (CSt); and A, b are constants obtained from Figure 29.5.

VACUUM

Level of pressure	Effect
Moderate vacuum to 10^{-4} torr	Liquid lubricants tend to evaporate
High vacuum below 10^{-5} torr	Surface films are lost, and metals in contact can seize

Lubricant loss by evaporation

Effect	Method of control
As the pressure is reduced and the vapour pressure of a liquid lubricant is approached, its rate of evaporation increases At very low pressure the lubricant may evaporate too quickly to be usable	1 Use a lubricant with a very low vapour pressure.* The rate of loss can then be very low 2 If the pressure is below 10^{-6} torr lubricant evaporation can be reduced by a simple labyrinth seal because the mean free path of its vapour molecules will be greater than the labyrinth paths 3 If the space around the lubricated component can be sealed, the local pressure will stabilise at the lubricant vapour pressure

Sealing Method	Vacuum obtainable
Single lip seal†	10 torr
Double or triple lip seals†	10^{-3} torr

* For example, the materials in Table 29.1.
† Seals should be lubricated with a smear of vacuum grease.

Table 29.1 Lubricants and coatings which have been used in high vacuum

Material	Vacuum	Temperature, °C	Speed	Load	Remarks
Versilube F50	10^{-6}	160	High	Medium	Ball bearing and gear
Versilube G–300 Grease	10^{-6}	160	High	Medium	Ball bearing
Apiezon 'T'	10^{-6}	160	High	Medium	Ball bearing
Barium film	10^{-6}	200	High	Light	Ball bearings
PbO, PbI_2 or other halides in graphite	10^{-6}	750	High	Medium	Brushes
Everlube 811 (phenolic bonded MoS_2)	10^{-6}–10^{-7}	40–85	High	Medium	(Remove initial wear debris)
BR2S Grease	10^{-7}	Ambient	Medium	Medium	Ball bearing
MoS_2–graphite–silicate	10^{-7}	Max. 200	Low	Medium	Ball bearings, pin-on-disc, gears
15% tin in nickel	10^{-7}	Ambient	Medium	Medium	Slider
Apiezon 'L'	10^{-7}–10^{-8}	<50	Low	High radial	100 mm ball bearings, no atmospheric contamination
Cu–PTFE–WSe_2	10^{-7}–10^{-8}	−50 to +150	Medium	Medium	Rolling/sliding
Gold plating, silver plating	10^{-7}–10^{-8}	<50	Low	Low	Gears
AeroShell Grease 15	10^{-8}	<120	8000 rev/min	Low	Instrument bearings (sensitive to mis-alignment)
MoS_2–burnished 24 ct Au film	10^{-8}	Ambient	Low	Medium	Ball bearings
In situ MoS_2	10^{-9}	Ambient	Medium	Medium	Slider
Lead film	10^{-9}	−20 to +80	Medium	Low	Ball bearing, gear
Silver–copper–MoS_9	10^{-9}	Ambient	Medium	Low	Brushes
MoS_2 or Cu in polyimide	10^{-10}	Ambient	Medium	Medium	Slider

Loss of surface films in high vacuum

Surface contaminant films of soaps, oils and water, etc., and surface layers of oxides, etc., enable components to rub together without seizure under normal atmospheric conditions. Increasing vacuum causes the films to be lost, and reduces the rate at which oxide layers reform after rubbing. The chance of seizure is therefore increased.

Seizure can be minimised by using pairs of metals which are not mutually soluble, and Table 29.2 shows some compatible common metals under high vacuum conditions, but detailed design advice should usually be obtained.

Vacuum level	Effect on surfaces
0.5 bar (pressure)	Minimum pressure for there to be sufficient water vapour in average room air to enable graphite to work successfully
6 torr	Minimum pressure for graphite to work successfully in pure water vapour
10^{-1} torr	Water lost from surface
10^{-5} torr	Most soaps and oils lost from surface
10^{-5} torr	Oxide films become difficult to replace after rubbing
10^{-8} torr	Oxide films no longer replaced after rubbing
Below 10^{-10} torr	Oxide film may be lost without rubbing

Table 29.2 Some compatible metal pairs for vacuum use

Material	Satisfactory partner
(a) Stainless steel (martensitic)*	Polyimide + 20% Cu fibre
(b) Stainless steel (austenitic)*†	Rhenium; cobalt (below 300°C), cobalt + 25% molybdenum (up to 700°C)
Tool steel*†	Tool steel, 700 VPN; nickel alloy–MoS_2 composite
Mild steel*† Soft irons	Silver; lead
Cast irons	Assuming grey irons, graphite on its own is not recommended for vacuum work and this may apply to structures containing free graphite
Copper‡	Molybdenum; chromium; tungsten
Tin	Iron; nickel; cobalt; chromium
Lead	Chromium; cobalt; nickel; iron; copper; zinc; aluminium
Tungsten	Silver; copper
Molybdenum	Copper
Aluminium	Indium; lead; cadmium
Cadmium	Aluminium; iron; nickel
Nickel§	Tin; silver; lead
Chromium	Copper; lead; tin; silver
Gold	Rhenium; lead
Silver	Plain carbon steel; chromium; cobalt; nickel

* High sulphur contents of these will help reduction of wear under vacuum.
† Generally, for steels, PTFE-based composites are advised. PTFE is the film former + lamellar solids such as MoS_2, WS_2, $CdCl_2$, CdI_2, $CdBr_2$ or selenides but not carbons, graphites or BN. A binder phase of Ag or Cu can also be included. At very high vacuum all plastics will out-gas.
‡ Unlike normal atmospheric conditions, copper and copper alloys give high wear and friction against ferrous materials in a vacuum.
§ If sintered, dispersed oxide in the nickel will be beneficial.

HIGH TEMPERATURE

Temperature limitations of liquid lubricants

The chief properties of liquid lubricants which impose temperature limits are, in usual order of importance, (1) oxidation stability; (2) viscosity; (3) thermal stability; (4) volatility; (5) flammability.

Oxidation is the most common cause of lubricant failure. Figure 30.1 gives typical upper temperature limits when oxygen supply is unrestricted.

Compared with mineral oils most synthetic lubricants, though more expensive, have higher oxidation limits, lower volatility and less dependence of viscosity on temperature (i.e. higher viscosity index).

For greases (oil plus thickener) the usable temperature range of the thickener should also be considered (Figure 30.2).

Temperature limitations of solid lubricants

All solid lubricants are intended to protect surfaces from wear or to control friction when oil lubrication is either not feasible or undesirable (e.g. because of excessive contact pressure, temperature or cleanliness requirements).

There are two main groups of solid lubricant, as given in Table 30.1.

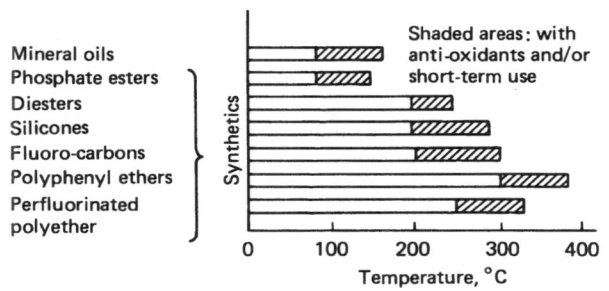

Figure 30.1

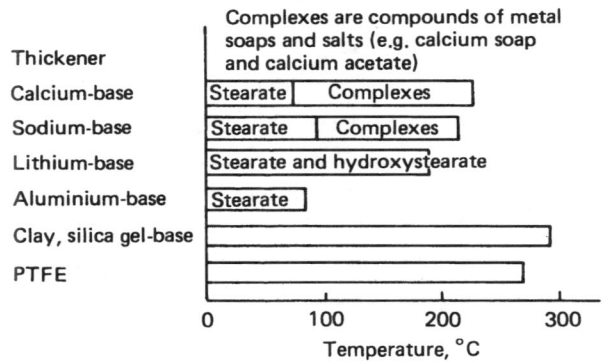

Figure 30.2

Table 30.1

		Example	Practical temperature limit, °C*	Common usages
1	Boundary lubricants and 'extreme pressure' additives (surface active)	Metal soap (e.g. stearate)	150 ⎫	Metal cutting, drawing and shaping. Highly-loaded gears
		Chloride (as Fe Cl$_3$)	300 ⎬	
		Sulphide (as FeS)	750 ⎭	
		Phthalocyanine (with Cu and Fe)	550	Anti-seizure
2	Lamellar solids and/or low shear strength solids	Graphite	600 ⎫	General, metal working, anti-seizure and anti-scuffing
		Molybdenum disulphide	350 ⎭	
		Tungsten disulphide	500	
		Lead monoxide†	650	
		Calcium fluoride	1000	
		Vermiculite	900	Anti-seizure
		PTFE	250	Low friction as bonded film or reinforced composite

* The limit refers to use in air or other oxidising atmospheres.
† Bonded with silica to retard oxidation.

Dry wear

When oil, grease or solid lubrication is not possible, some metallic wear may be inevitable but oxide films can be beneficial. These may be formed either by high ambient temperature or by high 'hot spot' temperature at asperities, the latter being caused by high speed or load.

Examples of ambient temperature effects are given in Figures 30.3 and 30.4, and examples of asperity temperature effects are given in Figures 30.5 and 30.6.

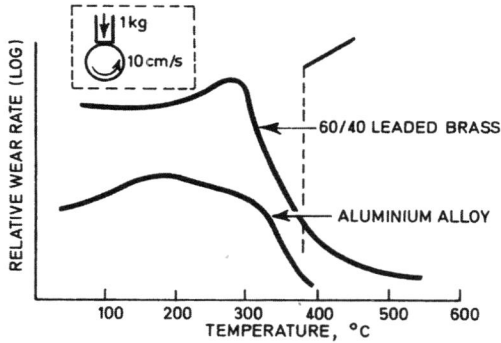

Figure 30.3 Wear of brass and aluminium alloy pins on tool steel cylinder, demonstrating oxide protection (negative slope region). Oxide on aluminium alloy breaks down at about 400°C, giving severe wear

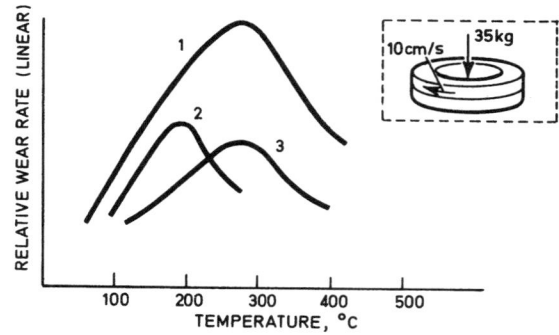

Figure 30.4 Wear of (1) nitrided EN41A; (2) high carbon tool steel; (3) tungsten tool steel. Oxides: αFe_2O_3 below maxima; αFe_3O_4-type above maxima

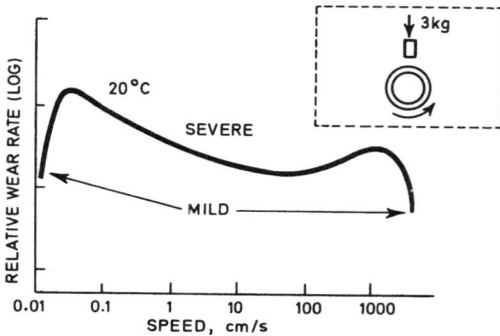

Figure 30.5 Wear of brass pin on tool steel-ring. At low speed wear is mild because time is available for oxidation. At high speed wear is again mild because of hot-spot temperatures inducing oxidation

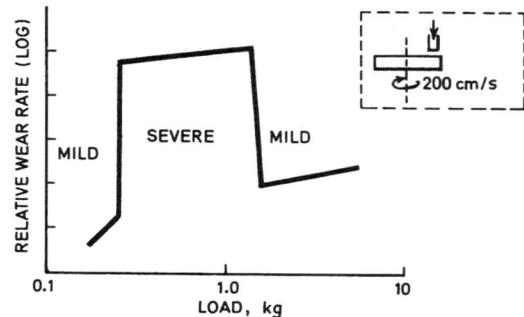

Figure 30.6 Transition behaviour of 3% Cr steel. Mild wear region characterised by oxide debris: severe wear region characterised by metallic debris

Bearing materials for high temperature use

When wear resistance, rather than low friction, is important, the required properties (see Table 30.2) of bearing materials depend upon the type of bearing.

Table 30.2

Rolling contact	Sliding contact	Gas bearings
High hot hardness (> 600 VPN). Dimensional stability, resistance to: oxidation, phase change, residual stress and creep. Thermal shock resistance	Moderate hot hardness. Good thermal conductivity and shock resistance. Resistance to oxidation and scaling	Extreme dimensional stability. Low thermal expansion and porosity. High elastic modulus. Capable of fine surface finish

Hot hardness, particularly in rolling contact bearings, is of high importance and Figure 30.7 shows maximum hardness for various classes of material.

Some practical bearing materials for use in oxidising atmospheres are shown in Table 30.3.

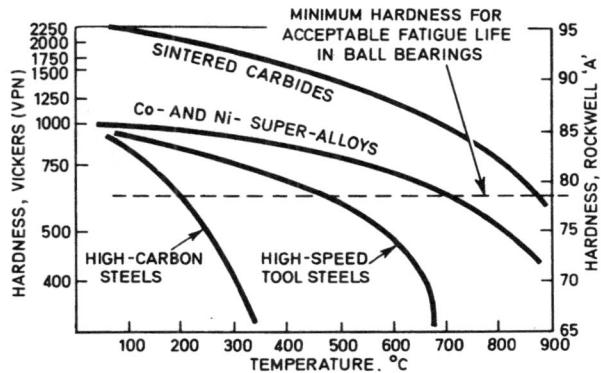

Figure 30.7

Table 30.3

Rolling contact	Sliding contact	Approximate temp. limit, °C
High-speed tool steel (Mo and W types)	PH stainless steel Nitrided steel	500
	Hastelloy (Ni super-alloy)	750
Stellite (Co super alloy) Titanium carbide	Stellite	850
Dense α-alumina Zirconia	Alumina–Cr–W cermet Silicon carbide Silicon nitride	1000 or above

LOW TEMPERATURE

General

'Low temperature' may conveniently be subdivided into the three classes shown in Table 30.4. In Class 1, oils are usable depending upon the minimum temperature at which they will flow, or the 'pour point'. Some typical values are given in Table 30.5. Classes 2 and 3 of Table 30.4 embrace most industrially important gases (or cryogenic fluids) with the properties shown in Table 30.6.

Because of their very low viscosity (compare to $7 \times 10^{-2}\,Ns/m^2$ for SAE 30 oil at 35°C) these fluids are impractical as 'lubricants' for hydrodynamic journal bearings. (Very high speed bearings are theoretically possible but the required dimensional stability and conductivity are severe restrictions.)

Table 30.5

Type of lubricant	Pour point, °C
Mineral oil	−57 (min.)
Diester	−60
Phosphate ester	−57
Silicate ester	−65
Di-siloxane	−70
Silicone	−70
Polyphenyl-ether	−7
Perfluorinated polyether	−75 to −90

Table 30.4

Class	Temperature range	Usages involving tribology
1	0°C to −80°C (190 K)	Domestic refrigeration, transport over snow and ice
2	−80°C to −196°C (77 K)	Liquefaction and handling of industrial gases, rocket propulsion (turbo-pumps, seals)
3	−196°C to −273°C (0 K)	Space exploration, liquid hydrogen and helium systems

Table 30.6

Fluid	Boiling point K	Viscosity at boiling point Ns/m²
Oxygen	90.2	1.9×10^{-4}
Nitrogen	77.4	1.6×10^{-4}
Argon	87.3	—
Methane	111.7	—
Hydrogen	20.4	1.3×10^{-5}
Helium	4.2	0.5×10^{-5}

Unlubricated metals

In non-oxidising fluids, despite low temperature, metals show adhesive wear (galling, etc.) but in oxygen the wear is often less severe because oxide films may be formed. Where there is condensation on shafts, seals or ball bearings (dry lubricated) a corrosion-resistant hard steel (e.g. 440°C) is preferable.

Plain bearing materials

As bushes and thrust bearings, filled PTFE/metal and filled graphite/metal combinations are often used – see Table 30.8.

Safety

Aspects of safety are summarised in Table 30.7.

Table 30.7

Hazard	Precaution
Heat generated at bearings or seals may cause local boiling of liquid or ignition	Design adequate venting system. For fuel liquids (e.g. methane, hydrogen) and oxygen particularly; ensure total compatibility of bearing materials under extreme conditions
Fine wear debris or grease residues	Thorough check on ignition aspects and/or extreme cleanliness in installation, particularly for liquid oxygen

Table 30.8 Some successful plain bearing materials for cryogenic fluids

Bush or face	Suitable journal or counterface	Remarks
Bronze/graphite-filled PTFE PTFE/lead-filled bronze (steel backed)	18/8 Stainless steel Martensitic steel (e.g. S.80) Chromium plate	Suitable for all fluids
Copper/lead-filled graphite	Chromium plate	Soft stainless steel is scored. Combination is best in liquid oxygen
Phenolic-impregnated carbon	Carbon	—
Pure PTFE	Duralumin or bronze	Thermal conductivity of counterface important

Ball bearings and seals for cryogenic temperatures

Table 30.9 Recommended tribological practice at cryogenic temperatures

Component	Recommended material
High speed ball bearings (> 10 000 RPM)	The raceway coating should include MoS_2 or PTFE, and the cage should be woven glass fibre reinforced PTFE
Low speed ball bearings	Either the cage should be PTFE filled with MoS_2 and chopped glass fibre, or a film of magnetron-sputtered MoS_2 (or ion-plated lead) should be present on the raceways and balls
Reciprocating seals	Use a seal manufactured from PTFE filled with chopped glass fibre or from a PTFE and bronze composite
Rotary seals	Use a carbon-graphite face loaded against a tungsten carbide or hard chromium plated face

This section is restricted to chemical effects on metals.

Chemical effects can arise whenever metals are in contact with chemicals, either alone or as a contaminant in a lubricant.

Wear in the presence of a corrosive liquid can lead to accelerated damage due to corrosive wear. This problem is complex, and specialist advice should be taken.

Corrosion or corrosive wear are often caused by condensation water in an otherwise clean system.

Table 31.1 Some specific contamination situations

	Problem	Typical occurrence	Solution
1	Contamination by reactive gases	Internal combustion engines	Sophisticated lubricants
2	Dilution by petroleum gases	Gas compressors, valves, flowmeters	Use of higher viscosity or non petroleum oil.
3	Inert or reducing gases	Gas compressors, valves, flowmeters	Use of self-lubricating materials

Table 31.2 Typical corrosion resistant materials

Material	Nominal composition	Reference number in next table
Corrosion resisting steel[1]	0.15 C max. 12 Cr 1 Ni max.	1
Corrosion resisting steel[1]	0.08 C max. 19 Cr 9 Ni	2
Corrosion resisting steel[1]	0.1 C max. 18 Cr 10 Ni 2 Mo	3
Corrosion resisting steel[1]	0.07 C max. 22 Cr 29 Ni 3 Mo	4
Corrosion resisting steel[1]	0.35 C max. 27 Cr 3 Ni max.	5
Grey cast irons[2]		6
High silicon irons	14 Si min.	7
Austenitic cast irons	Total Ni Cr Cu 22 min.	8
Bronzes		9
Nickel, chrome/molybdenum alloys	Fe 20 max. Some Cu, W, Si Mn	10
Monel metal		11
Nickel		12
Lead		13
Non metallic materials		14

Notes

The resistance of these materials to corrosion will be substantially affected by the velocity of corrosive fluids in contact with them. High fluid velocities increase corrosion rates.

1. Stainless steel materials are prone to seizure if allowed to rub together. The risk may be reduced by maintaining the highest practical hardness of both surfaces, or by MoS_2 lubrication of the interface.

2. With grey cast iron precautions must be taken against corrosion in storage or when temporarily out of action.

SELECTION OF CORROSION RESISTANT MATERIALS FOR CONTACT WITH VARIOUS LIQUIDS

Liquid	Condition	Materials suitable under the majority of conditions	Liquid	Condition	Materials suitable under the majority of conditions
Acetaldehyde		6	Acid, Chromic	Aqueous sol.	2 3 4 7 10
Acetate Solvents		2 3 4 6 9 10	Acid, Citric	Aqueous sol.	2 3 4 7 9 10
Acetone		6	Acids, Fatty (Oleic, Palmitic, Stearic, etc.)		
Acid, Acetic and Anhydride	Cold	2 3 4 7 10			2 3 4 9 10
Acid, Acetic	Boiling	3 4 7 10	Acid, Formic		3 4 10
Acid, Arsenic		2 3 4 7 10	Acid, Fruit		2 3 4 9 10 11
Acid, Benzoic		2 3 4 10	Acid, Hydrochloric	Dil. cold	4 7 10 11 12 14
Acid, Boric	Aqueous sol.	2 3 4 7 9 10	Acid, Hydrochloric	Dil. hot or conc.	7 10 14
Acid, Butyric	Conc.	2 3 4 10	Acid, Hydrocyanic		2 3 4 6 10
Acid, Carbolic	Conc. (m.p.106°F)	2 3 4 6 10	Acid, Hydroflouric	Anhydrous, with hydro carbon	11
Acid, Carbolic	Aqueous sol.	2 3 4 10			
Acid, Carbonic	Aqueous sol.	9	Acid, Hydroflouric	Aqueous sol.	9 11

Liquid	Condition	Materials suitable under the majority of conditions
Acid, Hydrofluosilicic		9 11
Acid, Lactic		2 3 4 7 9 10
Acid, Mixed		2 3 4 6 7 10
Acid, Naphthenic		1 2 3 4 6 10
Acid, Nitric	Conc. boiling	4 5 7
Acid, Nitric	Dilute	1 2 3 4 5 7
Acid, Oxalic	Cold	2 3 4 7 10
Acid, Oxalic	Hot	4 7 10
Acid, Ortho-Phosphoric		3 4 10
Acid, Picric		2 3 4 7 10
Acid, Pyrogallic		2 3 4 10
Acid, Pyroligneous		2 3 4 9 10
Acid, Sulphuric	Cold	4 7 10
Acid, Sulphuric	Hot	7 10
Acid, Sulphuric (Oleum)	Fuming	4 10
Acid, Sulphurous		2 3 4 9 10 13
Acid, Tannic		2 3 4 9 10 11
Acid, Tartaric	Aqueous sol.	2 3 4 9 10 11
Alcohols		9
Alum	(See Aluminum Sulphate and Potash Alum)	
Aluminum Sulphate	Aqueous sol.	4 7 10 11 13
Ammonia, Aqua		6
Ammonium Bicarbonate	Aqueous sol.	6
Ammonium Chloride	Aqueous sol.	3 4 7 10 11
Ammonium Nitrate	Aqueous sol.	2 3 4 6 10 11
Ammonium Phosphate	Aqueous sol.	2 3 4 6 10 11
Ammonium Sulphate	Aqueous sol.	2 3 4 6 10
Ammonium Sulphate	With H_2SO_4	3 4 7 9 10 13
Aniline		6
Aniline Hydrochloride	Aqueous sol.	7 10 14
Asphalt	Hot	1 6
Barium Chloride or Nitrate	Aqueous sol.	2 3 4 6 10
Beer or Beer Wort		2 9
Beet Juice or Pulp		2 9
Benzene (Benzol)		6
Blood		9
Brine, Calcium Chloride	pH > 8	6
Brine, Calcium Chloride	pH < 8	4 8 9 10 11
Brine, Calcium and Magnesium Chlorides	Aqueous sol.	4 8 9 10 11
Brine, Calcium and Sodium Chloride	Aqueous sol.	4 8 9 10 11
Brine, Sodium Chloride	Under 3% Salt, cold	6 8 9
Brine, Sodium Chloride	Over 3% Salt, Cold	2 3 4 8 9 10 11
Brine, Sodium Chloride	Over 3% Salt, hot	3 4 7 10 11
Brine, Sea Water		6 9
Butane		6
Calcium Bisulphite	Paper mill	3 4 10 13
Calcium Chlorate	Aqueous sol.	4 7 10 14
Calcium Hypochlorite		4 6 7 10
Cane Juice		8 9
Carbon Bisulphide		6
Carbon Tetrachloride	Anhydrous	6
Carbon Tetrachloride	Plus water	2 9
Caustic Potash	(see Potassium Hydroxide)	
Caustic Soda	(see Sodium Hydroxide)	
Cellulose Acetate		3 4 10
Chloride Water	(Depending on conc.)	3 4 7 10 13 14
Chlorobenzene		2 9
Chloroform		2 3 4 9 10 11
Chrome Alum	Aqueous sol.	4 7 10
Condensate	(see Water Distilled)	
Copper Ammonium Acetate	Aqueous sol.	2 3 4 6 10
Copper Chloride (Cupric)	Aqueous sol.	7 10 14
Copper Nitrite		2 3 4 10
Copper Sulphate, Blue Vitriol	Aqueous sol.	2 3 4 7 10 13
Creosote		6
Cresol, Meta		4 6 10
Cyanogen	In water	6
Diphenyl		6
Enamel		6
Ethanol	(see Alcohols)	
Ethylene Chloride (Dichloride)	Cold	2 3 4 9 10 11
Ferric Chloride	Aqueous sol.	7 10 14
Ferric Sulphate	Aqueous sol.	2 3 4 7 10
Ferrous Chloride	Cold. aqueous	7 10 14
Ferrous Sulphate (Green Copperas)	Aqueous sol.	3 4 7 10 11 13
Formaldehyde		2 3 4 9 10
Fruit Juices		2 3 4 9 10 11
Furfural		2 3 4 6 9 10
Glaubers Salt	(see Sodium Sulphate)	
Glucose		9
Glue	Hot	6
Glue Sizing		9
Glycerol (Glycerin)		6 9
Heptane		6
Hydrogen Peroxide	Aqueous sol.	2 3 4 10
Hydrogen Sulphide	Aqueous sol.	2 3 4 10
Hydrosulphite of Soda	(see Sodium Hydrosulphite)	
Hyposulphite of Soda	(see Sodium Thiosulphate)	
Kaolin Slip	Suspension in water	6
Kaolin Slip	Suspension in acid	4 7 10
Lard	Hot	6
Lead Acetate (Sugar of Lead)	Aqueous sol.	3 4 10 11
Lead	Molten	6
Lime Water (Milk of Lime)		6
Liquors—Pulp Mill:		3 4 6 8 10 11
Sulphite		3 4 10 13
Lithium Chloride	Aqueous sol.	6
Magnesium Chloride	Aqueous sol.	4 7 10 14
Magnesium Sulphate (Epsom Salts)	Aqueous sol.	2 3 4 6 10

Liquid	Condition	Materials suitable under the majority of conditions
Manganese Chloride	Aqueous sol.	2 3 4 7 9 10
Manganese Sulphate	Aqueous sol.	2 3 4 6 9 10
Mash		2 9
Mercuric Chloride	Aqueous sol.	7 10
Mercuric Sulphate	In H$_2$SO$_4$	4 7 10 14
Mercurous Sulphate	In H$_2$SO$_4$	4 7 10 14
Methyl Chloride		6
Methylene Chloride		2 6
Milk		2
Molasses		9
Mustard		2 3 4 7 9 10
Nicotine Sulphate		4 7 10 11
Nitre	(see Potassium Nitrate)	
Nitre Cake	(see Sodium Bisulphate)	
Nitro Ethane		6
Nitro Methane		6
Oil, Coal Tar		2 3 4 6 10
Oil, Coconut		2 3 4 6 9 10 11
Oil, Crude	Cold	6
Oil, Crude	Hot	6
Oil, Linseed		2 3 4 6 9 10 11
Oil, Mineral (including kerosene, fuel, naptha, lubricating, paraffin)		6
Oil, vegetable (including olive, palm, soya, turpentine)		6
Oil, quenching		6
Oil, rapeseed		2 3 4 9 10 11
Perhydrol	(see Hydrogen Peroxide)	
Petrol (Petroleum Ether)		6
Phenol	(see Acid, Carbolic)	
Photographic Developers		2 3 4 10 14
Potash	Plant liquor	2 3 4 8 9 10 11
Potash Alum	Aqueous sol.	3 4 7 8 9 10 11
Potassium Bichromate	Aqueous sol.	6
Potassium Carbonate	Aqueous sol.	6
Potassium Chlorate	Aqueous sol.	2 3 4 7 10
Potassium Chloride	Aqueous sol.	2 3 4 9 10 11
Potassium Cyanide	Aqueous sol.	6
Potassium Hydroxide	Aqueous sol.	1 2 3 4 6 8 10 11 12
Potassium Nitrate	Aqueous sol.	1 2 3 4 6 10
Potassium Sulphate	Aqueous sol.	2 3 4 9 10
Propane		6
Pyridine		6
Pyridine Sulphate		4 7 13
Resin (Colophony)	Paper Mill	6
Salt Cake	Aqueous sol.	2 3 4 7 9 10
Sea Water	(see Brines)	
Sewage		6 9
Shellac		9
Silver Nitrate	Aqueous sol.	2 3 4 7 10
Soap Liquor		6
Soda Ash	Cold	6
Soda Ash	Hot	2 3 4 8 10 11
Sodium Bicarbonate	Aqueous sol.	2 3 4 6 8 10
Sodium Bisulphate	Aqueous sol.	4 7 10 13
Sodium Carbonate	(see Soda Ash)	
Sodium Chlorate	Aqueous sol.	2 3 4 7 10

Liquid	Conditions	Materials suitable under the majority of conditions
Sodium Chloride	(see Brines)	
Sodium Cyanide	Aqueous sol.	6
Sodium Hydroxide	Aqueous sol.	1 2 3 4 6 8 10 11 12
Sodium Hydrosulphite	Aqueous sol.	2 3 4 10 13
Sodium Hypochlorite		4 7 10 14
Sodium Hyposulphite	(see Sodium Thiosulphate)	
Sodium Meta Silicate		6
Sodium Nitrate	Aqueous sol.	1 2 3 4 6 10
Sodium Phosphate:		
Monobasic, Dibasic	Aqueous sol.	2 3 4 9 10
Tribasic	Aqueous sol.	6
Meta	Aqueous sol.	2 3 4 9 10
Hexameta	Aqueous sol.	2 3 4 10
Sodium Plumbite	Aqueous sol.	6
Sodium Sulphate	Aqueous sol.	2 3 4 9 10
Sodium Sulphide	Aqueous sol.	2 3 4 6 10 13
Sodium Sulphite	Aqueous sol.	2 3 4 9 10 13
Sodium Thiosulphate	Aqueous sol.	2 3 4 10 14
Stannic Stannous Chloride	Aqueous sol.	7 10 13 14
Starch		9
Strontium Nitrate	Aqueous sol.	2 6
Sugar	Aqueous sol.	2 3 4 8 9 10
Sulphite Liquors	(see Liquors, Pulp Mill)	
Sulphur	In water	6 8 9
Sulphur	Molten	6
Sulphur Chloride	Cold	6 13
Syrup	(see Sugar)	
Tallow	Hot	6
Tanning Liquors		2 3 4 7 9 10 11
Tar	Hot	6
Tar and Ammonia	In water	6
Tetraethyl Lead		6
Toluene (Toluol)		6
Trichloroethylene		2 6 9
Varnish		2 6 9 11
Vegetable Juices		2 3 4 9 10 11
Vinegar		2 3 4 7 9 10
Vitriol, Blue	(see Copper Sulphate)	
Vitriol, Green	(see Ferrous Sulphate)	
Vitriol, Oil of	(see Acid, Sulphuric)	
Water, Boiler Feed High Makeup	Not evaporated	6
Water, Boiler Feed Low Makeup	Evaporated	1 2 11
Water, Distilled	High purity	2 9
	Condensate	9
Water, Fresh		6
Wine		2 9
Wood Pulp (Stock)		6 9
Wood Vinegar	(see Acid Pyroligneous)	
Xylol (Xylene		2 3 4 6 10
Yeast		2 9
Zinc Chloride	Aqueous sol.	3 4 7 10
Zinc Sulphate	Aqueous sol.	3 4 9 10

(*Data in table courtesy*: American Hydraulic Institute)

The purpose of maintenance is to preserve plant and machinery in a condition in which it can operate, and can do so safely and economically.

Table 1.1 Situations requiring maintenance action

Situation	Basis of decision	Decision mechanism
Loss of operating efficiency	Recognition of the problem associated with a study of the costs	The economic loss arising from the reduced efficiency is greater than the effective cost of the repair.
Loss of function	Safety risk	This must be avoided by prior maintenance action. There must therefore be a method of predicting the approach of a failure condition.
	Downtime cost Repair cost	The maintenance method that is selected needs to be chosen to keep total costs to a minimum, while maximising the profit and market opportunities. The ongoing cost of maintenance versus total replacement also needs to be considered.

It is the components of plant and machinery which fail individually, and can lead to the loss of function of the whole unit or system. Maintenance activity needs therefore to be concentrated on those components that are critical.

Table 1.2 Factors in the selection of critical components

Important factor	Typical examples	Guide to selection
Likelihood of failure	Components subject to: Occasional overload Fatigue loading Wear Corrosion or other environmental effects	Can be identified by a review of the design of the system or by an analysis of relevant operating experience. Components which are more likely to fail should be selected for maintenance attention.
Effect of the component failure on the system	Components which, when they fail, cause a failure of the whole system, possibly with some consequential damage	These components need to be selected for special attention. They must be selected if there are safety implications.
	Components which, when they fail, may allow other components to be overloaded, and then cause failure of the system	An analysis of the effects of the failure needs to be carried out to detect any safety implications.
	Components which, when they fail, only produce a reduction in the performance of the system	The likely economic effects of the failure need to be analysed to decide whether these components merit special attention.
The time required to replace a failed component or to rectify the effect of its failure	Components that can be replaced quickly such as bulbs, fuses and some printed circuit boards, or components whose function can be taken over immediately by a standby system	If there are no safety implications, no particular maintenance action is required other than component replacement or repair after failure
	Components which take a long time to replace or repair	Their condition needs to be monitored and their maintenance planned in advance

Table 1.3 Maintenance methods which can be used

Maintenance method	Advantages	Disadvantages
Allow the equipment to break down and then repair it	Can be the cheapest solution if the repair is easy and there is no safety risk or possibility of consequential damage	The cost of consequential damage can be high and there might be safety risks. The plant may be of a type which cannot be allowed to stop suddenly because of product solidification or deterioration, etc. The failure may occur at an inconvenient time or, if the plant is mobile, at an inconvenient place. The necessary staff and replacement components may not then be available.
Preventive maintenance carried out at regular intervals	Reduces the risk of failure in service Enables the work and availability of special tools and spare parts to be planned well in advance Is particularly appropriate for components which need changing because of capacity absorption such as filter elements	Some failures will continue to occur in service because components do not fail at regular intervals as shown in Figure 10.2. The failures can be unexpected and inconvenient. During the overhaul many components in good condition will be stripped and inspected unnecessarily. Mistakes can be made on re-assembly so that the plant can end up in a worse condition after the overhaul. The overhaul process can take a considerable time resulting in a major loss of profitable production.
Opportunistic maintenance carried out when the plant or equipment happens to become available	The maintenance can be carried out with no loss of effective operating time.	Staff and spare parts have to be kept available which may cause increased costs. Is only practical in combination with condition based maintenance or some preventive maintenance, because otherwise any work which needs to be done is not known about. Breakdowns are still likely to occur in service.
Condition-based maintenance	Enables the plant to be left in service until a failure is about to occur. It can then be withdrawn from service in a planned manner and repaired at minimum cost. Consequential damage from a failure can be avoided.	Can only detect failures which show a progression to failure, which enables the incipient failure to be detected. Components with a sudden failure mode cannot provide the required advanced detection. Failures of components due to an unexpected random overload in service cannot be avoided, even by this method.
Maintenance/replacement based on expected component life	Is the only safe method usable for components which have a sudden failure mode	Requires monitoring of the operating conditions of the component so that its expected life can be estimated. If all failures are to be avoided many components will need to be changed when individually they have a useful life still remaining, e.g. in Figure 10.2 the working life has to be kept below the range in which failures may occur.

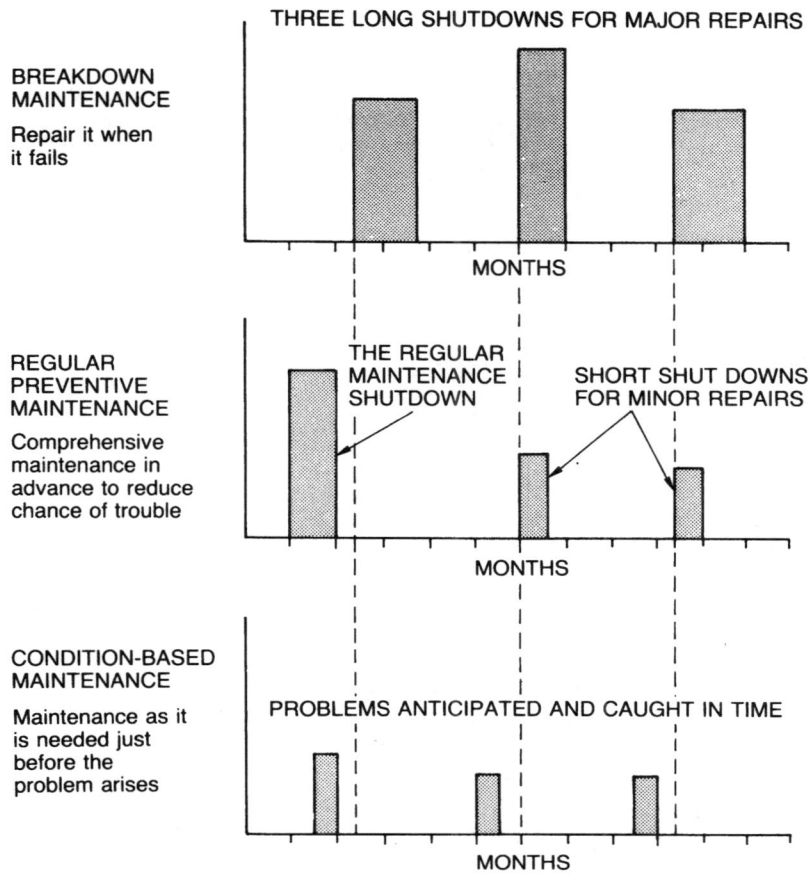

Figure 1.1 The results of maintaining the same industrial plant in different ways. The height of the bars indicates the amount of maintenance effort required

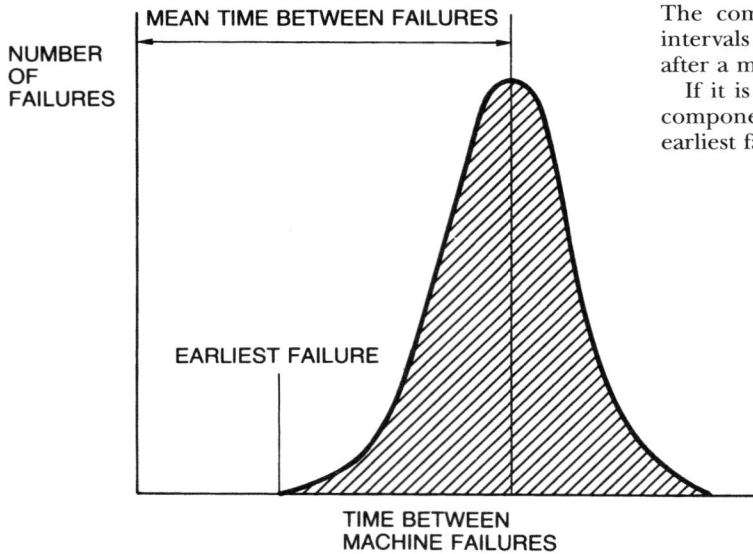

The components of machines do not fail at regular intervals but show a range of times to failure before and after a mean time.

If it is essential that no failures occur in service, the components must be changed within the time that the earliest failure may be expected.

Figure 1.2 The distribution of the time to failure for a typical component

Table 1.4 Feedback from maintenance to equipment design

Opportunity	Remarks
The maintenance activity provides an opportunity to record and analyse the problems and failures which occur during the service life of equipment. In particular, records need to be kept of the operating time to failure, the failure mode and any particular operating conditions which may have contributed to the failure.	This is a unique source of information which is of very high value to the process of design and development of improved equipment. It is a profit centre of the maintenance activity.
An intelligent review of maintenance operations can enable repair times to be determined for various components. The nature and cause of accessibility problems can be determined together with the definition of the design principles which need to be used to improve it.	This is useful for maintenance planning but it is particularly important as a method of getting new equipment that is designed to make maintenance more simple and of reduced cost.
The maintenance activity can be reviewed to identify opportunities in which modular design can be used to reduce downtime caused by in service faults. It can also often simplify the process of equipment manufacture.	If sub-systems can be designed to be self-contained and readily removable, this can enable the sub-system to be exchanged when a fault occurs. The actual problem can then be corrected off-line at a comprehensive test facility and the sub-system returned for further use as an exchange unit.

Table 1.5 Maintenance management

Technique	Description	Remarks
A separate maintenance department	Called in by the operating departments when something goes wrong	Inherently operates on a breakdown basis. Has little chance to contribute to a major uplift in plant performance. An outdated technique.
Terotechnology	Viewing plant and equipment in terms of their total life cost and thus placing more emphasis on cost effective maintainability as distinct from first cost alone	A level of management philosophy which prepares the way for efficient maintenance systems
Computerised maintenance information	Systems covering: Plant inventory and data Operating history Spares usage and storage On-going condition data Repair work records Failure analysis and reliability data Cost records	Provides the basis for an effective management system that is able to concentrate on problem areas, plan ahead, reduce costs and contribute to the specification of improved future plant
Total productive maintenance	A technique developed in Japan in which the production operators carry out the first line of problem detection and maintenance. There is intentionally no clear interface between the production and maintenance departments.	An effective way of involving all the people in a company and achieving recognition of the importance of process plant availability. It can reduce in-service failures by 25%.
Reliability centred maintenance	A technique developed in the airlines. A system which looks at component functions, failure modes and effects in order to concentrate on key issues. Safety in operation is paramount followed closely by the control of downtime, quality and customer service.	An effective means of concentrating maintenance effort cost effectively, with an ability to identify problems and any need for design improvements
Mobile equipment maintenance	The basic objective is to ensure that the equipment does not break down away from its base. Also, when visiting its base, any essential work that is needed should be carried out.	It is particularly important to monitor the condition of such plant and its expected component lives. It is sometimes necessary to carry out some maintenance work before it becomes essential in order to match the time of access to the equipment, and the overhaul department's workload.

Condition monitoring is a technique used to monitor the condition of equipment in order to give an advanced warning of failure. It is an essential component of condition-based maintenance in which equipment is maintained on the basis of its condition.

MONITORING METHODS

The basic principle of condition monitoring is to select a physical measurement which indicates that deterioration is occurring, and then to take readings at regular intervals. Any upward trend can then be detected and taken as an indication that a problem exists. This is illustrated in Figure 2.1 which shows a typical trend curve and the way in which this provides an alert that an incipient failure is approaching. It also gives a lead time in which to plan and implement a repair.

Since failures occur to individual components, the monitoring measurements need to focus on the particular failure modes of the critical components.

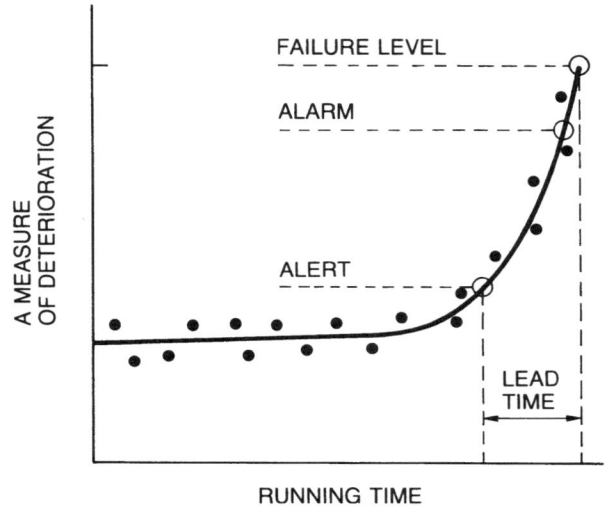

Figure 2.1 The principle of condition monitoring measurements which give an indication of the deterioration of the equipment

Table 2.1 Monitoring methods and the components for which they are suitable

Method	Principle	Application examples
Wear debris monitoring	The collection and analysis of wear debris derived from component surfaces, and carried away in the lubricating oil	Components such as bearings or other rubbing parts which wear, or suffer from surface pitting due to fatigue
Vibration monitoring	The detection of faults in moving components, from the change in the dynamic forces which they generate, and which affect vibration levels at externally accessible points	Rotating components such as gears and high speed rotors in turbines and pumps
Performance monitoring	Checking that the machine components and the complete machine system are performing their intended functions	The temperature of a bearing indicates whether it is operating with low friction.

The pressure and flow rate of a pump indicate whether its internal components are in good condition. |

The monitoring measurements give an indication of the existence of a problem as shown in Figure 2.1. More detailed analysis can indicate the nature of the problem so that rectification action can be planned. Other sections of this handbook give more details about these methods of monitoring.

In wear debris monitoring, the amount of the debris and its rate of generation indicate when there is a problem. The material and shape of the debris particles can indicate the source and the failure mechanism.

The overall level of a vibration measurement can indicate the existence of a problem. The form and frequency of the vibration signal can indicate where the problem is occurring and what it is likely to be.

Introducing condition monitoring

If an organisation has been operating with breakdown maintenance or regular planned maintenance, a change over to condition-based maintenance can result in major improvements in plant availability and in reduced costs. There are, however, up front costs for organisation and training and for the purchase of appropriate instrumentation. There are operational circumstances which can favour or retard the potential for the introduction of condition-based maintenance.

Table 2.2 Factors which can assist the introduction of condition-based maintenance

Factor	Mechanism of action
Where a safety risk is particularly likely to arise from the breakdown of machinery	Typical examples are plant handling dangerous materials, and machines for the transport of people.
Where accurate advanced planning of maintenance is essential	Typical examples are equipment situated in a remote place which is visited only occasionally for maintenance, and mobile equipment which makes only occasional visits to its base.
Where plant or equipment is of recent design, and may have some residual development problems	Condition monitoring enables faults to be detected early while damage is still slight, thus providing useful evidence to guide design improvements. It also improves the negotiating position with the plant manufacturer.
Where relatively insensitive operators use expensive equipment whose breakdown may result in serious damage	Condition monitoring enables a fault to be detected in sufficient time for an instruction to be issued for the withdrawal of the equipment before expensive damage is done.
Where the manufacturer can offer a condition monitoring service to several users of his equipment	The cost to each user can be reduced in this way, and the manufacturer gets a useful feed-back to guide his product design and development.
Where instruments or other equipment required for condition monitoring can be used, or is already being used, for another purpose	Other applications of the instruments or equipment may be process control or some servicing activity such as rotor balancing.

Table 2.3 Factors which can retard the introduction of condition-based maintenance

Factor	Mechanism of action
Where an industry is operating at a low level of activity, or operates seasonally, so that plant and machinery is often idle	If the plant is only operating part of the time, there is generally plenty of opportunity for inspection and maintenance during idle periods.
Where there is too small a number of similar machines or components being monitored by one engineer or group of engineers to enable sufficient experience to be built up for the effective interpretation of readings and for correct decisions on their significance	To gain experience in a reasonable time, the minimum number of machines tends to vary between 4 and 10 depending on the type of machine or component. The problem may be overcome by pooling monitoring services with other companies, or by involving machine manufacturers or external monitoring services.
Where skilled operators have close physical contact with their machines, and can use their own senses for subjective monitoring	Machine tools and ships can be examples of this situation, but any trends towards the use of less skilled operators or supervisory engineers, favours the application of condition monitoring.

Table 2.4 A procedure for setting up a plant condition monitoring activity

Activity	Remarks
1. Check that the plant is large enough to justify having its own internal system.	If the total plant value is less than £2M it may be worth sub-contracting the activity.
2. Consider the cost of setting up.	For most plant, a setting up cost of 1% of the plant value can be justified. If there is a major safety risk, up to 5% of the plant value may be appropriate.
3. Select the machines in the plant that should be monitored.	The important machines for monitoring will tend to be those which: (a) Are in continuous operation. (b) Are involved in single stream processes. (c) Have minimum parallel or stand-by capacity. (d) Have the minimum product storage capacity on either side of them. (e) Handle dangerous or toxic materials. (f) Operate to particularly high pressures or speeds.
4. Select the components of the critical machines on which the monitoring needs to be focussed.	The important components will be those where: (a) A failure is possible. (b) The consequences of the failure are serious in terms of safety or machine operation. (c) If a failure is allowed to occur the time required for a repair is likely to be long.
5. Choose the monitoring method or methods to be used.	List the possible techniques for each critical component and try to settle for two or at the most three techniques for use on the plant.

Table 2.5 Problems which can arise

Problem	Solution
Regular measurements need to be taken, often for months or years before a critical situation arises. The operators can therefore get bored.	The management need to keep the staff motivated by stressing the importance of their work. The use of portable electronic data collectors partially automates the collection process, provides a convenient interface with a computer for data analysis, and can also monitor the tour of duty of the operators.
One of the measurements indicates that an alert situation has arisen and a decision has to be made on whether to shut down the plant and incur high costs from loss of use, or whether it is a false alarm.	To avoid this situation install at least two physically different systems for monitoring really critical components. e.g. measure bearing temperature and vibration. In any event always recheck deviant readings and re-examine past trends.
The operators take a long time to acquire the necessary experience in detection and diagnosis, and can create false alarms.	Start taking the measurements while still operating a planned regular maintenance procedure. Take many measurements just prior to shut down and then check the components to see whether the diagnosis was correct.

Table 2.6 The benefits that can arise from the use of condition monitoring

Benefit	Mechanism
1. Increased plant availability resulting in greater output from the capital invested. 2. Reduced maintenance costs.	Machine running time can be increased by maximising the time between overhauls. Overhaul time can be reduced because the nature of the problem is known, and the spares and men can be ready. Consequential damage can be reduced or eliminated.
3. Improved operator and passenger safety.	The lead time given by condition monitoring enables machines to be stopped before they reach a critical condition, especially if instant shut-down is not permitted.
4. More efficient plant operation, and more consistent quality, obtained by matching the rate of output to the plant condition.	The operating load and speed on some machines can be varied to obtain a better compromise between output, and operating life to the next overhaul.
5. More effective negotiations with plant manufacturers or repairers, backed up by systematic measurements of plant condition.	Measurements of plant when new, at the end of the guarantee period, and after overhaul, give useful comparative values.
6. Better customer relations following from the avoidance of inconvenient breakdowns which would otherwise have occurred.	The lead time given by condition monitoring enables such breakdowns to be avoided.
7. The opportunity to specify and design better plant in the future.	The recorded experience of the operation of the present machinery is used for this purpose.

The temperatures in Table 3.1 are indicative of design limits. In practice it may be difficult to measure the contact temperature. Table 3.2 indicates practical methods of measuring temperatures and the limits that can be accepted.

Table 3.1 Maximum contact temperatures for typical tribological components

Component	Maximum temperature	Reason for limitation
White metal bearing	200°C at 1.5 MN/m² to 130°C at 7 MN/m²	Failure by incipient melting at low loading (1.5 MN/m²); by plastic deformation at high loading (7 MN/m²)
Rolling bearing	125°C	Normal tempering temperature (special bearings are available for higher temperature operation)
Steel gear	150–250°C	Scuffing; the temperature at which scuffing occurs is a function of both the lubricant and the steel and cannot be defined more closely

Table 3.2 Temperature as an indication of component failure

Component	Method of temperature measurement	Comments	Action limits [1] [4]
White metal bearing	Thermocouple in contact with back of white metal in thrust pad or at load line in journal bearings[5]	Extremely sensitive, giving immediate response to changes in load. Failure is indicated by rapid temperature rise	Alarm at rise of 10°C above normal running temperature. Trip at rise of 20°C
	Thermometer/thermocouple in oil bleed from bearing (viz. through hole drilled in bearing land)	Reasonably sensitive, may be preferable for journal bearings where there is difficulty in fitting a thermocouple into the back of the bearing in the loaded area	Alarm at rise of 10°C above normal running temperature. Trip at rise of 20°C
	Thermometer in bearing pocket or in drain oil	Relatively insensitive as majority of heat is carried away in oil that passes through bearing contact and this is rapidly cooled by excess oil that is fed to bearing. Can be useful in commissioning or checking replacements	Normal design 60°C Acceptable limit 80°C
Rolling bearing	Thermocouple or thermometer in contact with outer race (inner race rotating)	Two failure mechanisms cause temperature rise[2]	
		(a) breakdown of lubrication	Slow rise of temperature from steady value is indicative of deterioration of lubrication: Alarm at 10°C rise. Acceptable limit 100°C[3]
		(b) loss of internal clearance	Failure occurs so rapidly that there is insufficient time for warning of failure to be obtained from temperature indication of outer race
	Thermometer in oil		Acceptable limit 100°C
Gears	Thermometer in oil		Acceptable limit 80°C above ambient
Metallic packing	Contact thermometer on rod		Acceptable limit 80°C

(1) Temperature rise above normal value is more useful as an indication of trouble than the absolute value. The more the running value is below the acceptable limit the greater the margin of safety.

(2) Failure by fatigue or wear of raceways does not give temperature rise. They may be detected by an increase in noise level.

(3) Temperature in grease-packed bearing will rise to peak value until grease clears into housing and then fall to normal running value. Peak value may be 10–20°C above normal and attainment of equilibrium may take up to six hours. With bearing with grease relief valve a similar cycle will occur on each re-lubrication.

(4) Running-in. Higher than normal temperatures may occur during the initial running. Equilibrium temperatures can be expected after about twenty-four hours. The acceptable limits given should not be exceeded; if the limit is reached the machine should be stopped and and allowed to cool before proceeding with the run-in.

(5) Care must be taken to avoid deforming the bearing surface as this will result in a falsely high reading.

Vibration analysis

B4

PRINCIPLES

Vibration analysis uses vibration measurements taken at an accessible position on a machine, and analyses these measurements in order to infer the condition of moving components inside the machine.

Table 4.1 The generation and transmission of vibration

The signal	Mechanism	Examples
Generation of the signal	The mass centres of moving parts move during machine operation, generally in a cyclic manner. This gives rise to cyclic force variations.	Unbalanced shafts. Bent shafts and resonant shafts Rolling elements in rolling bearings moving unevenly. Gear tooth meshing cycles. Loose components. Cyclic forces generated by fluid interactions.
Transmission of the signal	From the moving components via their supporting bearing components to the machine casing	Ideally, there should be a relatively rigid connecting path between the area where the vibration is transmitted internally to the machine casing, and the points on the outside of the machine where the measurement is taken.
Transmission problems	If the moving parts are very light and the machine casing is very heavy and rigid, the signal measured externally may be too small for accurate analysis and diagnosis.	High speeds rotors in high pressure machines with rigid barrel casings can have this problem. A solution is to take a direct measurement of the cyclic movement of the shaft, relative to the casing at its supporting bearings.

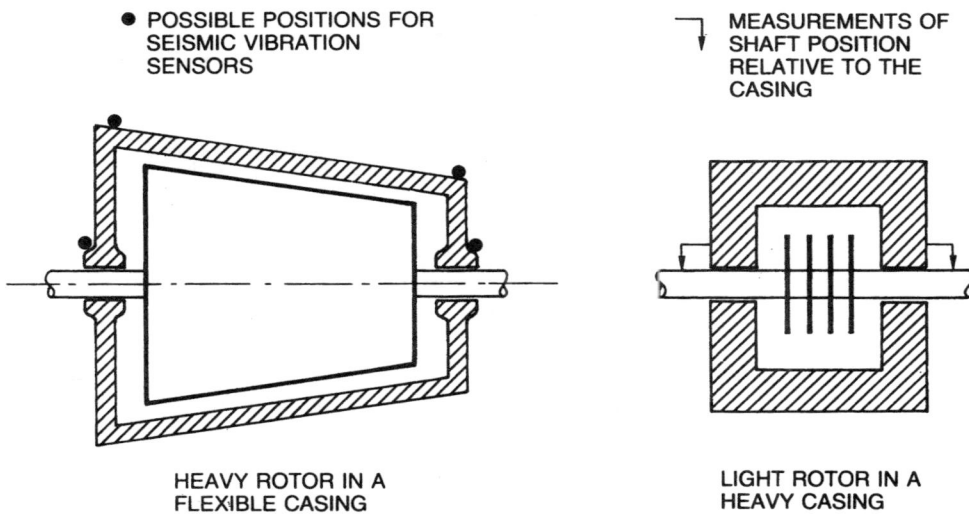

Figure 4.1 Vibration measurements on machines

Table 4.2 Categories of vibration measurement

Measurement	The principle behind the technique	Applications
Overall level of vibration (see subsequent section)	The general level of vibration over a wide frequency band. It determines the degree to which the machine may be running roughly. It is a means of quantifying the technique of feeling a machine by hand.	All kinds of rotating machines but with particular application to higher speed machines Not usually applicable to reciprocating machines
Spectral analysis of vibration (see subsequent section on vibration frequency monitoring)	The vibration signal is analysed to determine any frequencies where there is a substantial component of the vibration level. It is equivalent to scanning the frequency bands on a radio receiver to see if any station is transmitting.	From the value of the frequencies where there is a signal peak, the likely source of the vibration can be determined. Such a frequency might be the rotational speed of a particular shaft, or the tooth meshing frequency of a particular pair of gear wheels.
Discrete frequency monitoring	A method of monitoring a particular machine component by measuring the vibration level generated at the particular frequency which that component would be expected to generate	If a particular shaft in a machine is to be examined for any problems, the monitoring would be tuned to its rotational speed.
Shock pulse monitoring	Using a vibration probe, with a natural resonant frequency that is excited by the shocks generated in rolling element bearings, when they operate with fatigue pits in the surfaces of their races	The monitoring of rolling element bearings with a simple hand held instrument
Kurtosis measurement	This is a technique that looks at the 'spikyness' of a vibration signal, i.e. the number of sharp peaks as distinct from a smoother sinusoidal profile.	The monitoring of fatigue development in rolling bearings with a simple portable instrument, that is widely applicable to all types and sizes of bearing
Signal averaging (see subsequent section)	The accumulation over a few seconds of the parts of a cyclic vibration signal, which contain a particular frequency. Parts of the signal at other frequencies are averaged out. By matching the particular frequency to, for example, the rotational speed of a particular machine component, the resulting diagram will show the characteristics of that component.	The monitoring of a gear by signal averaging, relative to its rotational speed, will show the cyclic action of each tooth. A tooth with a major crack could be detected by its increased flexibility.
Cepstrum analysis	If two vibration frequencies are superimposed in one signal, sideband frequencies are generated on either side of the higher frequency peak, with a spacing related to that of the lower frequency involved. Cepstrum analysis looks at these sidebands in order to understand the underlying frequency patterns and their relative effects.	Interactions between the rotational frequency of bladed rotors and the blade passing frequency Also between gear tooth meshing frequencies and gear rotational speeds

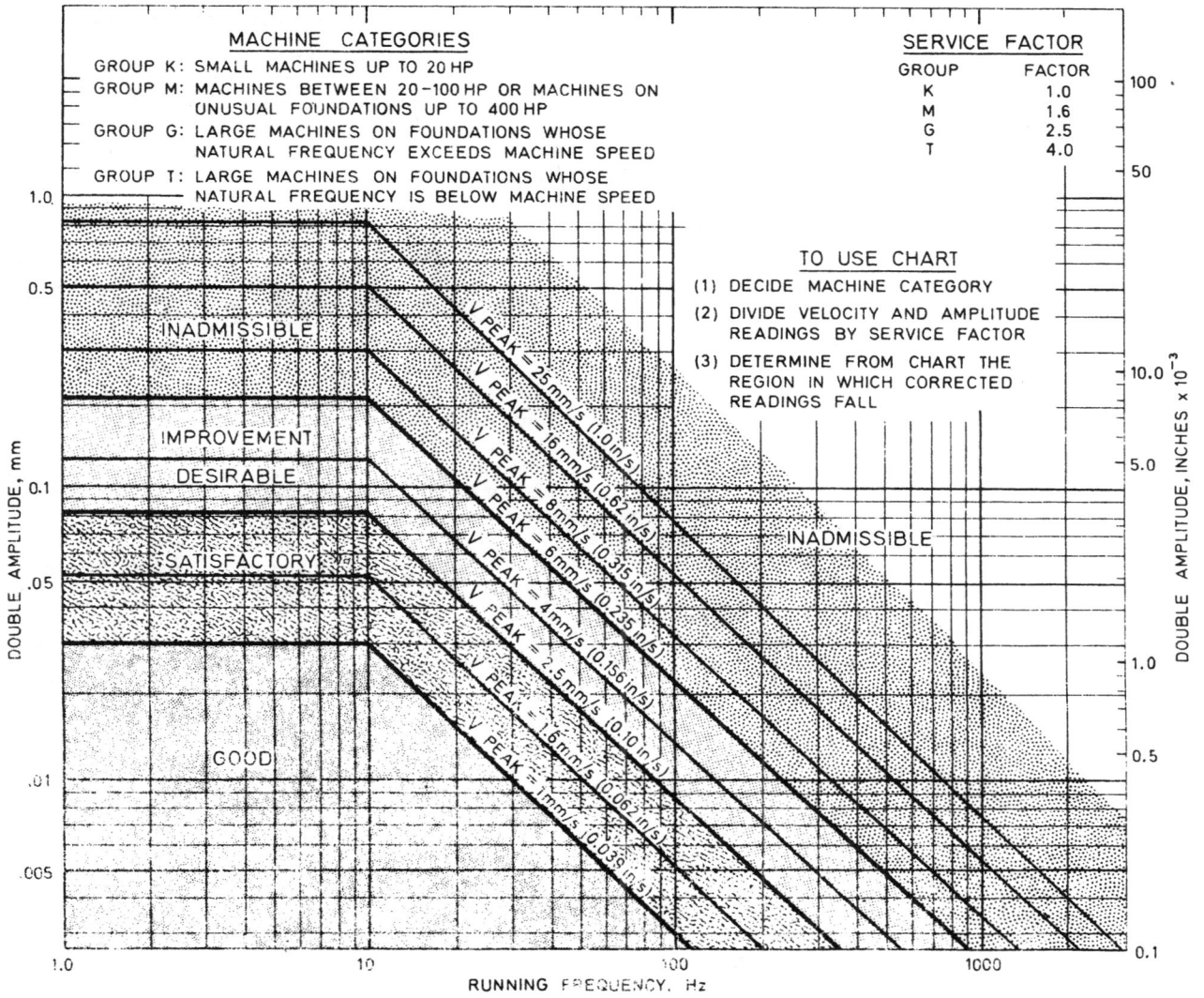

Figure 4.2 Guidance on the levels of overall vibration of machines

OVERALL LEVEL MONITORING

This is the simplest method for the vibration monitoring of complete machines. It uses the cheapest and most compact equipment. It has the disadvantage however that it is relatively insensitive, compared with other methods, which focus more closely on to the individual components of a machine.

The overall vibration level can be presented as a peak to peak amplitude of vibration, as a peak velocity or as a peak acceleration. Over the speed range of common machines from 10 Hz to 1000 Hz vibration velocity is probably the most appropriate measure of vibration level. The vibration velocity combines displacement and frequency and is thus likely to relate to fatigue stresses.

The normal procedure is to measure the vertical, horizontal and axial vibration of a bearing housing or machine casing and take the largest value as being the most significant.

As in all condition monitoring methods, it is the trend in successive readings that is particularly significant. Figure 4.2, however, gives general guidance on acceptable overall vibration levels allowing for the size of a machine and the flexibility of its mounting arrangements.

For machine with light rotors in heavy casings, where it is more usual to make a direct measurement of shaft vibration displacement relative to the bearing housing, the maximum generally acceptable displacement is indicated in the following table.

Table 4.3 Allowable vibrational displacements of shafts

Ratio Vibration displacement / Diametral clearance	Speed rev/min
0.5	300
0.25	3000
0.1	12000

VIBRATION FREQUENCY MONITORING

The various components of a machine generate vibration at characteristic frequencies. If a vibration signal is analysed in terms of its frequency content, this can give guidance on its source, and therefore on the cause of any related problem. This spectral analysis is a useful technique for problem diagnosis and is often applied, when the overall level of vibration of a machine exceeds normal values.

In spectral analysis the vibration signal is converted into a graphical plot of signal strength against frequency as shown in Figure 4.3, in this case for a single reduction gearbox.

In Figure 4.3 there are three particular frequencies which contribute to most of the vibration signal and, as shown in Figure 4.4, they will usually correspond to the shaft speeds and gear tooth meshing frequencies.

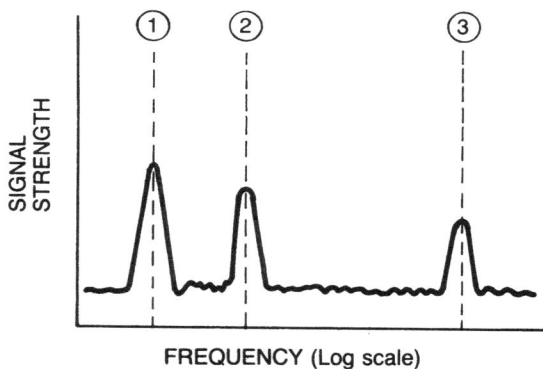

Figure 4.3 The spectral analysis of the vibration signal from a single reduction gearbox

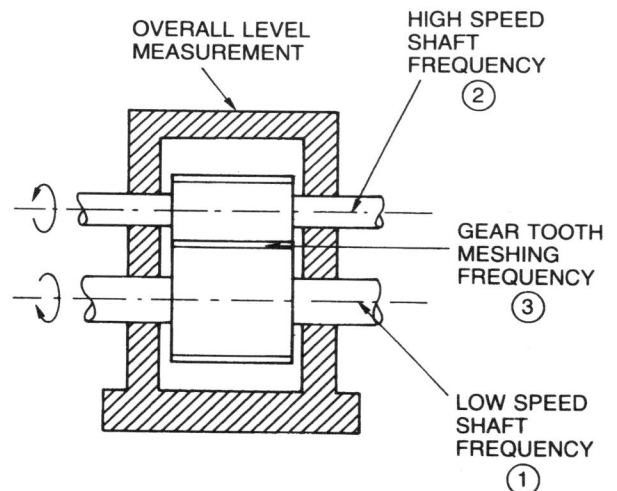

Figure 4.4 An example of the sources of discrete frequencies observable in a spectral analysis

Discrete frequency monitoring

If it is required to monitor a particular critical component the measuring system can be turned to signals at its characteristic frequency in order to achieve the maximum sensitivity. This discrete frequency monitoring is particularly appropriate for use with portable data collectors, particularly if these can be preset to measure the critical frequencies at each measuring point. The recorded values can then be fed into a base computer for conversion into trends of the readings with the running time of the machine.

Table 4.4 Typical discrete frequencies corresponding to various components and problems

Component/problem	Frequency	Characteristics
Unbalance in rotating parts	Shaft speed	Tends to increase with speed and when passing through a resonance such as a critical speed
Bent shaft	Shaft speed	Usually mainly axial vibration
Shaft misalignment	Shaft speed or 2 × shaft speed	Often associated with high levels of axial vibration
Shaft rubs	Shaft speed and 2 × shaft speed	Can excite higher resonant frequencies. May vary in level between runs.
Oil film whirl	0.45 to 0.5 × shaft speed	Only on machines with lubricated sleeve bearings
Gear tooth problems	Tooth meshing frequency	Generally also associated with noise
Reciprocating components	Running speed and 2 × running speed.	Inherent in reciprocating machinery
Rolling element bearing fatigue damage	Shock pulses at high frequency	Caused by the rolling elements hitting the fatigue pits
Cavitation in fluid machines	High frequency similar to shock pulses	Can be mistaken for rolling element bearing problems

SIGNAL AVERAGING

If a rotating component carries a number of similar peripheral sub-units, such as the teeth on a gear wheel or the blades on a rotor which interact with a fluid, then signal averaging can be used as an additional monitoring method.

A probe is used to measure the vibrations being generated and the output from this is fed to a signal averaging circuit, which extracts the components of the signal which have a frequency base corresponding to the rotational speed of the rotating component which is to be monitored. This makes it possible to build up a diagram which shows how the vibration forces vary during one rotation of the component. Some typical diagrams of this kind are shown in Figure 4.5 which indicates the contribution to the vibration signal that is made by each tooth on a gear. An outline of the technique for doing this is shown in Figure 4.6.

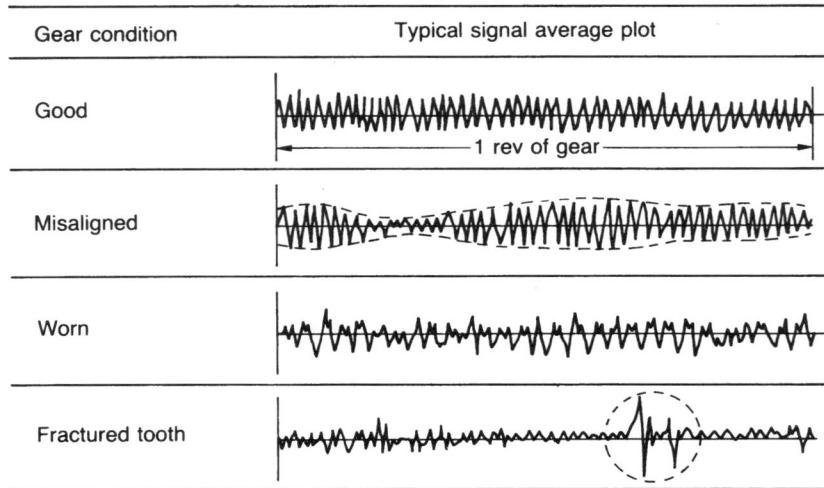

Figure 4.5 *Signal average plots used to monitor a gear and showing the contribution from each tooth*

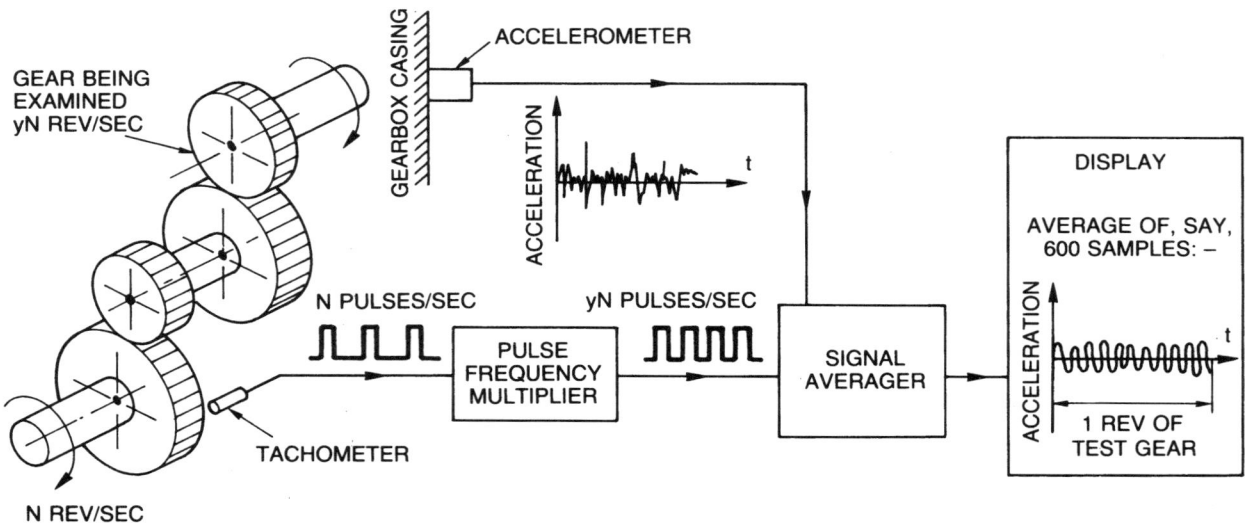

Figure 4.6 *A typical layout of a signal averaging system for monitoring a particular gear in a transmission system*

In wear debris analysis machine lubricants are monitored for the presence of particles derived from the deterioration of machine components. The lubricant itself may also be analysed, to indicate its own conditon and that of the machine.

WEAR DEBRIS ANALYSIS

Table 5.1 Wear debris monitoring methods

Type	Mechanism of operation
IN LINE	
Monitoring the main flow of oil through the machine	Magnetic plugs or systems which draw ferro-magnetic particles from the oil flow for inspection, or on to an inductive sensor, that produces a signal indicating the mass of material captured
	Inductive systems using measuring coils to assess the amount of ferrous material in circulation
	Measurement of pressure drop across the main full flow filter
ON LINE	
Monitoring a by-passed portion of the main oil flow	Optical measurement of turbidity as an indicator of particle concentration
	Pressure drop across filters of various pore sizes to indicate particle size distribution
	Discoloration of a filter strip after the passage of a fixed sample volume
	Resistance change between the grid wires of a filter to indicate the presence of metallic particles
OFF LINE	
Extracting a representative sample from the oil volume and analysing it remotely from the machine	Spectrometric analysis of the elemental content of the wear debris in order to determine its source
	Magnetic gradient separation of wear particles from a sample to determine their relative size, as a measure of problem severity
	Microscopic examination of the shape and size of the particles to determine the wear mechanisms involved
	Inductive sensor to give a direct numerical measurement of the level of ferrous debris in a sample of oil
	Optical particle counting on a diluted oil sample

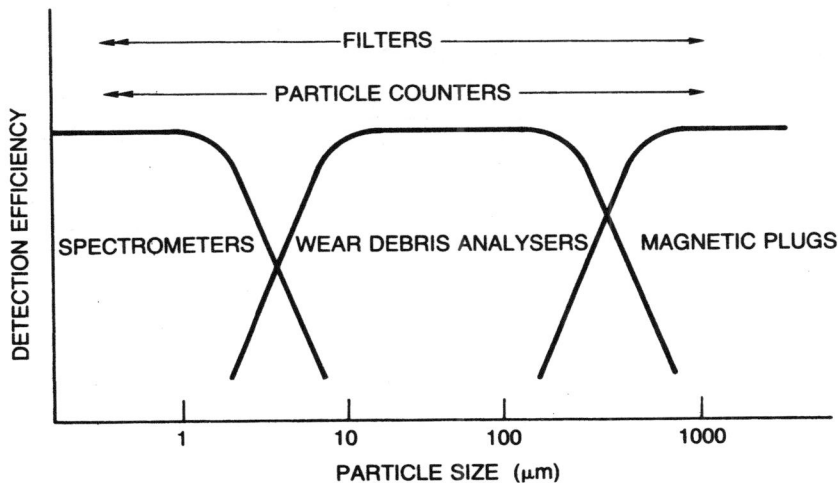

Figure 5.1 The relative efficiency of various wear debris monitoring methods

Table 5.2 Off-line wear debris analysis techniques

Technique	Description	Application
Atomic absorption spectroscopy	Oil sample is burnt in a flame and light beams of a wave length characteristic of each element are passed through the flame. The amount of light absorbed is a measure of the amount of the element that is present in the oil sample	Detects most common engineering metals. Detects particles smaller than $10\mu m$ only. Accurate at low concentrations of less than 5 ppm.
Atomic emission spectroscopy	Oil sample is burnt in an electric arc and the spectral colours in the arc are analysed for intensity by a bank of photomultipliers. Gives a direct reading of the content of many elements in the oil.	Detects most common engineering metals. Detects wear particles smaller than $10\mu m$ only. Accuracy is poor below 5 ppm.
Ferrography	A diluted oil sample is flowed across a glass slide above a powerful magnet. The particles deposit out on the slide with a distribution related to their size.	The size distribution can indicate the severity of the wear. The particle shapes indicate the wear mechanism.
Rotary particle depositor	A diluted oil sample is placed on a glass cover above rotating magnets. Wear particles are deposited as a set of concentric rings.	As for ferrography with the added advantage that it can be linked directly to a particle quantifier
X-ray fluorescence	When the sample is exposed to a radioactive beam, X-rays characteristic of the material content are emitted.	Detects most engineering metals. Accurate to only ± 5 ppm.
Inductively coupled plasma emission spectroscopy	The oil sample is sprayed into an argon plasma torch. The spectral colours of the emitted light and its intensity are then measured to indicate the amount of various elements that are present.	Detects most engineering metals. Accurate down to parts per billion.

Table 5.3 Problems with wear debris analysis

Problem	Solution
Poor quality or variable samples	Ensure that sample bottles are clean and properly labelled. Use sampling valves or suction syringes. Hydraulic fluid sampling methods defined in ISO 3722.
Unrecorded oil change or a large oil addition	Often indicated by a sudden drop in contaminant levels. The importance of recording oil top-ups needs to be emphasised to the operators.
Addition of the wrong oil to the machine detected by an increase of elements commonly used in oil additives	Take a second sample and discuss the problem with the machine operator in order to avoid a recurrence

Table 5.4 Sources of materials found in wear debris analysis

Material	Likely worn component or other source	Material	Likely worn component or other sources
Aluminium	Light alloy pistons Aluminium tin crankshaft bearings Components rubbing on aluminium casings	Lead Magnesium	Plain bearings Wear of plastic components with talc fillers Seawater intrusion
Antimony	White metal plain bearings	Nickel	Valve seats Alloy steels
Boron	Coolant leaks Can be present as an oil additive	Potassium	Coolant leaks
Chromium	Piston rings or cylinder liners Valve seats	Silicon Silver	Mineral dust intrusion Silver-plated bearing surfaces Fretting of silver soldered joints
Cobalt	Valve seats Hard coatings		
Copper	Copper-lead or bronze bearings Rolling element bearing cages	Sodium Tin	Coolant leakage Seawater intrusion Plain bearings
Indium	Crankshaft bearings	Vanadium	Intrusion of heavy fuel oil
Iron	Gears Shafts Cast iron cylinder bores	Zinc	A common oil additive

Table 5.5 Quick tests for metallic debris from filters

Metal	Test method	Method	Test method
Steel and nickel	Can be attracted by a permanent magnet	Silver	Dissolves to form a white fog in nitric acid
Tin	Fuses with tin solder on a soldering iron	Copper and Bronze	Dissolves to produce a blue/green cloud in nitric acid
Aluminium	Dissolves rapidly in sodium or potassium hydroxide solution to form a white cloud	Chromium	Dissolves to produce a green cloud in hydrochloric acid

Physical characteristics of wear debris

Rubbing wear

The normal particles of benign wear of sliding surfaces. Rubbing wear particles are platelets from the shear mixed layer which exhibits super-ductility. Opposing surfaces are roughly of the same hardness. Generally the maximum size of normal rubbing wear is 15 μm.

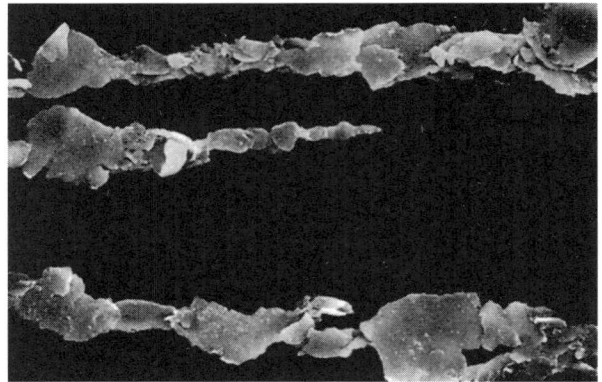

Break-in wear particles are typical of components having a ground or machined surface finish. During the break-in period the ridges on the wear surface are flattened and elongated platelets become detached from the surface often 50 μm long.

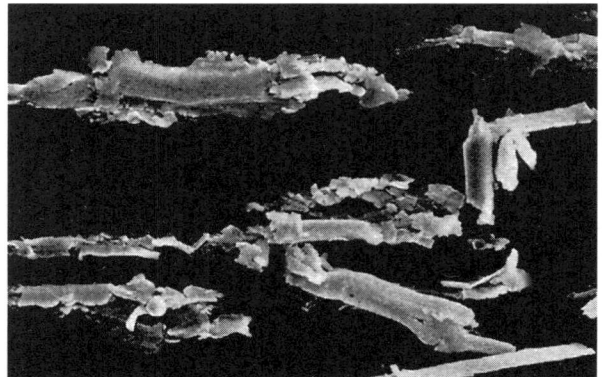

Cutting wear

Wear particles which have been generated as a result of one surface penetrating another. The effect is to generate particles much as a lathe tool creates machining swarf. Abrasive particles which have become embedded in a soft surface, penetrate the opposing surface generating cutting wear particles. Alternatively a hard sharp edge or a hard component may penetrate the softer surface. Particles may range in size from 2–5 μm wide and 25 to 100 μm long.

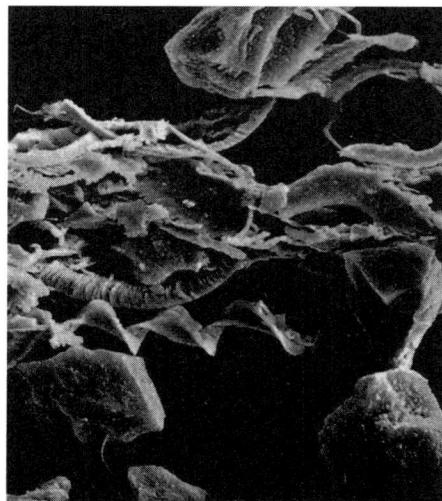

Rolling fatigue wear

Fatigue spall particles are released from the stressed surface as a pit is formed. Particles have a maximum size of 100 μm during the initial microspalling process. These flat platelets have a major dimension to thickness ratio greater than 10:1.

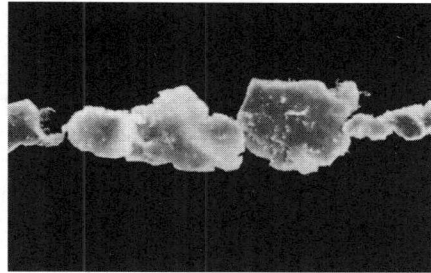

Spherical particles associated with rolling bearing fatigue are generated in the bearing fatigue cracks. The spheres are usually less than 3 μm in diameter.

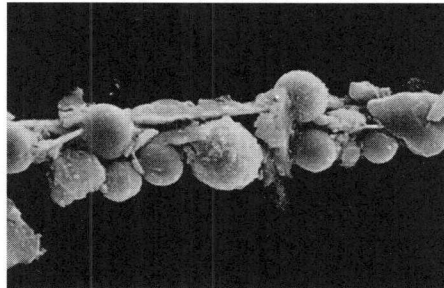

Laminar particles are very thin free metal particles between 20–50 μm major dimension with a thickness ratio approximately 30:1. Laminar particles may be formed by their passage through the rolling contact region.

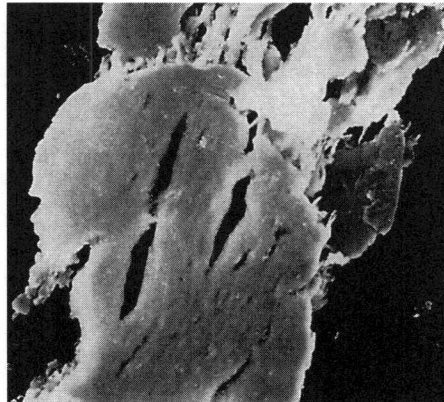

Combined rolling and sliding (gear systems)

There is a large variation in both sliding and rolling velocities at the wear contacts; there are corresponding variations in the characteristics of the particles generated. Fatigue particles from the gear pitch line have similar characteristics to rolling bearing fatigue particles. The particles may have a major dimension to thickness ratio between 4:1 and 10:1. The chunkier particles result from tensile stresses on the gear surface causing fatigue cracks to propagate deeper into the gear tooth prior to pitting. A high ratio of large (20 μm) particles to small (2 μm) particles is usually evident.

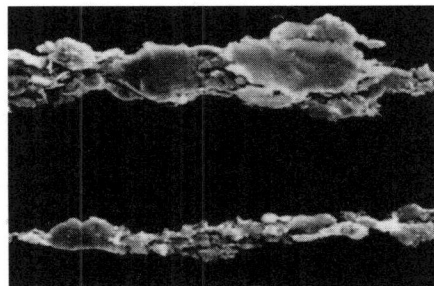

Severe sliding wear

Severe sliding wear particles range in size from 20 μm and larger. Some of these particles have surface striations as a result of sliding. They frequently have straight edges and their major dimension to thickness ratio is approximately 10:1.

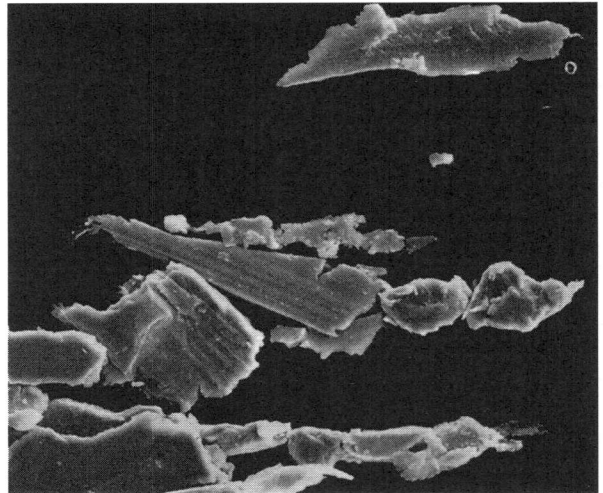

Crystalline material

Crystals appear bright and changing the direction of polarisation or rotating the stage causes the light intensity to vary. Sand appears optically active under polarised light.

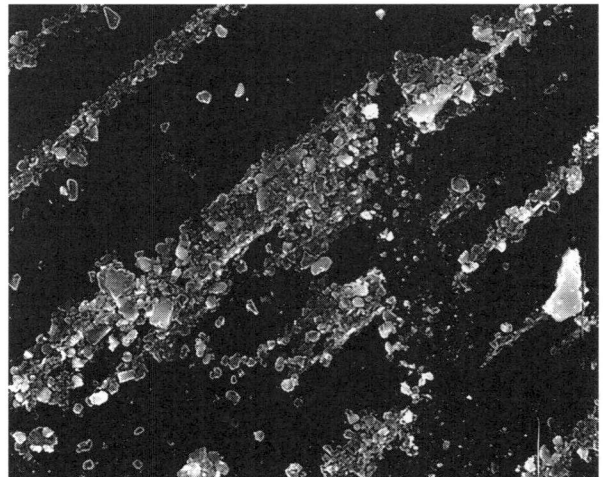

Weak magnetic materials

The size and position of the particles after magnetic separation on a slide indicates their magnetic susceptibility. Ferro-magnetic particles (Fe, Co, Ni) larger than 15 μm are always deposited at the entry or inner ring zone of the slide. Particles of low susceptibility such as aluminium, bronze, lead, etc, show little tendency to form strings and are deposited over the whole of the slide.

Polymers

Extruded plastics such as nylon fibres appear very bright when viewed under polarised light.

Examples of problems detected by wear debris analysis

Crankshaft bearings from a diesel engine

Rapid wear of the bearings occurred in a heavy duty cycle transport operation. The copper, lead and tin levels relate to a combination of wear of the bearing material and its overlay plating.

Sample no.	1	2	3	4
Iron ppm	35	40	42	40
Copper ppm	15	25	35	45
Lead ppm	20	28	32	46
Tin ppm	4	4	10	15

Grease lubricated screwdown bearing

The ratio of chromium to nickel, corresponding broadly to that in the material composition, indicated severe damage to the large conical thrust bearing.

Sample no.	1	2	3	4
Iron ppm	150	240	1280	1540
Chromium ppm	1	2	11	31
Nickel ppm	2	4	23	67

Differential damage in an intercity bus

Excessive iron and the combination of chromium and nickel resulted from the disintegration of a nose cone bearing

Sample no.	1	2	3	4
Iron ppm	273	383	249	71000
Chromium ppm	2	3	2	21
Nickel ppm	0	1	0	5

Large journal bearing in a gas turbine pumping installation

The lead based white metal wore continuously.

Sample no.	1	2	3	4
Iron ppm	0	2	2	2
Copper ppm	4	5	11	15
Lead ppm	10	24	59	82

Piston rings from an excavator diesel engine

Bore polishing resulted in rapid wear of the piston rings. The operating lands of the oil control rings were worn away.

Sample no.	1	2	3	4
Iron ppm	5	5	9	10
Chromium ppm	15	25	40	120

Engine cylinder head cracked

The presence of sodium originates from the use of a corrosion inhibitor in the cooling water. A crack was detected in the cylinder head allowing coolant to enter the lubricant system.

Sample no.	1	2	3	4
Iron ppm	60	84	104	203
Chromium ppm	7	12	12	53
Nickel ppm	7	10	12	27
Sodium ppm	19	160	330	209

LUBRICANT ANALYSIS

Table 5.6 Off-line lubricant analysis techniques

Technique	Description	Interpretation
Viscosity measurement	Higher viscosity than a new sample	Oxidation of the oil and/or heavy particulate contamination
	Lower viscosity than a new sample	Fuel dilution in the case of engine oils
Total acid number, TAN ASTM D974 D664 IP 139 177	Indicates the level of organic acidity in the oil	A measure of oil oxidation level For hydraulic oils the value should not exceed twice the level for the new oil
Strong acid number, SAN ASTM D974 D664 IP 139 177	Indicates the level of strong inorganic acidity in the oil	In engine oils indicates the presence of sulphur-based acids from fuel combustion. Synthetic hydraulic oils can show high values if they deteriorate
Total base number, TBN ASTM D664 D2896 IP 177 276	Indicates the reserves of alkalinity present in the oil	Running engines on higher sulphur fuels creates acids which are neutralised by the oil, as long as it continues to be alkaline
Infra-red spectroanalysis	Measures molecular compounds in the oil such as water, glycol, refrigerants, blow by gases, liquid fuels, etc. Also additive content.	A very versatile monitoring method for lubricant condition

Table 5.7 Analysis techniques for the oil from various types of machine

** essential * useful	Diesel engine	Gasoline engine	Gears	Hydraulic systems	Air compressor	Refrigeration compressor	Gas turbine	Steam turbine	Trans-formers	Heat transfer
Spectro-chemical analysis	**	**	**	**	**	**	**	**		
Infra-red analysis	**	**	*	*	*	**	*	*	*	*
Wear debris quantifier	**	**	**	*	*	*	*	*		
Viscosity at 40°C	**	**	**	**	**	**	**	**		*
Viscosity at 100°C	*	*								
Total base number	**	**								
Total acid number			*	**	*	*	**	*	*	**
Water %	**	**	*	**	*			**	**	
Total solids	*	*							*	*
Fuel dilution	**	**								
Particle counting				**			*			

THE NEED FOR LUBRICANT CHANGES

Reason for changing lubricant	Cause of the problem	Effect of the problem	Methods of extending lubricant life
DEGRADATION	 *Approximate life of refined mineral oils*	Alteration of the physical characteristics of the lubricant Formation of sludges that can block oil-ways	Where possible bulk oil temperature should be kept below 60°C and unnecessary aeration should be avoided
CONTAMINATION	Wear debris, external solids and water entering the system Access of working fluids to the lubricant	Solids and immiscible liquids interfere with lubrication and cause wear Soluble contaminants, such as hydrocarbons and organic solvents, alter the physical properties of the oil	Contamination can be reduced or eliminated by use of filters, centrifuges and regular draining of separated material from the bottom of the lubricant reservoir

CHANGE PERIODS

Systems containing less than 250 litre (50 gal)

Analytical testing is not justified and change periods are best based on experience. The following examples in the opposite column are typical of industrial practice:

System	Operating conditions	Change period years
Oil bath	All	1
Well-sealed systems (e.g. gearboxes, hydraulic systems, compressor crankcases)	40–50°C 60°C	2–3 1
Grease-packed ball and cylindrical roller bearings[1]	Up to $dn^{(2)} = 150\,000$ $dn = 200\,000$	2 1

Notes: (1) Horizontally mounted, medium series bearings lubricated with a rolling bearing quality grease operating in normal ambient temperature conditions.

(2) d = bore in mm, n = rev/min.

Systems containing more than 250 litre (50 gal)

Regular testing should be carried out to determine when the lubricant is approaching the end of its useful service life. A combination of visual examination and laboratory testing is recommended.

Visual examination	Laboratory tests
1 Carry out at moderately short intervals, say weekly	1 Longer term tests, say every 6–12 months
2 Can be carried out on the plant with simple equipment and little experience	2 If suitable facilities are not available, oil supplier may be able to help
3 Allows immediate action to be taken and stimulates interest on the plant	3 Takes some time before results are available, but gives quantitative values for assessment of lubricant condition
4 Samples should be kept for comparison with the next set	4 Records of laboratory tests enable more subtle changes in the oil to be followed and allow prediction of further useful life

The results obtained are only representative of the sample. This should preferably be taken when the system is running, and a clean container must be used. Guidance on interpreting the results is given in the following tables.

VISUAL EXAMINATION OF USED LUBRICATING OIL

1 Take sample of circulating oil in clean glass bottle (50–100 ml).
2 If dirty or opaque, stand for 1 h, preferably at 60°C (an office radiator provides a convenient source of heat).

Appearance of sample		Reason	Action to be taken	
When taken	After 1 h		System without filter[1] or centrifuge	System with filter[1] or centrifuge
Clear	—	—	None	None
Opaque[2]	Clear	Foaming	Cause of foaming to be sought[3]	Cause of foaming to be sought[3]
	Clear oil with separated water layer	Unstable emulsion[4]	Run off water (and sludge) from drain[5]	Check centrifuge[6]
	No change	Stable emulsion	Submit sample for analysis[7]	Check centrifuge;[6] if centrifuge fails to clear change oil
Dirty	Solids separated[8]	Contamination	Submit sample for analysis[7]	Check filter or centrifuge
Black (acrid smell)	No change	Oil oxidised	Submit sample for analysis[7]	Submit sample for analysis[7]

Notes: (1) The term filter is restricted to units able to remove particles less than 50 μm; coarse strainers, which are frequently fitted in oil pump suctions to protect the pump, do not remove all particles liable to damage bearings, etc.
(2) Both foams (mixtures of air and oil) and emulsions (mixtures of water and oil) render the oil opaque.
(3) Foaming is usually mechanical in origin, being caused by excessive churning, impingement of high-pressure return oil on the reservoir surface, etc. Foam can be stabilised by the presence of minor amounts of certain contaminants, e.g. solvents, corrosion preventives, grease. If no mechanical reason can be found for excessive foam generation, it is necessary to change the oil.
(4) Steps should be taken to remove the water as soon as possible. Not only is water liable to cause lubrication failure, but it will also cause rusting; the presence of finely divided rust tends to stabilise emulsions.
(5) Failure of water to separate from oil in service may be the result of inadequate lubricant capacity or the oil pump suction being too close to the lowest part of the reservoir. More commonly it results from re-entrainment of separated water from the bottom of the sump when, by neglect, it has been allowed to build up in the system.
(6) The usual reason for a centrifuge failing to remove water is that the temperature is too low. The oil should be heated to 80°C before centrifuging.
(7) It is not always possible to decide visually whether the oil is satisfactory or not. In doubtful cases it is necessary to have laboratory analysis (see next table).
(8) In a dark oil, solids can be seen by inverting the bottle and examining the bottom.

LABORATORY TESTS FOR USED MINERAL LUBRICATING OILS

Property	Test method[1]	Type of oil	Action level
Viscosity	IP71/ASTM D445	All	15% increase from new oil value (cSt at 40°C)[2]
Acidity (neutralisation value)	IP1	Straight oil	2–3 mg KOH/g[3]
	IP139/ASTM D947	Turbine oil	1 mg KOH/g[4]
	or	hd hydraulic oil	Discuss with oil supplier[5]
	IP177/ASTM D664	ep gear oil	
Ash (check for contamination by solids)	IP4	Straight oil / Turbine oil	> 0.2%[6]
	ASTM D893 or IP4	hd hydraulic oil / ep gear oil	> 0.2%[7]
Water level	ASTM D1744	All	200 ppm[8]
Additive level (checks for additive loss)		Turbine oil / hd hydraulic oil / ep gear oil / Engine oil	Discuss with oil supplier[9]

Notes: (1) *I.P. Standards for Petroleum and Its Products*, Institute of Petroleum and also ASTM Standards.
(2) Change oil if viscosity has increased by 15% as a result of degradation. (Where increase or decrease is caused by contamination with miscible liquid or topping up with the wrong grade of oil, much greater changes can be tolerated and limits have to be chosen to suit particular cases.)
(3) Acidity develops as a result of oxidation; the oxidation reaction is auto-catalytic so that once it has started it proceeds at an increasing rate. The acids formed are not corrosive to materials of construction, but measurement of acidity gives a useful guide to the condition of the oil. The exact value at which the oil should be changed is not critical, but experience shows that when a value of 2–3 mg KOH/g is reached the useful life of the oil is limited.
(4) In oils containing oxidation inhibitors there is a long induction period before acidity develops; once this occurs it develops rapidly and the figure given provides a useful indication of the end of the service life of the oil.
(5) Additives in the oil interfere with the determination of acidity.
(6) This is an arbitrary value and particular cases may require individual judgement.
(7) Oil-soluble ash-forming additives render the simple ash test inapplicable in these cases.
(8) This figure applies to systems with critical components, e.g. rolling bearings, hydraulic control valves, gears. In less critical systems, e.g. steam-engine crankcases, much higher levels can be tolerated.
(9) In circulation systems with a large inventory of oil it may be worth checking the depletion of additives in additive-containing oils. No general tests for these are available, but oil suppliers should be able to help in providing test methods for the particular additives they use.

NOTES ON GOOD MAINTENANCE PRACTICE

Attention to detail will give improved performance of oils in lubrication systems. The following points should be noted:

1 Oil systems should be checked weekly and topped up as necessary. Systems should not be over-filled as this may lead to overheating through excessive churning.
2 Oil levels in splash-lubricated gearboxes may be different when the machine is running from when it is stationary. For continuously running machines the correct running level should be marked to avoid the risk of over- or under-filling.
3 Degradation is a function of temperature. Where possible the bulk oil temperature in systems should not exceed 60°C. The outside of small enclosed systems should be kept clean to promote maximum convection cooling.
4 Care must be exercised to prevent the ingress of dirt during topping up.

The ability of micro-organisms to use petroleum products as nourishment is relatively common. When they do so in very large numbers a microbiological problem may arise in the use of the petroleum product. Oil emulsions are particularly prone to infection, as water is essential for growth, but problems also arise in straight oils.

CHARACTERISTICS OF MICROBIAL PROBLEMS

1 They are most severe between 20°C and 40°C.
2 They get worse.
3 They are 'infectious' and can spread from one system to another.
4 Malodours and discolorations occur, particularly after a stagnation period.
5 Degradation of additives by the organisms may result in changes in viscosity, lubricity, stability and corrosiveness.
6 Masses of organisms agglomerate as 'slimes' and 'scums'.
7 Water is an essential requirement.

Factors affecting level of infection of emulsions

The severity of a problem is related to the numbers and types of organisms present. Most of the factors in the following table also influence straight oil infections.

Increase infection level	Decrease infection level
1 High initial 'inoculum' from dregs of previous charge and equipment slimes	1 Temperatures consistently outside optimum range for growth
2 Oil formulation containing complete diet for organisms	2 Local temperatures in sterilising range (above 70°C), for example in a purifier
3 Growth stimulation from added sandwiches, urine, etc.	3 High pH (above 9.5)
4 Leaks or carry-over into the system from other infected systems	4 Organisms removed as slimes in centrifuges and filters by adhering to metal swarf and machined parts
5 Poor quality water used in preparing charge	5 Anti-microbial inhibitors present in oil formulation
6 Over diluted emulsion in use for prolonged period	6 Variety of oil formulations used in successive charges

Characteristics of principal infecting organisms (generalised scheme)

Organism	pH Relationship	Products of growth	Type of growth
Aerobic bacteria (use oxygen)	Prefer neutral to alkaline pH	Completely oxidised products (CO_2 and H_2O) and some acids. Occasionally generate ammonia	Separate rods, forming slime when agglomerated. Size usually below 5 μm in length
Anerobic bacteria (grow in absence of oxygen)	Prefer neutral to alkaline pH	Incompletely oxidised and reduced products including CH_4, H_2, and H_2S	As above. Often adhere to steel surfaces, particularly swarf
Yeasts	Prefer acid pH	Oxidised and incompletely oxidised products. pH falls	Usually separate cells, 5–10 μm, often follow bacterial infections or occur when bacteria have been inhibited. Sometimes filamentous
Fungi (moulds)	Prefer acid pH but some flourish at alkaline pH in synthetic metal working fluids	Incompletely oxidised products—organic acids accumulate	Filaments of cells forming visible mats of growth. Spores may resemble yeasts. Both yeasts and moulds grow more slowly than bacteria

Comparison of microbial infections in oil emulsions and straight oils

Oil emulsions	Straight oils
Organisms live in the continous (water) phase	Organisms live in a separate discrete water phase or a dispersed water phase
Alkaline environment, usually changing *slowly* to neutral or even slightly acidic	Water often becoming rapidly acidic
Micro-organisms predominantly bacteria except in synthetic formulations	Bacteria, fungi and yeasts commonly occur
Oil-in-water emulsion becomes unstable due to degradation of emulsifiers. Effect is less pronounced in bio-stable formulations	Water-in-oil emulsions become stabilised due to surface activity of micro-organisms
Anti-microbials (biocides) often included in formulation and give some initial protection	Anti-microbials rarely included, but can be added in use
Temperature (just above ambient) favours growth	Temperature often so high that systems are self-sterilising. Heated purifiers also pasteurise the oil
Anti-microbial treatment usually involves water-soluble biocides	Biocides must have some oil solubility

ECONOMICS OF INFECTION

The total cost of a problem is rarely concerned with the cost of the petroleum product infected but is made up from some of the following components:

1 Direct cost of replacing spoiled oil or emulsion.
2 Loss of production time during change and consequential down-time in associated operations.
3 Direct labour and power costs of change.
4 Disposal costs of spoiled oil or emulsion.
5 Deterioration of product performance particularly:
   (*a*) surface finish and corrosion of product in machining;
   (*b*) staining and rust spotting in steel rolling;
   (*c*) 'pick-up' in aluminium rolling.
6 Cost of excessive slime accumulation, e.g. overloading centrifuges, 'blinding' grinding wheels, blocking filters.
7 Wear and corrosion of production machinery, blocked pipe-lines, valve and pump failure.
8 Staff problems due to smell and possibly health.

ANTI-MICROBIAL MEASURES

These may involve:

1 Cleaning and sterilising machine tools, pipework, etc., between charges.
2 Addition of anti-microbial chemicals to new charges of oil or emulsion.
3 Changes in procedures, such as:
   (*a*) use of clean or even de-ionised water;
   (*b*) continuous aeration or circulation to avoid malodours;
   (*c*) prevention of cross-infection;
   (*d*) re-siting tanks, pipes and ducts, eliminating dead legs;
   (*e*) frequent draining of free water from straight oils;
   (*f*) change to less vulnerable formulations
4 Continuous laboratory or on-site evaluation of infection levels.

Physical methods of controlling infection (heat, u.v., hard irradiation) are feasible but chemical methods are more generally practised for metal working fluids. Heat is sometimes preferred for straight oils. There is no chemical 'cure-all', but for any requirement the following important factors will affect choice of biocide.

1 Whether water or oil solubility is required – or both.
2 Speed of action required. Quick for a 'clean-up', slow for preventing re-infection.
3 pH of system – this will affect the activity of the biocide and, conversely, the biocide may affect the pH of the system (most biocides are alkaline).
4 Identity of infecting organisms.
5 Ease of addition – powders are more difficult to measure and disperse than liquids.
6 Affect of biocide on engineering process; e.g. reaction with oil formulation, corrosive to metals present.
7 Toxicity of biocide to personnel – most care needed where contact and inhalation can occur – least potential hazard in closed systems, e.g. hydraulic oils.
8 Overall costs over a period.
9 Environmental impact on disposal.

Most major chemical suppliers can offer one or more industrial biocides and some may offer an advisory service.

A useful condition monitoring technique is to check the performance of components, to check that they are performing their intended function correctly.

COMPONENT PERFORMANCE

The technique for selecting a method of monitoring a component is to decide what function it is required to perform and then to consider the various ways in which that function can be measured.

Table 8.1 Methods of monitoring the performance of fixed components for fault detection

Component	Function	Monitoring method
Casings and frameworks	Rigid support and transfer of loads to foundations	Crack detection by: Visual inspection Dye penetrants Ultrasonic tests Eddy current probes Magnetic flux Radiography Tests for deflection under a known applied load. Visual checks for material loss by corrosion
Cold pressure vessels	The containment of fluid under pressure	Detection of external surface cracks by: Visual inspection Dye penetrants Eddy current probes Magnetic flux Detection of internal cracks by: Ultrasonic tests Radiography Boroscopes (when out of service) Detection of loss of wall thickness by corrosion: Ultrasonic tests Electrochemical probes Sacrificial coupons Small sentinel holes (where permissible) Detection of strain growth by acoustic emission
Boilers and thermal reactors	The heating of fluids and containment of pressure	Detection of surface cracks and leakage by visual inspection Detection of hidden cracks by ultrasonic tests Hammer testing of the shell Chemical check of feed water and boiler water samples, to indicate likely corrosion or deposit build up
Nozzle blades	The profiled flow of fluids and transmission of forces	Checking for profile changes and integrity by boroscopes (when out of service)
Ducts	Guiding the flow of air or other gases	Detection of leaks by gas sniffer detectors If gas is hot and carrying fine solids, partial blockages can be detected by: Infrared thermography Thermographic paints
Pipes	Guiding the flow of fluids under pressure	Checking for reduction in wall thickness by: Ultrasonics Corrosion coupons Electrochemical probes Sentinel holes

Table 8.2 Methods of monitoring the performance of moving components for fault detection

Component	Function	Monitoring method
Journal bearings	Locating rotating shafts radially with the minimum friction	Checking low friction operation by temperature measurement
Thrust bearings	Locating rotating shafts axially with the minimum friction	Checking low friction operation by temperature measurement. Checking location by shaft position probes
Shafts and couplings	Smooth rotation while transmitting torques	Stroboscopes for visual inspection while rotating
Seals	Allowing rotating shafts to enter pressurised fluid containments with minimum leakage	Checking for leakage: Visually via deposits Gas sniffer detectors Liquid leakage pools
Pistons and cylinders	The interchange of fluid pressure and axial force with minimum leakage	Listening for leakage at ultrasonic frequencies
I.C. engine combustion	The provision of regular power pulses to drive the crankshaft	Toothed wheel fitted to the free end of the crankshaft to check for even rotation and pulses
Belt drives	The smooth transmission of power by differential tension	Stroboscopes for visual inspection while rotating. Proximity probes to detect excess slackness
Springs	Allowing controlled deflection with increasing loads	Deflection measurement at known loads. Crack detection by visual inspection and dye penetrants

In addition to monitoring the performance of components, it is also useful to monitor the performance of complete machines and systems.

Table 8.3 Methods of monitoring the performance of machines and systems for fault detection

Machine/System	Function	Monitoring method
Hydraulic power systems	Converting fluid pressure and flow into mechanical power	Measuring the relationship between pressure and flow in the system
Pumps	Converting mechanical power into fluid pressure and flow	Measuring the relationship between delivery pressure and flow, to detect any deterioration in the internal components
Textile machines	The production of fabric from threads	Checking the fabric for any patterns which indicate inconsistencies in machine operation
Coal or ore crushing mills	Reducing the particle size of materials	Checking changes in the size distribution of the output material
Motor vehicle	Consuming fuel to provide transportation	Measuring the distance covered per unit of fuel consumption
Heat exchangers	Producing temperature changes between two fluid flow streams	Monitoring the relationship between flow rate and temperature change
Thermodynamic and chemical process systems	Using pressure, temperature and volume/flow changes to interchange energy or change materials	The measurement and comparison against datums of pressure, temperature and flow relationships in the system
Systems with automatic control	The control of system variables to obtain a required output	The monitoring of the control actions taken by the system in normal operation to determine extreme values or combinations which indicate a system fault. Giving the system a control exercise designed to detect likely faults

BALL AND ROLLER BEARINGS

If there is evidence of pitting on the balls, rollers or races, suspect fatigue, corrosion or the passage of electrical current. Investigate the cause and renew the bearing.

If there is observable wear or scuffing on the balls, rollers or races, or on the cage or other rubbing surfaces, suspect inadequate lubrication, an unacceptable load or misalignment. Investigate the cause and renew the bearing.

ALL OTHER COMPONENTS

Wear weakens components and causes loss of efficiency. Wear in a bearing may also cause unexpected loads to be thrown on other members such as seals or other bearings due to misalignment. No general rules are possible because conditions vary so widely. If in doubt about strength or efficiency, consult the manufacturer. If in doubt about misalignment or loss of accuracy, experience of the particular application is the only sure guide.

Bearings as such are considered in more detail below.

JOURNAL BEARINGS, THRUST BEARINGS, CAMS, SLIDERS, etc.

Debris

If wear debris is likely to remain in the clearance spaces and cause jamming, the volume of material worn away in intervals between cleaning should be limited to $\frac{1}{5}$ of the available volume in the clearance spaces.

Surface treatments

Wear must not completely remove hardened or other wear resistant layers.

Note that some bearing materials work by allowing lubricant to bleed from the bulk to the surface. No wear is normally detectable up to the moment of failure. In these cases follow the manufacturer's maintenance recommendations strictly.

Some typical figures for other treatments are:

Treatment	Allowable wear depth
Good quality carburising	2.5 mm (0.1 in)
Gas nitriding	0.25 mm (0.01 in)
Salt bath nitriding	None
Cyanide hardening (shallow)	0.025 mm (0.001 in)
Cyanide hardening (deep)	0.25 mm (0.01 in)
Graphite or MoS_2 films	None
White metal, etc.	up to 50% of original thickness

Surface condition

Roughening (apart from light scoring in the direction of motion) usually indicates inadequate lubrication, overloading or poor surfaces. Investigate the cause and renew the bearing.

Pitting usually indicates fatigue, corrosion, cavitation or the passage of electrical current. Investigate the cause. If a straight line can be drawn (by eye) across the bearing area such that 10% or more of the metal is missing due to pits, then renew the components.

Scoring usually indicates abrasives either in the lubricant or in the general surroundings.

Journal bearings with smoothly-worn surfaces

The allowable increase in clearance depends very much on the application, type of loading, machine flexibilities, importance of noise, etc., but as a general guide, an increase of clearance which more than doubles the original value may be taken as a limit.

Wear is generally acceptable up to these limits, subject to the preceding paragraphs and provided that more than 50% of the original thickness of the bearing material remains at all points.

Thrust bearings, cams, sliders, etc. with smoothly-worn surfaces

Wear is generally acceptable, subject to the preceding paragraphs, provided that no surface features (for example jacking orifices, oil grooves or load generating profiles) are significantly altered in size, and provided that more than 50% of the original thickness of the bearing material remains at all points.

CHAINS AND SPROCKETS

For effects of wear on efficiency consult the manufacturers. Some components may be case-hardened in which case data on surface treatments will apply.

CABLES AND WIRES

For effects of wear on efficiency consult the manufacturer. Unless there is previous experience to the contrary any visible wear on cables, wires or pulleys should be investigated further.

METAL WORKING AND CUTTING TOOLS

Life is normally set by loss of form which leads to unacceptable accuracy or efficiency and poor surface finish on the workpiece.

THE SIGNIFICANCE OF FAILURE

Failure is only one of three ways in which engineering devices may reach the end of their useful life.

The way in which the end of useful life is reached		Typical devices which can end their life this way
Failure	Slow	Seals leaking progressively
	Sudden	Electric lamp bulbs
Obsolescence		Gas lamps Steam locomotives
Completion		Bullets and bombs Packaging

In the design process an attempt is usually made to ensure that failure does not occur before a specified life has been reached, or before a life limit has been reached by obsolescence or completion. The occurrence of a failure, without loss of life, is not so much a disaster, as the ultimate result of a design compromise between perfection and economics.

When a limit to operation without failure is accepted, the choice of this limit depends on the availability required from the device.

Availability is the average percentage of the time that a device is available to give satisfactory performance during its required operating period. The availability of a device depends on its reliability and maintainability.

Reliability is the average time that devices of a particular design will operate without failure.

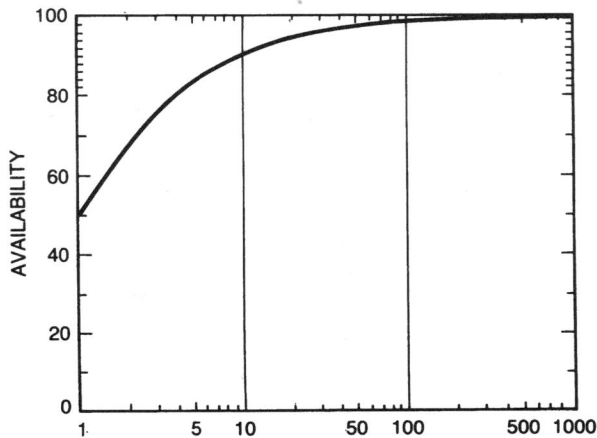

$$\frac{\text{RELIABILITY}}{\text{MAINTAINABILITY}} = \frac{\text{AVERAGE RUNNING TIME TO FAILURE}}{\text{AVERAGE TIME TO REPAIR}}$$

Figure 10.1 The relationship between availability, reliability and maintainability. *High availabilities can only be obtained by long lives or short repair times or both*

Maintainability is measured by the average time that devices of a particular design take to repair after a failure

The availability required, is largely determined by the application and the capital cost.

FAILURE ANALYSIS

The techniques to be applied to the analysis of the failures of tribological components depend on whether the failures are isolated events or repetitive incidents. Both require detailed examination to determine the primary cause, but, in the case of repeated failures, establishing the temporal pattern of failure can be a powerful additional tool.

Investigating failures

When investigating failures it is worth remembering the following points:

(a) Most failures have several causes which combine together to give the observed result. A single cause failure is a very rare occurrence.
(b) In large machines tribological problems often arise because deflections increase with size, while in general oil film thicknesses do not.
(c) Temperature has a very major effect on the performance of tribological components both directly, and indirectly due to differential expansions and thermal distortions. It is therefore important to check:

    Temperatures
    Steady temperature gradients
    Temperature transients

Causes of failure

To determine the most probable causes of failure of components, which exist either in small numbers, or involve mass produced items the following procedure may be helpful:

1 Examine the failed specimens using the following sections of this Handbook as guidance, in order to determine the probable mode of failure.
2 Collect data on the actual operating conditions and double check the information wherever possible.
3 Study the design, and where possible analyse its probable performance in terms of the operating conditions to see whether it is likely that it could fail by the mode which has been observed.
4 If this suggests that the component should have operated satisfactorily, examine the various operating conditions to see how much each needs to be changed to produce the observed failure. Investigate each operating condition in turn to see whether there are any factors previously neglected which could produce sufficient change to cause the failure.

Repetitive failures

Two statistics are commonly used:-

1 MTBF (mean time between failures)

$$= \frac{L_1 + L_2 + \ldots + L_n}{n}$$

where L_1, L_2, etc., are the times to failure and n the number of failures.

2 L_{10} Life, is the running time at which the number of failures from a sample population of components reaches 10%. (Other values can also be used, e.g. L_1 Life, viz the time to 1% failures, where extreme reliability is required.)

MTBF is of value in quantifying failure rates, particularly of machines involving more than one failing component. It is of most use in maintenance planning, costing and in assessing the effect of remedial measures.

L_{10} Life is a more rigorous statistic that can only be applied to a statistically homogeneous population, i.e. nominally identical items subject to nominally identical operating conditions.

Failure patterns

Repetitive failures can be divided by time to failure according to the familiar 'bath-tub' curve, comprising the three regions: early-life failures (infantile mortality), 'mid-life' (random) failures and 'wear-out'.

Early-life failures are normally caused by built-in defects, installation errors, incorrect materials, etc.

Mid-life failures are caused by random effects external to the component, e.g. operating changes, (overload) lightning strikes, etc.

Wear-out can be the result of mechanical wear, fatigue, corrosion, etc.

The ability to identify which of these effects is dominant in the failure pattern can provide an insight into the mechanism of failure.

As a guide to the general cause of failure it can be useful to plot failure rate against life to see whether the relationship is falling or rising.

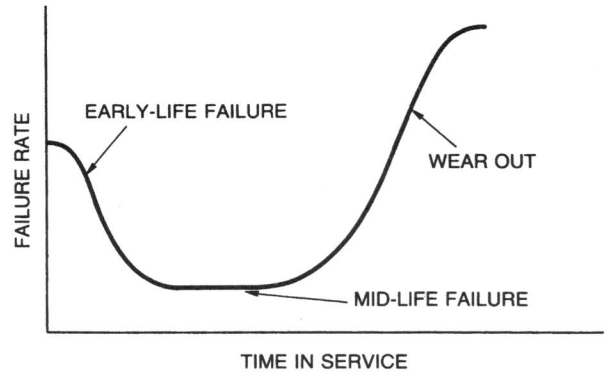

Figure 10.2 *The failure rate with time of a group of similar components*

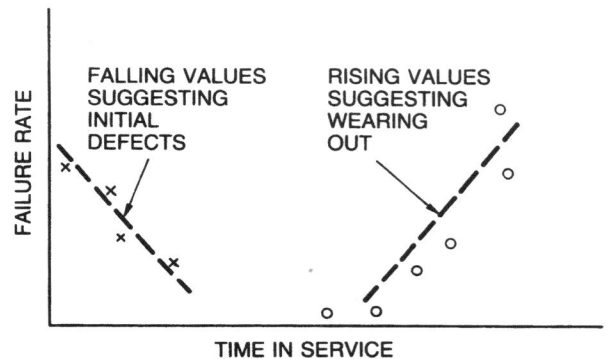

Figure 10.3 *The failure rate with time used as an investigative method*

Weibull analysis

Weibull analysis is a more precise technique. Its power is such that it can provide useful guidance with as few as five repeat failures. The following form of the Weibull probability equation is useful in component failure analysis:

$$F(t) = 1 - \exp[\alpha(t - \gamma)^{\beta}]$$

where $F(t)$ is the cumulative percentage failure, t the time to failure of individual items and the three constants are the scale parameter (α), the Weibull Index (β) and the location parameter (γ).

For components that do not have a shelf life, i.e. there is no deterioration before the component goes into service, $\gamma = 0$ and the expression simplifies to:

$$F(t) = 1 - \exp[\alpha t^{\beta}].$$

The value of the Weibull Index depends on the temporal pattern of failure, viz:

early-life failures $\beta = 0.5$
random failures $\beta = 1$
wear out $\beta = 3.4$

Weibull analysis can be carried out simply and quickly as follows:

1 Obtain the values of $F(t)$ for the sample size from Table 10.1
2 Plot the observed times to failure against the appropriate value of $F(t)$ on Weibull probability paper (Figure 10.5).
3 Draw best fit straight line through points.

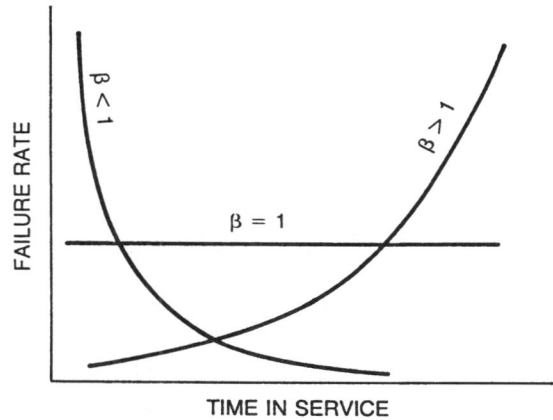

Figure 10.4 The relationship between the value of β and the shape of the failure rate curve

4 Drop normal from 'Estimation Point' to the best fit straight line.
5 Read off β value from intersection on scale.

For $n > 20$ – Calculate approximate values of $F(t)$ from

$$\frac{100(i - 0.3)}{n + 0.4}$$

where: i is the ith measurement in a sample of n arranged in increasing order.

Sample size																			
5	12.9	31.3	50.0	68.8	87.0														
6	10.9	26.4	42.1	57.8	73.5	89.0													
7	9.4	22.8	36.4	50.0	63.5	77.1	90.5												
8	8.3	20.1	32.0	44.0	55.9	67.9	79.8	91.7											
9	7.4	17.9	28.6	39.3	50.0	60.6	71.3	82.0	92.5										
10	6.6	16.2	25.8	35.5	45.1	54.8	64.4	74.1	83.7	93.3									
11	6.1	14.7	23.5	32.3	41.1	50.0	58.8	67.6	76.4	85.2	93.8								
12	5.6	13.5	21.6	29.7	37.8	45.9	54.0	62.1	70.2	78.3	86.4	94.3							
13	5.1	12.5	20.0	27.5	35.0	42.5	50.0	57.4	64.9	72.4	79.9	87.4	94.8						
14	4.8	11.7	18.6	25.6	32.5	39.5	46.5	53.4	60.4	67.4	74.3	81.3	88.2	95.1					
15	4.5	10.9	17.4	23.9	30.4	36.9	43.4	50.0	56.5	63.0	69.5	76.0	82.5	89.0	95.4				
16	4.2	10.2	16.3	22.4	28.5	34.7	40.8	46.9	53.0	59.1	65.2	71.4	77.5	83.6	89.7	95.7			
17	3.9	9.6	15.4	21.1	26.9	32.7	38.4	44.2	50.0	55.7	61.5	67.2	73.0	78.8	84.5	90.3	96.0		
18	3.7	9.1	14.5	20.0	25.4	30.9	36.3	41.8	47.2	52.7	58.1	63.6	69.0	74.5	79.9	85.4	90.8	96.2	
19	3.5	8.6	13.8	18.9	24.1	29.3	34.4	39.6	44.8	50.0	55.1	60.3	65.5	70.6	75.8	81.0	86.1	91.3	96.4
20	3.4	8.2	13.1	18.0	22.9	27.8	32.7	37.7	42.6	47.5	52.4	57.3	62.2	67.2	72.1	77.0	81.9	86.8	91.7	96.5

Table 10.1 Values of the cumulative per cent failure F(t) for the individual failures in a range of sample sizes

⊕ ESTIMATION POINT

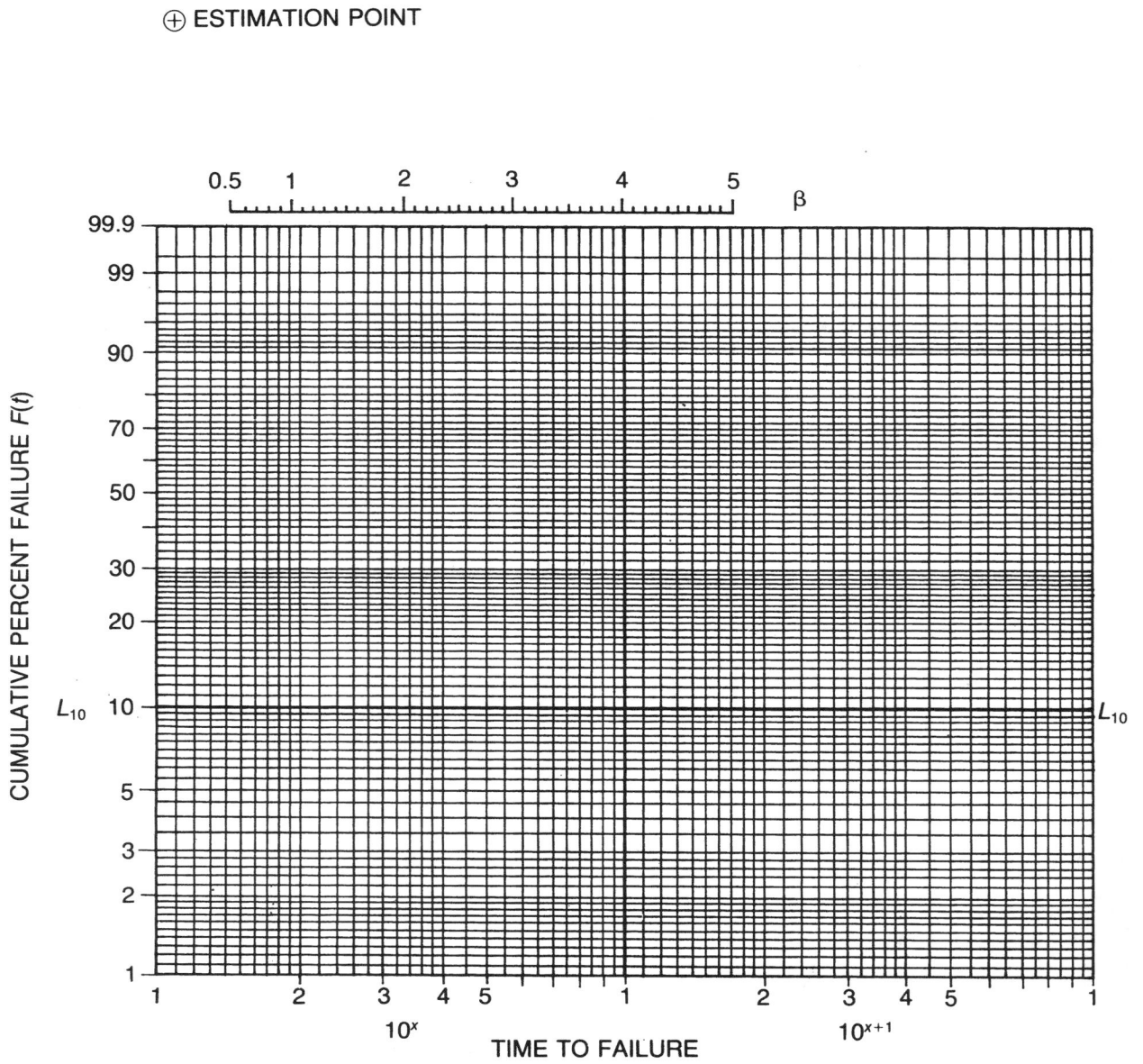

Figure 10.5 Weibull probability graph paper

Figure 10.6 gives an example of 9 failures of spherical roller bearings in an extruder gearbox. The β value of 2.7 suggests wear-out (fatigue) failure. This was confirmed by examination of the failed components. The L_{10} Life corresponds to a 10% cumulative failure. L_{10} Life for rolling bearings operating at constant speed is given by:

$$L_{10} \text{ Life (hours)} = \frac{10^6 \ C^x}{n \ P}$$

Where n = speed (rev/min), C = bearing capacity, P = equivalent radial load, x = 3 for ball bearings, 10/3 for roller bearings.

Determination of the L_{10} Life from the Weibull analysis allows an estimate to be made of the actual load. This can be used to verify the design value. In this particular example, the exceptionally low value of L_{10} Life (2500 hours) identified excessive load as the cause of failure.

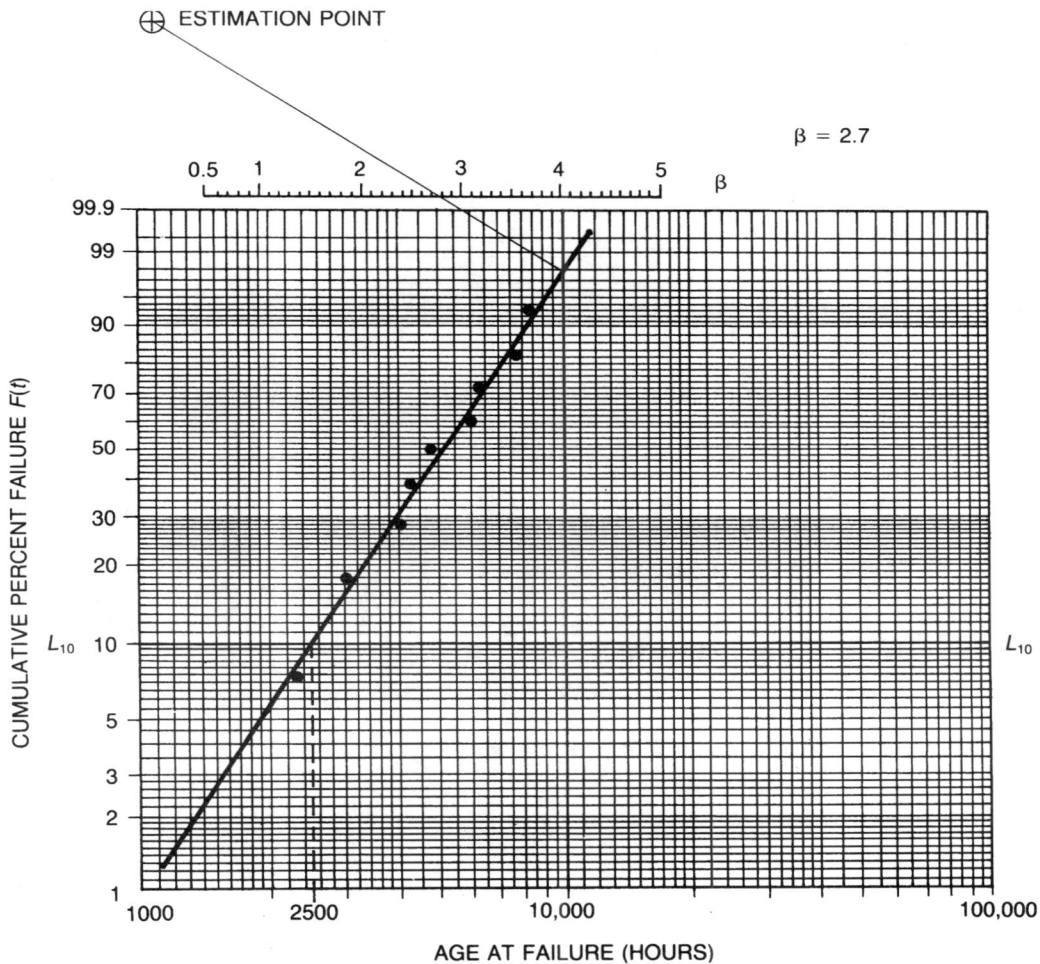

Figure 10.6 Thrust rolling bearing failures on extruder gearboxes

Figure 10.7 gives an example for 17 plain thrust bearing failures on three centrifugal air compressors. The β value of 0.7 suggests a combination of early-life and random failures. Detailed examination of the failures showed that they were caused in part by assembly errors, in part of machine surges.

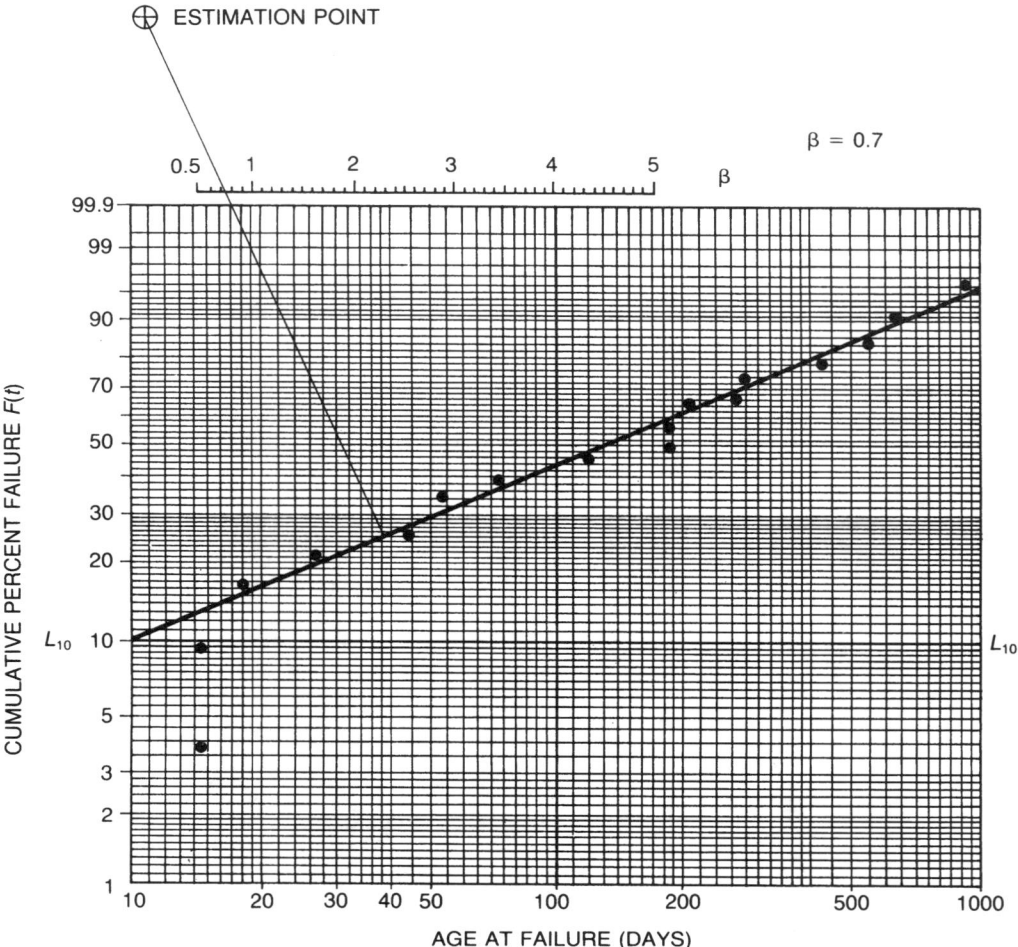

Figure 10.7 Plain thrust bearing failures on centrifugal air compressors

Foreign matter

Characteristics

Fine score marks or scratches in direction of motion, often with embedded particles and haloes.

Causes

Dirt particles in lubricant exceeding the minimum oil film thickness.

Wiping

Characteristics

Surface melting and flow of bearing material, especially when of low-melting point, e.g. whitemetals, overlays.

Causes

Inadequate clearance, overheating, insufficient oil supply, excessive load, or operation with a non-cylindrical journal.

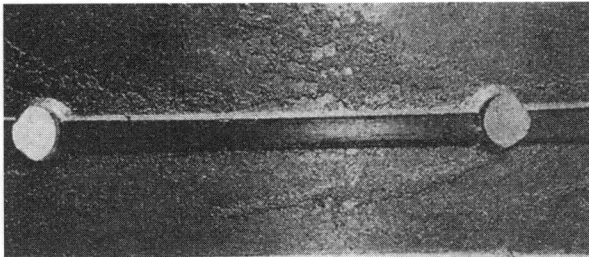

Foreign matter

Characteristics

Severe scoring and erosion of bearing surface in the line of motion, or along lines of local oil flow.

Causes

Contamination of lubricant by excessive amounts of dirt particularly non-metallic particles which can roll between the surfaces.

Fatigue

Characteristics

Cracking, often in mosaic pattern, and loss of areas of lining.

Causes

Excessive dynamic loading or overheating causing reduction of fatigue strength; overspeeding causing imposition of excessive centrifugal loading.

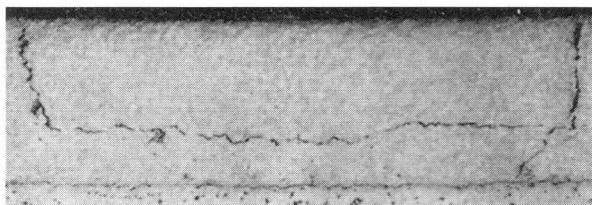

Fatigue

Characteristics

Loss of areas of lining by propagation of cracks initially at right angles to the bearing surface, and then progressing parallel to the surface, leading to isolation of pieces of the bearing material.

Causes

Excessive dynamic loading which exceeds the fatigue strength at the operating temperature.

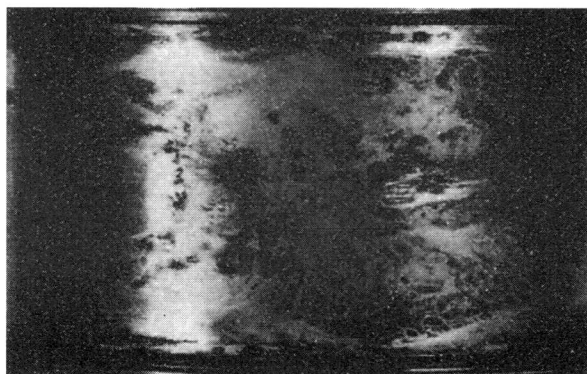

Fretting

Characteristics

Welding, or pick-up of metal from the housing on the back of bearing. Can also occur on the joint faces. Production and oxidation of fine wear debris, which in severe cases can give red staining.

Causes

Inadequate interference fit; flimsy housing design; permitting small sliding movements between surfaces under operating loads.

Excessive interference

Characteristics

Distortion of bearing bore causing overheating and fatigue at the bearing joint faces.

Causes

Excessive interference fit or stagger at joint faces during assembly.

Misalignment

Characteristics

Uneven wear of bearing surface, or fatigue in diagonally opposed areas in top and bottom halves.

Causes

Misalignment of bearing housings on assembly, or journal deflection under load.

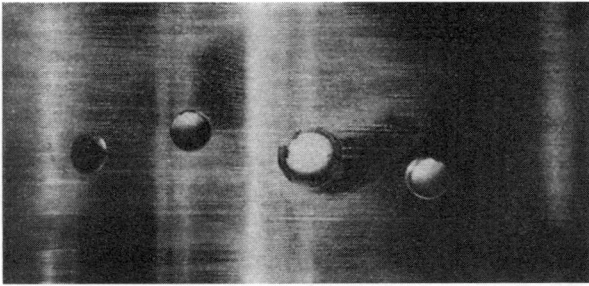

Dirty assembly

Characteristics

Localised overheating of the bearing surface and fatigue in extreme cases, sometimes in nominally lightly loaded areas.

Causes

Entrapment of large particles of dirt (e.g. swarf), between bearing and housing, causing distortion of the shell, impairment of heat transfer and reduction of clearance (see next column).

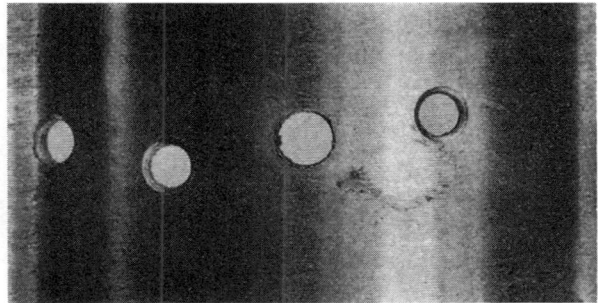

Dirty assembly

Characteristics

Local areas of poor bedding on the back of the bearing shell, often around a 'hard' spot.

Causes

Entrapment of dirt particles between bearing and housing. Bore of bearing is shown in previous column illustrating local overheating due to distortion of shell, causing reduction of clearance and impaired heat transfer.

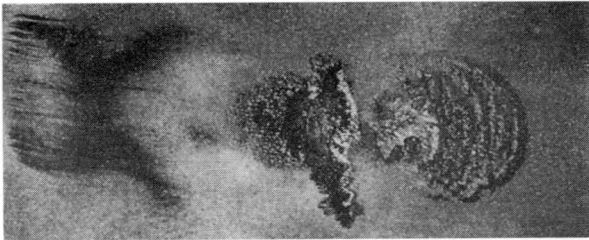

Cavitation erosion

Characteristics

Removal of bearing material, especially soft overlays or whitemetal in regions near joint faces or grooves, leaving a roughened bright surface.

Causes

Changes of pressure in oil film associated with interrupted flow.

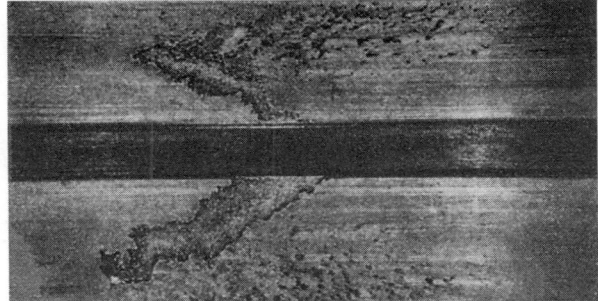

Discharge cavitation erosion

Characteristics

Formation of pitting or grooving of the bearing material in a V-formation pointing in the direction of rotation.

Causes

Rapid advance and retreat of journal in clearance during cycle. It is usually associated with the operation of a centrally grooved bearing at an excessive operating clearance.

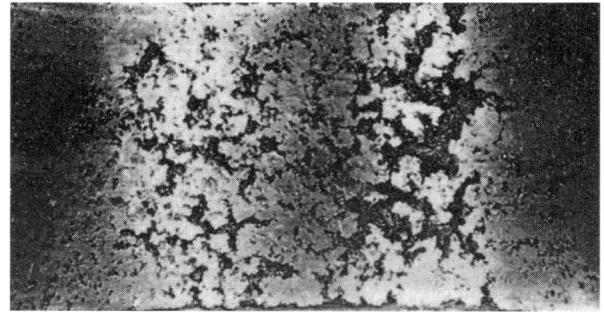

Cavitation erosion

Characteristics

Attack of bearing material in isolated areas, in random pattern, sometimes associated with grooves.

Causes

Impact fatigue caused by collapse of vapour bubbles in oil film due to rapid pressure changes. Softer overlay (Nos 1, 2 and 3 bearings) attacked. Harder aluminium −20% tin (Nos 4 and 5 bearings) not attacked under these particular conditions.

Corrosion

Characteristics

Removal of lead phase from unplated copper-lead or lead-bronze, usually leading on to fatigue of the weakened material.

Causes

Formation of organic acids by oxidation of lubricating oil in service. Consult oil suppliers; investigate possible coolant leakage into oil.

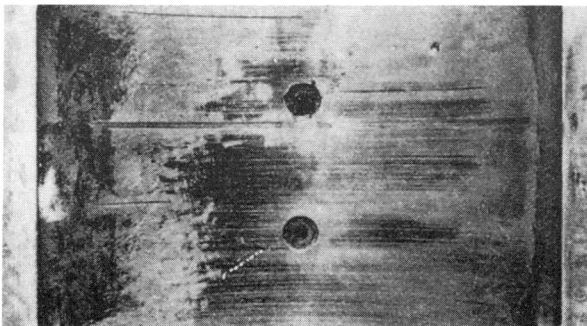

Tin dioxide corrosion

Characteristics

Formation of hard black deposit on surface of white-metal lining, especially in marine turbine bearings. Tin attacked, no tin-antimony and copper-tin constituents.

Causes

Electrolyte (sea water) in oil.

'Sulphur' corrosion

Characteristics

Deep pitting and attack or copper-base alloys, especially phosphor-bronze, in high temperature zones such as small-end bushes. Black coloration due to the formation of copper sulphide.

Causes

Attack by sulphur-compounds from oil additives or fuel combustion products.

'Wire wool' damage

Characteristics

Formation of hard black scab on whitemetal bearing surface, and severe machining away of journal in way of scab, as shown on the right.

Causes

It is usually initiated by a large dirt particle embedded in the whitemetal, in contact with journal, especially chromium steel.

'Wire wool' damage

Characteristics

Severe catastrophic machining of journal by 'black scab' formed in whitemetal lining of bearing. The machining 'debris' looks like wire wool.

Causes

Self-propagation of scab, expecially with 'susceptible' journals steels, e.g. some chromium steels.

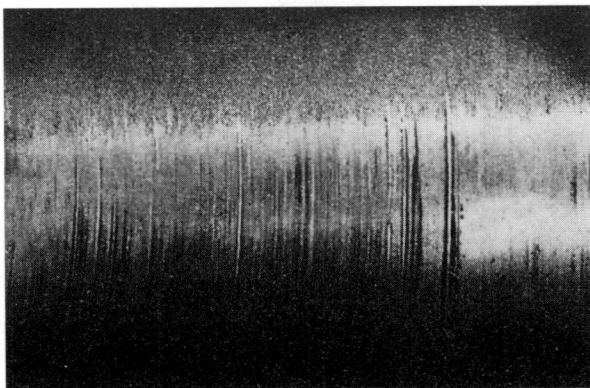

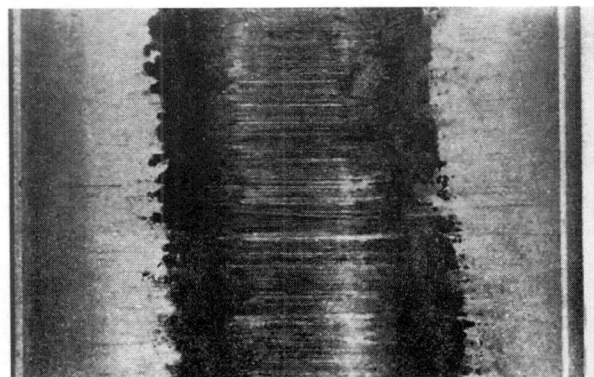

Electrical discharge

Characteristics

Pitting of bearing surface and of journal; may cause rapid failure in extreme cases.

Causes

Electrical currents from rotor to stator through oil film, often caused by faulty earthing.

Fretting due to external vibration

Characteristics

Pitting and pick-up on bearing surface.

Causes

Vibration transmitted from external sources, causing damage while journal is stationary.

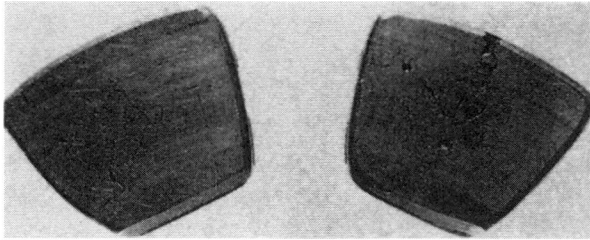

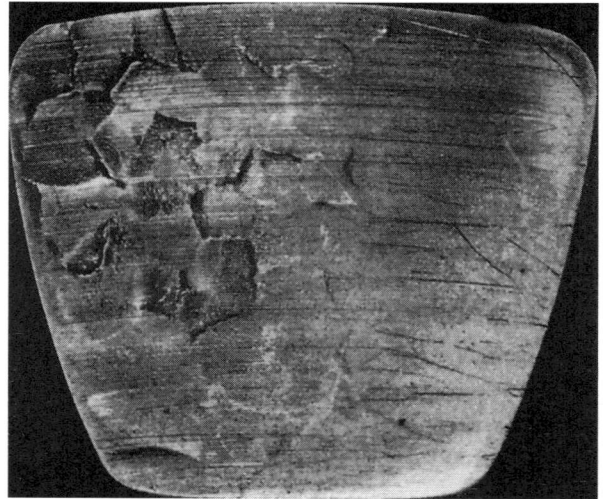

Overheating

Characteristics

Extrusion and cracking, especially of whitemetal-lined bearings.

Causes

Operation at excessive temperatures.

Thermal cycling

Characteristics

Surface rumpling and grain-boundary cracking of tin-base whitemetal bearings.

Causes

Thermal cycling in service, causing plastic deformation, associated with the non-uniform thermal expansion of tin crystals.

Faulty assembly

Characteristics

Localised fatigue or wiping in nominally lightly loaded areas.

Causes

Stagger at joint faces during assembly, due to excessive bolt clearances, or incorrect bolt disposition (bolts too far out).

Faulty assembly

Characteristics

Overheating and pick-up at the sides of the bearings.

Causes

Incorrect grinding of journal radii, causing fouling at fillets.

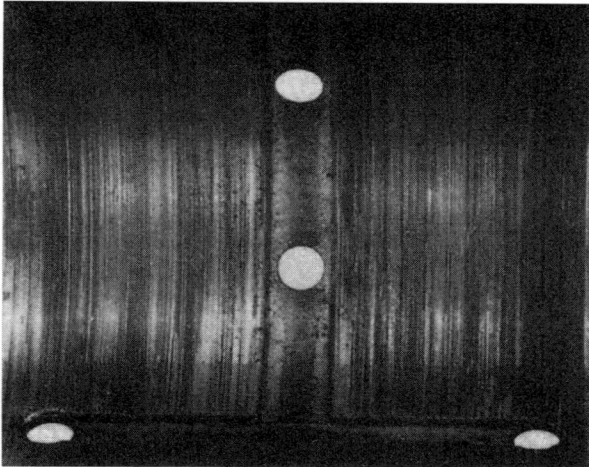

Incorrect journal grinding

Characteristics

Severe wiping and tearing-up of bearing surface.

Causes

Too coarse a surface finish, or in the case of SG iron shafts, the final grinding of journal in wrong direction relative to rotation in bearing.

Inadequate oil film thickness

Characteristics

Fatigue cracking in proximity of a groove.

Causes

Incorrect groove design, e.g. positioning a groove in the loaded area of the bearing.

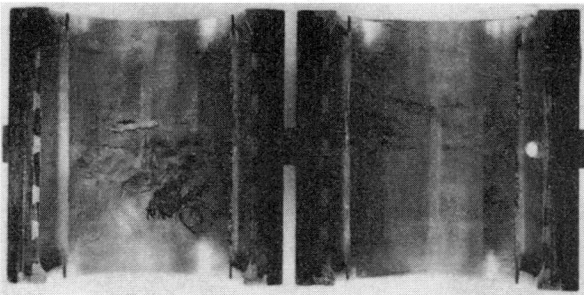

Inadequate lubrication

Characteristics

Seizure of bearing.

Causes

Inadequate pump capacity or oil gallery or oilway dimensions. Blockage or cessation of oil supply.

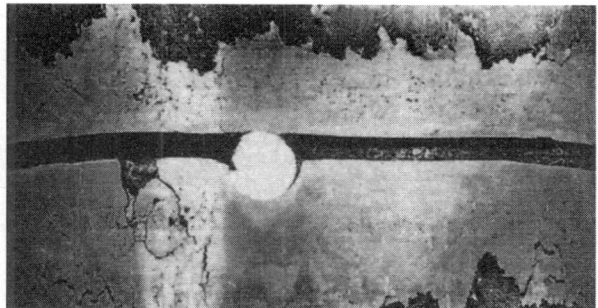

Bad bonding

Characteristics

Loss of lining, sometimes in large areas, even in lightly loaded regions, and showing full exposure of the backing material.

Causes

Poor tinning of shells; incorrect metallurgical control of lining technique.

All photographs courtesy of Glacier Metal Co. Ltd

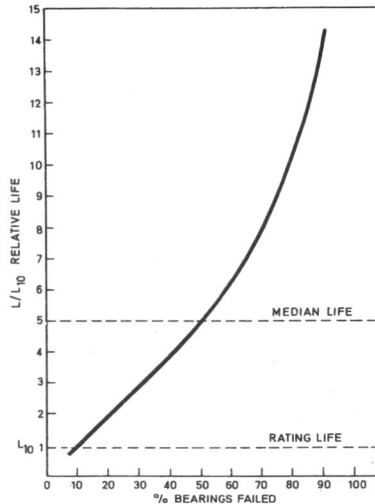

FATIGUE FLAKE

Characteristics
Flaking with conchoidal or ripple pattern extending evenly across the loaded part of the race.

Causes
Fatigue due to repeated stressing of the metal. This is not a fault condition but it is the form by which a rolling element bearing should eventually fail. The multitude of small dents are caused by the debris and are a secondary effect.

EARLY FATIGUE FLAKE

Characteristics
A normal fatigue flake but occurring in a comparatively short time. Appearance as for fatigue flake.

Causes
Wide life-expectancy of rolling bearings. The graph shows approximate distribution for all types. Unless repeated, there is no fault. If repeated, load is probably higher than estimated; check thermal expansion and centrifugal loads.

ATMOSPHERIC CORROSION

Characteristics
Numerous irregular pits, reddish brown to dark brown in colour. Pits have rough irregular bottoms.

Causes
Exposure to moist conditions, use of a grease giving inadequate protection against water corrosion.

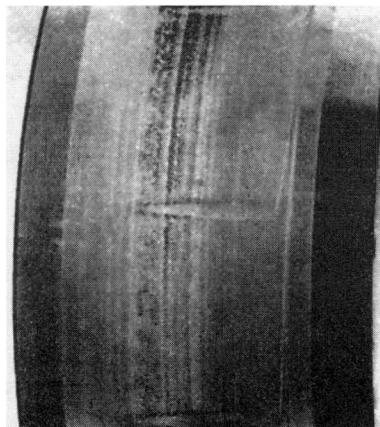

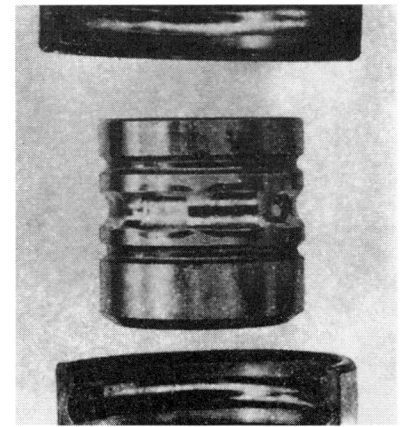

ROLLER STAINING

Characteristics
Dark patches on rolling surfaces and end faces of rollers in bearings with yellow metal cages. The patches usually conform in shape to the cage bars.

Causes
Bi-metallic corrosion in storage. May be due to poor storage conditions or insufficient cleaning during manufacture. Special packings are available for severe conditions. Staining, as shown, can be removed by the manufacturer, to whom the bearing should be returned.

BRUISING (OR TRUE BRINELLING)

Characteristics
Dents or grooves in the bearing track conforming to the shape of the rolling elements. Grinding marks not obliterated and the metal at the edges of the dents has been slightly raised.

Causes
The rolling elements have been brought into violent contact with the race; in this case during assembly using impact.

FALSE BRINELLING

Characteristics
Depressions in the tracks which may vary from shallow marks to deep cavities. Close inspection reveals that the depressions have a roughened surface texture and that the grinding marks have been removed. There is usually no tendency for the metal at the groove edges to have been displaced.

Causes
Vibration while the bearing is stationary or a small oscillating movement while under load.

B12.1

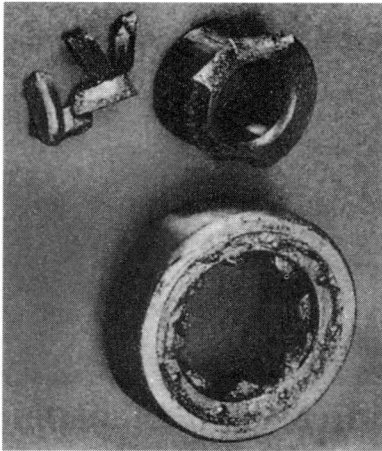

FRACTURED FLANGE

Characteristics

Pieces broken from the inner race guiding flange. General damage to cage and shields.

Causes

Bad fitting. The bearing was pressed into housing by applying load to the inner race causing cracking of the flange. During running the cracks extended and the flange collapsed. A bearing must never be fitted so that the fitting load is transmitted via the rolling elements.

OUTER RACE FRETTING

Characteristics

A patchy discoloration of the outer surface and the presence of reddish brown debris ('cocoa'). The race is not softened but cracks may extend inwards from the fretted zone.

Causes

Insufficient interference between race and housing. Particularly noticeable with heavily loaded bearings having thin outer races.

INNER RACE FRETTING

Characteristics

Heavy fretting of the shaft often with gross scalloping; presence of brown debris ('cocoa'). Inner race may show some fretting marks.

Causes

Too little interference, often slight clearance, between the inner race and the shaft combined with heavy axial clamping. Axial clamping alone will not prevent a heavily loaded inner race precessing slowly on the shaft.

INNER RACE SPINNING

Characteristics

Softening and scoring of the inner race and the shaft, overheating leading to carbonisation of lubricant in severe cases, may lead to complete seizure.

Causes

Inner race fitted with too little interference on shaft and with light axial clamping.

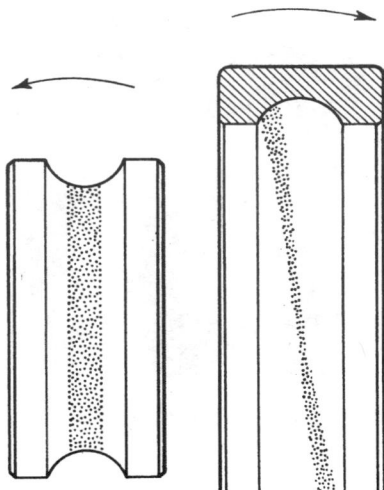

SKEW RUNNING MARKS

Characteristics

The running marks on the stationary race are not parallel to the faces of the race. In the figure the outer race is stationary.

Causes

Misalignment. The bearing has not failed but may do so if allowed to continue to run out of line.

UNEVEN FATIGUE

Characteristics

Normal fatigue flaking but limited to, or much more severe on, one side of the running track.

Causes

Misalignment.

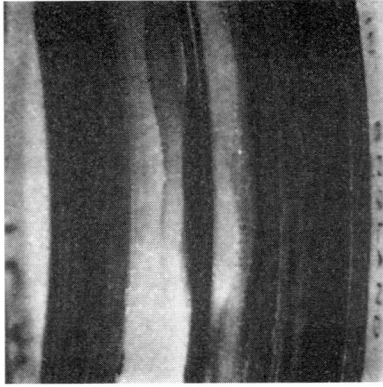

UNEVEN WEAR MARKS

Characteristics

The running or wear marks have an uneven width and may have a wavy outline instead of being a uniform dark band.

Causes

Ball skidding due to a variable rotating load or local distortion of the races.

ROLLER END COLLAPSE

Characteristics

Flaking near the roller-end radius at one end only. Microscopic examination reveals roundish smooth-bottomed pits.

Causes

Electrical damage with some misalignment. If the pits are absent then the probable cause is roller end bruising which can usually be detected on the undamaged shoulder. Although misalignment accentuates this type of damage it has rarely been proved to be the sole cause.

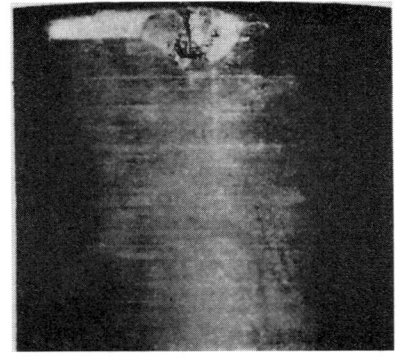

ROLLER END CHIPPING

Characteristics

A collapse of the material near the corner radii of the roller. In this instance chipping occurred simultaneously at opposite ends of the roller. A well-defined sub-surface crack can be seen.

Causes

Subcutaneous inclusions running the length of the roller. This type of failure is more usually found in the larger sizes of bearing.

Chipping at one end only may be caused by bruising during manufacture, or by electrical currents, and accentuated by misalignment.

ROLLER PEELING

Characteristics

Patches of the surface of the rollers are removed to a depth of about 0.0005 in.

Causes

This condition usually follows from an initial mild surface damage such as light electrical pitting; this could be confirmed by microscopic examination. It has also been observed on rollers which were slightly corroded before use.

If the cause is removed this damage does not usually develop into total failure.

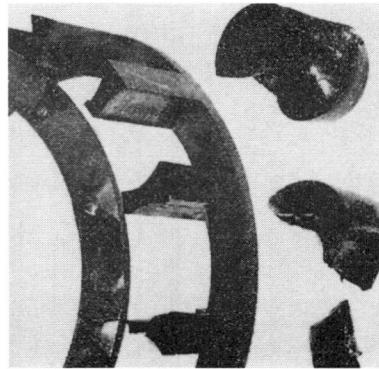

ROLLER BREAKAGE

Characteristics

One roller breaks into large fragments which may hold together. Cage pocket damaged.

Causes

Random fatigue. May be due to faults or inclusions in the roller material. Replacement bearing usually performs satisfactorily.

MAGNETIC DAMAGE

Characteristics

Softening of the rotating track and rolling elements leading to premature fatigue flaking.

Causes

Bearing has been rotating in a magnetic field (in this case, 230 kilolines (230 × 10^{-5} Wb), 300 rev/min, 860 h).

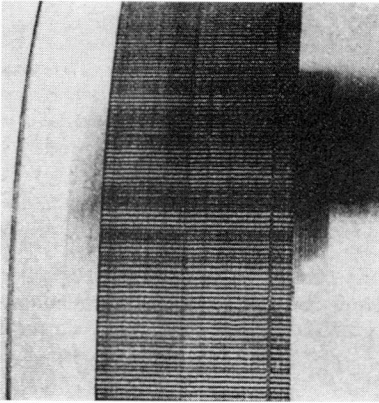

LADDER MARKING OR WASHBOARD EROSION

Characteristics

A regular pattern of dark and light bands which may have developed into definite grooves. Microscopic examination shows numerous small, almost round, pits.

Causes

An electric current has passed across the bearing; a.c. or d.c. currents will cause this effect which may be found on either race or on the rolling elements.

GREASE FAILURE

Characteristics

Cage pockets and rims worn. Remaining grease dry and hard; bearing shows signs of overheating.

Causes

Use of unsuitable grease. Common type of failure where temperatures are too high for the grease in use.

MOLTEN CAGE

Characteristics

Cage melted down to the rivets, inner race shows temper colours.

Causes

Lubrication failure on a high-speed bearing. In this case an oil failure at 26 000 rev/min. In a slower bearing the damage would not have been so localised.

OVERHEATING

Characteristics

All parts of the bearing are blackened or show temper colours. Lubricant either absent or charred. Loss of hardness on all parts.

Causes

Gross overheating. Mild overheating may only show up as a loss of hardness.

SMEARING

Characteristics

Scuff marks, discoloration and metal transfer on non-rolling surfaces. Usually some loss of hardness and evidence of deterioration of lubricant. Often found on the ends of rollers and the corresponding guide face on the flanges.

Causes

Heavy loads and/or poor lubrication.

ABRASIVE WEAR

Characteristics

Dulling of the working surfaces and the removal of metal without loss of hardness.

Causes

Abrasive particles in the lubricant, usually non-metallic.

B12.4

Gear failures rarely occur. A gear pair has not failed until it can no longer be run. This condition is reached when (*a*) one or more teeth have broken away, preventing transmission of motion between the pair or (*b*) teeth are so badly damaged that vibration and noise are unacceptable when the gears are run.

By no means all tooth damage leads to failure and immediately it is observed, damaged teeth should be examined to determine whether the gears can safely continue in service.

SURFACE FATIGUE

This includes case exfoliation in skin-hardened gears and pitting which is the commonest form of damage, especially with unhardened gears. Pitting, of which four types are distinguished, is indicated by the development of relatively smooth-bottomed cavities generally on or below the pitch line. In isolation they are generally conchoidal in appearance but an accumulation may disguise this.

Case exfoliation

Case exfoliation on a spiral bevel pinion

Characteristics

Appreciable areas of the skin on surface hardened teeth flake away from the parent metal in heavily loaded gears. Carburised and hardened, nitrided and induction hardened materials are affected.

Causes

Case exfoliation often indicates a hardened skin that is too thin to support the tooth load. Cracks sometimes originate on the plane of maximum Hertzian shear stress and subsequently break out to the surface, but more often a surface crack initiates the damage. Another possible reason for case exfoliation is the high residual stress resulting from too severe a hardness gradient between case and core. Exfoliation may be prevented by providing adequate case depth and tempering the gear material after hardening.

Initial or arrested pitting

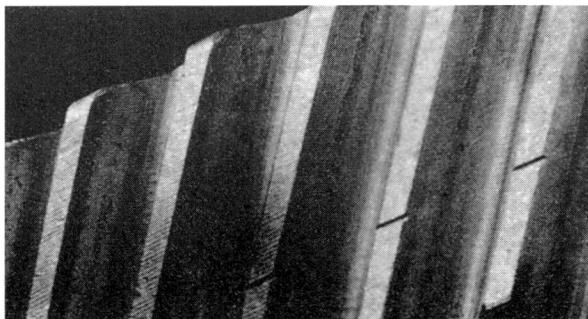

Initial or arrested pitting on a single helical gear

Characteristics

Initial pitting usually occurs on gears that are not skin hardened. It may be randomly distributed over the whole tooth flank, but more often is found around the pitch line or in the dedendum. Single pits rarely exceed 2 mm across and pitting appears in the early running life of a gear.

Causes

Discrete irregularities in profile or surface asperities are subjected to repeated overstress as the line of contact sweeps across a tooth to produce small surface cracks and clefts. In the dedendum area the oil under the high pressure of the contact can enter these defects and extend them little by little, eventually reaching the surface again so that a pit is formed and a small piece of metal is dislodged. Removal of areas of overstress in this way spreads the load on the teeth to a level where further crack or cleft formation no longer occurs and pitting ceases.

Progressive or potentially destructive pitting

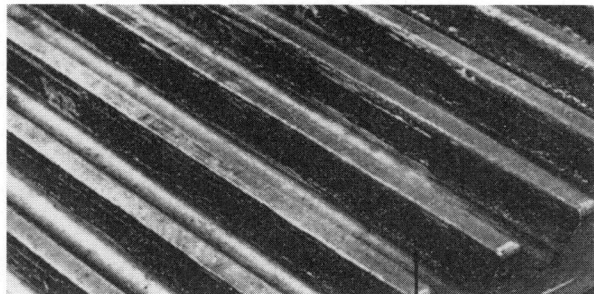

PITTING ON SOME TEETH
IS CONTINUOUS AND
QUITE DEEP

Progressive pitting on single helical gear teeth

Characteristics

Pits continue to form with continued running, especially in the dedendum area. Observation on marked teeth will indicate the rate of progress which may be intermittent. A rapid increase, particularly in the root area, may cause complete failure by increasing the stress there to the point where large pieces of teeth break away.

Causes

Essentially the gear material is generally overstressed, often by repeated shock loads. With destructive pitting the propagating cracks branch at about the plane of maximum Hertzian shear stress; one follows the normal initial pitting process but the other penetrates deeper into the metal.

Remedial action is to remove the cause of the overload by correcting alignment or using resilient couplings to remove the effect of shock loads. The life of a gear based on surface fatigue is greatly influenced by surface stress. Thus, if the load is carried on only half the face width the life will only be a small fraction of the normal value. In slow and medium speed gears it may be possible to ameliorate conditions by using a more viscous oil, but this is generally ineffective with high speed gears.

In skin-hardened gears pits of very large area resembling case exfoliation may be formed by excessive surface friction due to the use of an oil lacking sufficient viscosity.

Dedendum attrition

Dedendum attrition on a large single helical gear

Characteristics

The dedendum is covered by a large number of small pits and has a matt appearance. Both gears are equally affected and with continued running the dedenda are worn away and a step is formed at the pitch line to a depth of perhaps 0.5 mm. The metal may be detached as pit particles or as thin flakes. The wear may cease at this stage but may run in cycles, the dedenda becoming smooth before pitting restarts. If attrition is permitted to continue vibration and noise may become intolerable. Pitting may not necessarily be present in the addendum.

Causes

The cause of this type of deterioration is not fully understood but appears to be associated with vibration in the gear unit. Damage may be mitigated by the use of a more viscous oil.

Micro-pitting

Characteristics

Found predominantly on the dedendum but also to a considerable extent on the addendum of skin-hardened gears. To the naked eye affected areas have a dull grey, matt or 'frosted' appearance but under the microscope they are seen to be covered by a myriad of tiny pits ranging in size from about 0.03 to 0.08 mm and about 0.01 mm deep.

Depending on the position of the affected areas, micro-pitting may be corrective, especially with helical gears.

Causes

Overloading of very thin, brittle and super-hard surface layers, as in nitrided surfaces, or where a white-etching layer has formed, by normal and tangential loads. Coarse surface finishes and low oil viscosity can be predisposing factors. In some cases it may be accelerated by unsuitable load-carrying additives in the oil.

SMOOTH CHEMICAL WEAR

Can arise where gears using extreme pressure oil run under sustained heavy loads, at high temperatures.

Smooth chemical wear

Hypoid pinion showing smooth chemical wear

Characteristics

The working surfaces of the teeth, especially of the pinion, are worn and have a burnished appearance.

Causes

Very high surface temperatures cause the scuff resistant surface produced by chemical reaction with the steel to be removed and replaced very rapidly. The remedies are to reduce the operating temperatures, to reduce tooth friction by using a more viscous oil and to use a less active load-carrying additive.

SCUFFING

Scuffing occurs at peripheral speeds above about 3 m/s and is the result of either the complete absence of a lubricant film or its disruption by overheating. Damage may range from a lightly etched appearance (slight scuffing) to severe welding and tearing of engaging teeth (heavy scuffing). Scuffing can lead to complete destruction if not arrested.

Light scuffing

LIGHTLY SCUFFED AREAS

Light scuffing

Characteristics

Tooth surfaces affected appear dull and slightly rough in comparison with unaffected areas. Low magnification of a scuffed zone reveals small welded areas subsequently torn apart in the direction of sliding, usually at the tip and root of the engaging teeth where sliding speed is a maximum.

Causes

Disruption of the lubricant film occurs when the gear tooth surfaces reach a critical temperature associated with a particular oil and direct contact between the sliding surfaces permits discrete welding to take place. Low viscosity plain oils are more liable to permit scuffing than oils of higher viscosity. Extreme pressure oils almost always prevent it.

Heavy scuffing

Heavy scuffing on a case hardened hypoid wheel

Characteristics

Tooth surfaces are severely roughened and torn as the result of unchecked adhesive wear.

Causes

This is the result of maintaining the conditions that produced light scuffing. The temperature of the contacting surfaces rises so far above the critical temperature for the lubricant that continual welding and tearing of the gear material persists.

Spur, helical and bevel gears, may show so much displacement of the metal that a groove is formed along the pitch line of the driving gear and a corresponding ridge on that of the driven gear. It may be due to the complete absence of lubricant, even if only temporarily. Otherwise, the use of a more viscous oil, or one with extreme pressure properties is called for.

GENERAL COMMENTS ON GEAR TOOTH DAMAGE

Contact marking is the acceptance criterion for all toothed gearing, and periodic examination of this feature until the running pattern has been established, is the most satisfactory method of determining service performance. It is therefore advisable to look at the tooth surfaces on a gear pair soon after it has been run under normal working conditions. If any surface damage is found it is essential that the probable cause is recognised quickly and remedial action taken if necessary, before serious damage has resulted. Finding the principal cause may be more difficult when more than one form of damage is present, but it is usually possible to consider each characteristic separately.

The most prolific sources of trouble are faulty lubrication and misalignment. Both can be corrected if present, but unless scuffing has occurred, further periodic observation of any damaged tooth surfaces should be made before taking action which may not be immediately necessary.

ABRASIVE WEAR

During normal operation, engaging gear teeth are separated from one another by a lubricant film, commonly about 0.5 μm thick. Where both gears are unhardened and abrasive particles dimensionally larger than the film thickness contaminate the lubricant, especially if it is a grease, both sets of tooth surfaces are affected (three-body abrasion). Where one gear has very hard tooth surfaces and surface roughness greater than the film thickness, two-body abrasive wear occurs and the softer gear only becomes worn. For example, a rough case-hardened steel worm mating with a bronze worm wheel, or a rough steel pinion engaging a plastic wheel.

Foreign matter in the lubricant

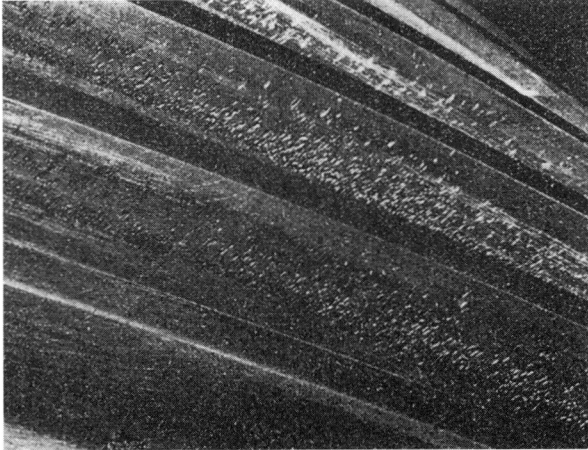

Effect of foreign matter in lubricant

Characteristics

Grooves are cut in the tooth flanks in the direction of sliding and their size corresponds to the size of the contaminant present. Displaced material piles up along the sides of a groove or is removed as a fine cutting. Usually scratches are short and do not extend to the tooth tips.

Causes

The usual causes of three-body abrasion are gritty materials falling into an open gear unit or, in an enclosed unit, inadequate cleaning of the gear case and oil supply pipes of such materials as casting sand, loose scale, shot-blast grit, etc.

Attrition caused by fine foreign matter in oil

Spur gear virtually destroyed by foreign matter in the oil

Characteristics

These are essentially similar to lapping. Very fine foreign matter suspended in a lubricant can pass through the gear mesh with little effect when normal film lubrication prevails. Unfavourable conditions permit abrasive wear; tooth surfaces appear dull and scratched in the direction of sliding. If unchecked, destruction of tooth profiles results from the lapping.

Causes

The size of the foreign matter permits bridging through the oil film. Most frequently, the origin of the abrasive material is environmental. Both gears and bearings suffer and systems should be cleaned, flushed, refilled with clean oil and protected from further contamination as soon as possible after discovery.

TOOTH BREAKAGE

If a whole tooth breaks away the gear has failed but in some instances a corner of a tooth may be broken and the gear can continue to run. The cause of a fracture should influence an assessment of the future performance of a gear.

Brittle fracture resulting from high shock load

Brittle fracture on spiral bevel wheel teeth

Characteristics

More than one tooth may be affected. With hard steels the entire fracture surface appears to be granular denoting a brittle fracture. With more ductile materials the surface has a fibrous and torn appearance.

Causes

A sudden and severe shock load has been applied to one or other member of a gear pair which has greatly exceeded the impact characteristics of the material. A brittle fracture may also indicate too low an Izod value in the gear material, though this is a very rare occurrence. A brittle fracture in bronze gears indicates the additional effect of overheating.

Tooth end and tip loading

LIGHT SCUFFING

Tooth end and tip loading

Characteristics

Spiral bevel and hypoid gears are particularly liable to heel end tooth breakage and other types of skin hardened gears may have the tooth tips breaking away. Fractured surfaces often exhibit rapid fatigue characteristics.

Causes

The immediate cause is excessive local loading. This may be produced by very high transmitted torque, incorrect meshing or insufficient tip relief.

Impact or excessive loading causing fatigue fracture

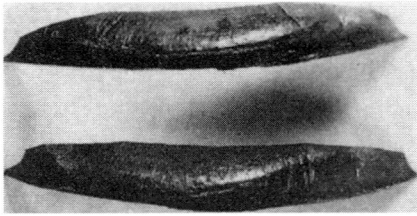

Slow fatigue on a through-hardened helical wheel

Characteristics

Often exhibit cracks in the roots on the loaded side of a number of teeth. If teeth have broken out the fracture surfaces show two phases; a very fine-grained, silky, conchoidal zone starting from the loaded side followed, where the final failure has suddenly occurred, by a coarse-grained brittle fracture.

Causes

The loading has been so intense as to exceed the tensile bending stress limit resulting in root cracking. Often stress-raisers in the roots such as blowholes, bruises, deep machining marks or non-metallic inclusions, etc. are involved. If the excessive loading continues the teeth will break away by slow fatigue and final sudden fracture.

Fatigue failure resulting from progressive pitting

LINE OF PITS THAT HAVE INITIATED FRACTURES

Fatigue failure from progressive pitting

Characteristics

Broken tooth surfaces exhibit slow fatigue markings, with the origin of the break at pits in the dedendum of the affected gear.

Causes

Progressive pitting indicates that the gears are being run with a surface stress intensity above the fatigue limit. Cracks originating at the surface continue to penetrate into the material.

PLASTIC DEFORMATION

Plastic deformation occurs on gear teeth due to the surface layers yielding under heavy loads through an intact oil film. It is unlikely to occur with hardness above HV 350.

Severe plastic flow in steel gears

Severe plastic flow in helical gears

Characteristics

A flash or knife-edge is formed on the tips of the driving teeth often with a hollow at the pitch cylinder and a corresponding swelling on the driven teeth. The ends of the teeth can also develop a flash and the flanks are normally highly burnished.

Causes

The main causes are heavy steady or repeated shock loading which raises the surface stress above the elastic limit of the material, the surface layers being displaced while in the plastic state, especially in the direction of sliding. Since a work-hardened skin tends to develop, the phenomenon is not necessarily detrimental, especially in helical gears, unless the tooth profiles are severely damaged. A more viscous oil is often advantageous, particularly with shock-loading, but the best remedy is to reduce the transmitted load, possibly by correcting the alignment.

CASE CRACKING

With correctly manufactured case hardened gears case cracking is a rare occurrence. It may appear as the result of severe shock or excessive overload leading to tooth breakage or as a condition peculiar to worm gears.

Heat/load cracking on worms

RADIAL CRACKS IN POLISHED CONTACT ZONE

Heat/load cracking on a worm wheel

Characteristics

On extremely heavily loaded worms the highly polished contact zone may carry a series of radial cracks. Spacing of the cracks is widest where the contact band is wide and they are correspondingly closer spaced as the band narrows. Edges rarely rise above the general level of the surface.

Causes

The cracks are thought to be the result of high local temperatures induced by the load. Case hardened worms made from high core strength material (En39 steel) resist this type of cracking.

FAILURES OF PLASTIC GEARS

Gears made from plastic materials are meshed with either another plastic gear or more often, with a cast iron or steel gear; non ferrous metals are seldom used. When applicable, failures generally resemble those described for metal gears.

Severe plastic flow, scoring and tooth fracture indicate excessive loading, possibly associated with inadequate lubrication. Tempering colours on steel members are the sign of unsatisfactory heat dispersal by the lubricant.

Wear on the metallic member of a plastic/metal gear pair usually suggests the presence of abrasive material embedded in the plastic gear teeth. This condition may derive from a dusty atmosphere or from foreign matter carried in the lubricant.

When the plastic member exhibits wear the cause is commonly attributable to a defective engaging surface on the metallic gear teeth. Surface texture should preferably not be rougher than 16 μin (0.4 μm) cla.

PISTON PROBLEMS

Piston problems usually arise from three main causes and these are:

1. Unsatisfactory rubbing conditions between the piston and the cylinder.
2. Excessive operating temperature, usually caused by inadequate cooling or possibly by poor combustion conditions.
3. Inadequate strength or stiffness of the piston or associated components at the loads which are being applied in operation.

Skirt scratching and scoring

Characteristics

The piston skirt shows axial scoring marks predominantly on the thrust side. In severe cases there may be local areas showing incipient seizure.

Causes

Abrasive particles entering the space between the piston and cylinder. This can be due to operation in a dusty environment with poor air filtration. Similar damage can arise if piston ring scuffing has occurred since this can generate hard particulate debris. More rarely the problem can arise from an excessively rough cylinder surface finish.

Piston skirt seizure

Characteristics

Severe scuffing damage, particularly on the piston skirt but often extending to the crown and ring lands. The damage is often worse on the thrust side.

Causes

Operation with an inadequate clearance between the piston and cylinder. This can be associated with inadequate cooling or a poor piston profile. Similar damage could also arise if there was an inadequate rate of lubricant feed up the bore from crankshaft bearing splash.

Piston crown and ring land damage

Characteristics

The crown may show cracking and the crown land and lands between the rings may show major distortion, often with the ring ends digging in to the lands.

Causes

Major overheating caused by poor cooling and in diesel engines defective injectors and combustion. The problem may arise from inadequate cylinder coolant flow or from the failure of piston cooling arising from blocked oil cooling jets.

Skirt scratching

Skirt seizure

Misaligned pistons

Characteristics

The bedding on the skirt is not purely axial but shows diagonal bedding.

Causes

Crankshaft deflections or connecting rod bending. Misalignment of rod or gudgeon pin bores.

Cracking inside the piston

Characteristics

Cracks near the gudgeon pin bosses and behind the ring grooves.

Causes

Inadequate gudgeon pin stiffness can cause cracking in adjacent parts of the piston, or parts of the piston cross section may be of inadequate area.

Diagonal skirt bedding

RING PROBLEMS

The most common problem with piston rings is scuffing of their running surfaces. Slight local scuffing is not uncommon in the first 20 to 50 hours of running from new when the rings are bedding in to an appropriate operating profile. However the condition of the ring surfaces should progressively improve and scuffing damage should not spread all round the rings.

Scuffing of cast iron rings

Characteristics

Local zones around the ring surface where there are axial dragging marks and associated surface roughening. Detailed examination often shows thin surface layers of material with a hardness exceeding 1000 Hv and composed of non-etching fine grained martensite (white layer).

Causes

Can arise from an unsuitable initial finish on the cylinder surface. It can also arise if the rings tend to bed at the top of their running surface due to unsuitable profiling or from thermal distortion of the piston.

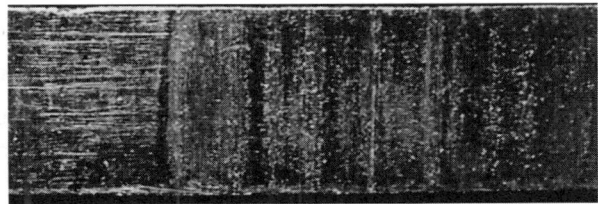

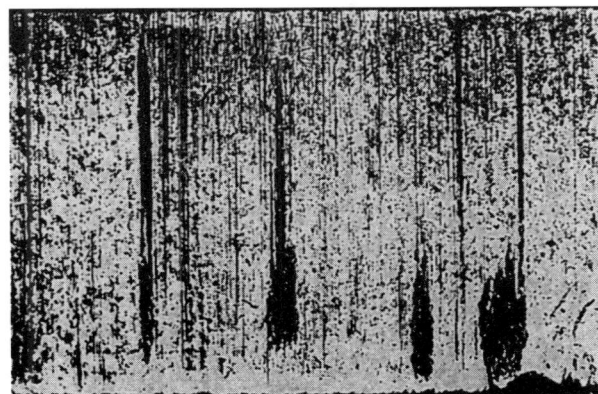

Scuffed cast iron rings

Scuffing of chromium plated piston rings

Characteristics

The presence of dark bands running across the width of the ring surface usually associated with transverse circumferential cracks. In severe cases portions of the chromium plating may be dragged from the surface.

Causes

Unsuitable cylinder surface finish or poor profiling of the piston rings. Chromium plated top rings need to have a barelled profile as installed to avoid hard bedding at the edges.

In some cases the problem can also arise from poor quality plating in which the plated surface is excessively rough or globular and can give local sharp areas on the ring edges after machining.

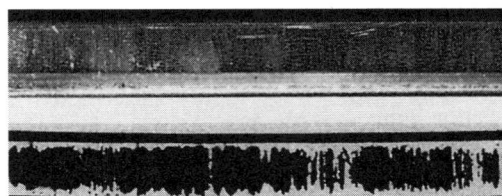

Scuffed chromium plated rings

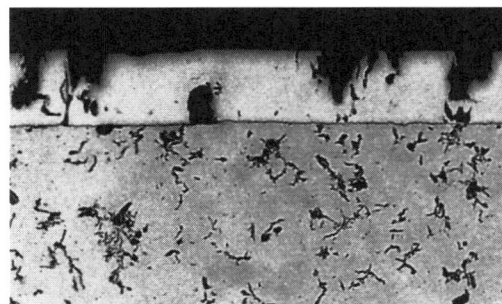

Severely damaged chromium plate

MACHINED SURFACE
OF THE RING →

CHAMFER AT EDGE
OF THE PLATING.
GLOBULAR FINISH CAN
CREATE LOCAL SHARP EDGES. →

The edge of a piston ring

Rings sticking in their grooves

Characteristics

The rings are found to be fixed in their grooves or very sluggish in motion. There may be excessive blow by or oil consumption.

Causes

The ring groove temperatures are too high due to conditions of operation or poor cooling. The use of a lubricating oil of inadequate quality can also aggravate the problem.

A stuck piston ring

CYLINDER PROBLEMS

Problems with cylinders tend to be of three types:

1. Running in problems such as bore polishing or in some cases scuffing.
2. Rates of wear in service which are high and give reduced life.
3. Other problems such as bore distortion arising from the engine design or cavitation erosion damage of the water side of a cylinder liner, which can penetrate through to the bore.

Bore polishing

Characteristics

Local areas of the bore surface become polished and oil consumption and blow by tend to increase because the piston rings do not then bed evenly around the bore. The polished areas can be very hard thin, wear-resistant 'white' layers.

Causes

The build up of hard carbon deposits on the top land of the piston can rub away local areas of the bore surface and remove the controlled surface roughness required to bed in the piston rings.

If there is noticeable bore distortion from structural deflections or thermal effects, the resulting high spots will be preferentially smoothed by the piston rings.

The chemical nature of the lubricating oil can be a significant factor in both the hard carbon build-up and in the polishing action.

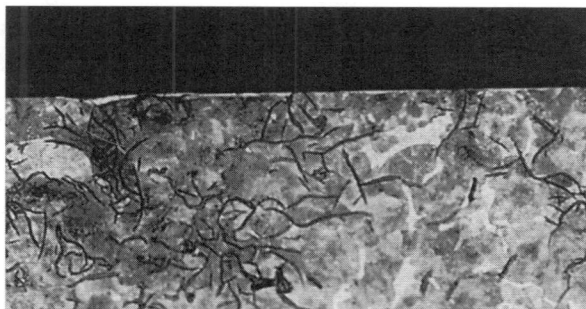

Bore polishing

High wear of cast iron cylinders

Characteristics

Cylinder liners wear in normal service due to the action of fine abrasive particles drawn in by the intake air. The greatest wear occurs near to the TDC position of the top ring.

Corrosion of a cast iron bore surface can however release hard flake-like particles of iron carbide from the pearlite in the iron. These give a greatly increased rate of abrasive wear.

Causes

Inadequate air filtration when engines are operated in dusty environments.

Engines operating at too low a coolant temperature, i.e. below about 80°C, since this allows the internal condensation of water vapour from the combustion process, and the formation of corrosion pits in the cylinder surface.

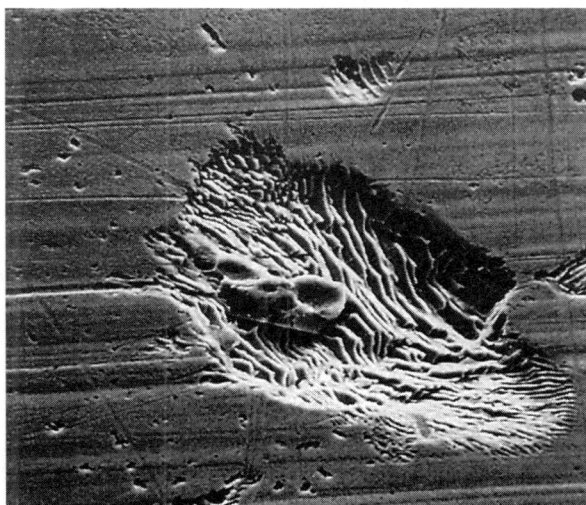

Corrosion of a cast iron bore

High wear of chromium plated cylinders

Characteristics

An increasing rate of wear with operating time associated with the loss of the surface profiling which provides a dispersed lubricant supply. The surface becomes smooth initially and then scuffs because of the unsatisfactory surface profile. This then results in a major increase in wear rate.

Causes

High rates of abrasive particle ingestion from the environment can cause this problem. A more likely cause may be inadequate quality of chromium plating and its finishing process aimed at providing surface porosity. Some finishing processes can leave relatively loose particles of chromium in the surface which become loose in service and accelerate the wear process.

Abbrasive turn round marks at TDC

Bore scuffing

Characteristics

Occurs in conjunction with piston ring scuffing. The surface of the cylinder shows areas where the metal has been dragged in an axial direction with associated surface roughening.

Causes

The same as for piston ring scuffing but in addition the problem can be accentuated if the metallurgical structure of the cylinder surface is unsatisfactory.

In the case of cast iron the material must be pearlitic and should contain dispersed hard constituents derived from phosphorus, chromium or vanadium constituents. The surface finish must also be of the correct roughness to give satisfactory bedding in of the piston rings.

In the case of chromium plated cylinder liners it is essential that the surface has an undulating or grooved profile to provide dispersed lubricant feeding to the surface.

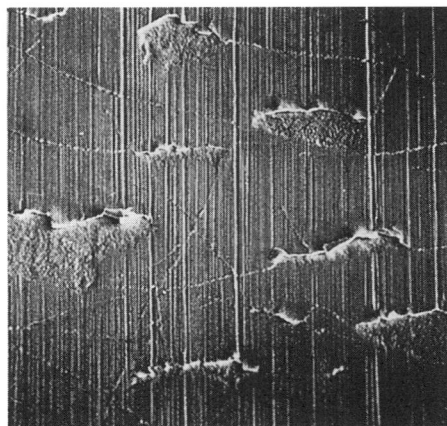

A chromium plated liner which has scuffed after losing its surface profiling by wear

Cavitation erosion of cylinder liners

Characteristics

If separate cylinder liners are used with coolant in contact with their outside surface, areas of cavitation attack can occur on the outside. The material removal by cavitation continues and eventually the liner is perforated and allows the coolant to enter the inside of the engine.

Causes

Vibration of the cylinder liner under the influence of piston impact forces is the main cause of this problem but it is accentuated by crevice corrosion effects if the outside of the liner has dead areas away from the coolant flow.

ROTARY MECHANICAL SEALS

Table 15.1 Common failure mechanisms of mechanical seals

Special conditions	Likely symptoms			Seizure	Failure mechanism	Remedy
	High leakage	High friction	High wear			
OPERATING CONDITIONS						
Speed of sliding high		X	X	X	Excessive frictional heating, film vaporises	Provide cooling
"	X				Thermal stress cracking of the face (Figure 15.1)	Use material with higher conductivity or higher tensile strength
"	X	X	X	X	Thermal distortion of seal (Figure 15.2)	Provide cooling
Speed of sliding low		X	X	X	Poor hydrodynamic lubrication, solid contact	Use face with good boundary lubrication capacity
Appreciable vibration present	X		X	X	Face separation unstable	Try to reduce vibration, avoid bellows seals, fit damper
Low pressure differential	(X)				Fluid pumped by seal against pressure	Try reversing seal to re-direct flow
High pressure differential		X	X	X	Hydrodynamic film over-loaded	Modify area ratio of seal to reduce load
"	X	X	X	X	Seal or housing distorting (Figure 15.2)	Stiffen seal and/or housing
Sterilisation or cleaning cycle used	X				High temperature or solvents incompatible with seal materials, especially rubbers	Use compatible materials
Exposure to sunlight, ozone, radiation	X				Seal materials (rubber) fail	Protect seal from exposure, consider other materials
FLUID						
Viscosity of fluid high		X	X	X	Excessive frictional heating, film vaporises	Provide cooling
	X	X	X	X	Excessive frictional heating, seal distorts	Provide cooling
Viscosity of fluid low		X	X	X	Poor hydrodynamic lubrication, solid contact	Use faces with good boundary lubrication capacity
Lubricity of fluid poor	X	X	X	X	Surfaces seize or 'pick-up'	Use faces with good boundary lubrication capacity
Abrasives in fluid	X		X		Solids in interface film	Circulate clean fluid round seal
Crystallisable fluid	X		X		Crystals form at seal face	Raise temperature or flush fluid outside seal
Polymerisable fluid	X		X		Solids form at seal face	Raise temperature or flush fluid outside seal
Ionic fluid, e.g. salt solutions	X	X	X	X	Corrosion damages seal faces	Select resistant materials
Non-Newtonian fluid, e.g. suspension, colloids, etc.	(X)				Fluid behaves unpredictably, leakage may be reversed	Try reversing seal to re-direct flow in acceptable direction
DESIGN						
Auxiliary cooling, flushing, etc.	X	X	X	X	Stoppage in auxiliary circuit	Overhaul auxiliaries
Double seals	X	X	X	X	Pressure build-up between seals if there is no provision for pressure control	Provide pressure control
Housing flexes due to pressure or temperature changes	X	X	X	X	Seal faces out of alignment, non-uniform wear	Stiffen housing and/or mount seal flexibly
Seal face flatness poor	X				Excessive seal gap (Figure 15.3)	Lap faces flatter
Seal faces rough	X	X	X	X	Asperities make solid contact	Lap or grind faces
Bellows type seal	X				Floating seal member vibrates	Fit damping device to bellows

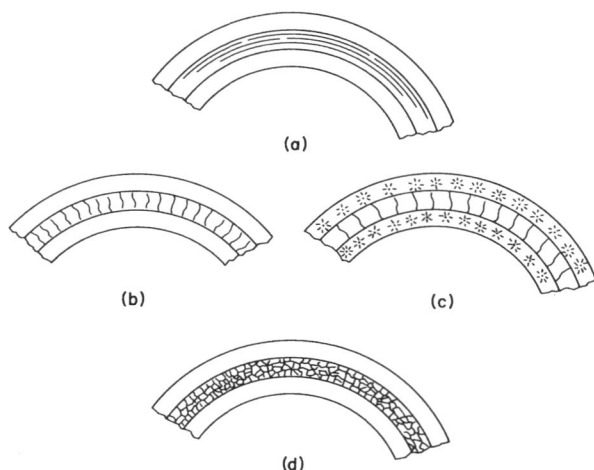

Figure 15.1 Mechanical seal faces after use:
(a) normal appearance, some circumferential scoring;
(b) parallel radial cracks; (c) radial cracks with blisters;
(d) surface crazing; (b)–(d) are due to overheating,
particularly characteristic of ceramic seal faces

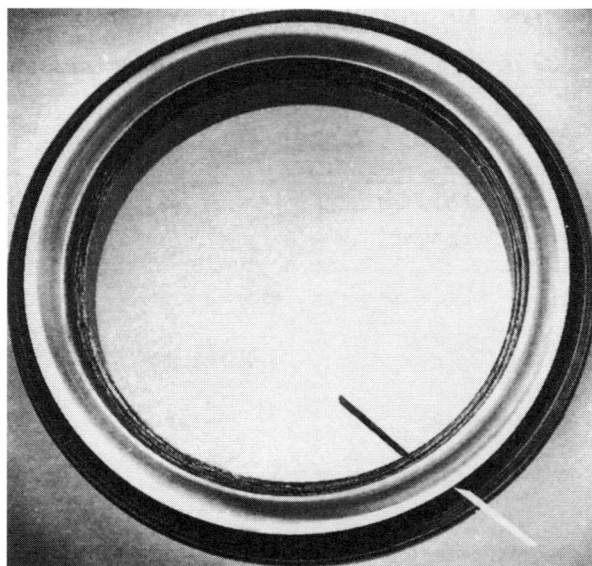

**Figure 15.2 Tungsten carbide mechanical seal face
showing symmetrical surface polishing
characteristic of mild hydraulic or thermal
distortion**

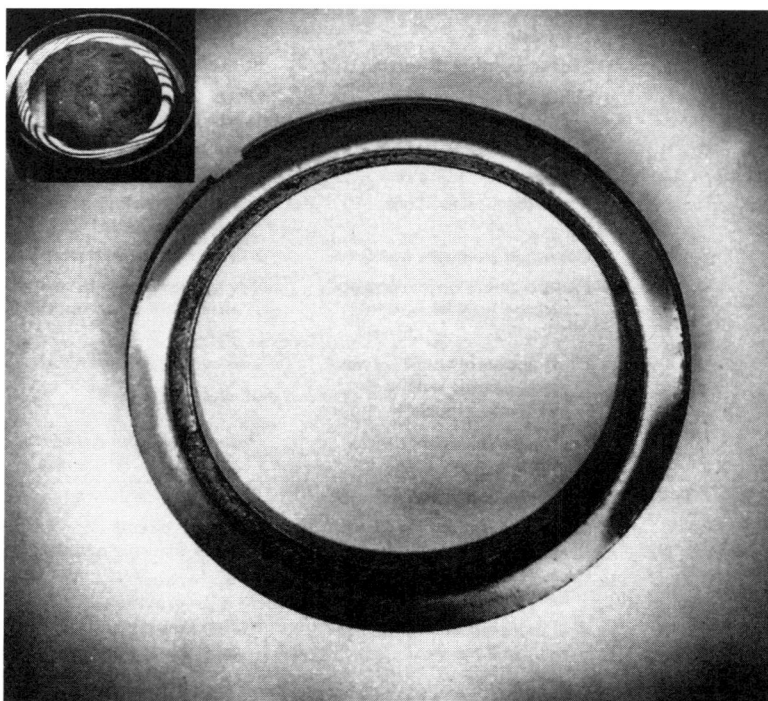

**Figure 15.3 Tungsten carbide mechanical seal face showing
localised polishing due to lack of flatness; this seal leaked badly.**
The inset illustrates a typical non-flat seal face viewed in sodium light
using an optical flat to give contour lines at 11 micro-inch increments of
height

RUBBER SEALS OF ALL TYPES

Table 15.2 Common failure mechanisms of rubber seals

Symptoms	Cause	Remedy
Rubber brittle, possibly cracked, seal leaks	Rubber ageing. Exposure to ozone/sunlight. Overheated due to high fluid temperature or high speed (Figure 15.4)	Renew seal, consider change of rubber compound; consider improving seal environmental or operating conditions
Rubber softened, possibly swollen	Rubber incompatible with sealed fluid (Figure 15.5)	Change rubber compound or fluid
Seal motion irregular, jerky, vibration, or audible squeal	'Stick-slip' (Figure 15.6)	Higher or lower speed may avoid problem; change fluid temperature; change rubber compound
Seal friction very high on starting	'Stiction', i.e. static friction is time dependent and much higher than kinetic friction (Figure 15.7)	Probably inevitable in some degree, time effect slowed by softer rubber and/or more viscous fluid
Seal permanently deformed	'Permanent set', a characteristic of rubbers, some more than others (Figure 15.4)	Change rubber compound

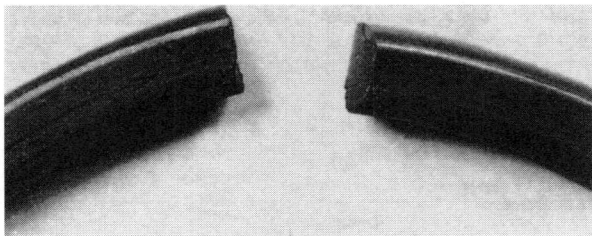

Figure 15.4 Rubber O-ring failure due to overheating. The brittle fracture is due to hardening of the originally soft rubber and the flattened appearance is a typical example of compression set

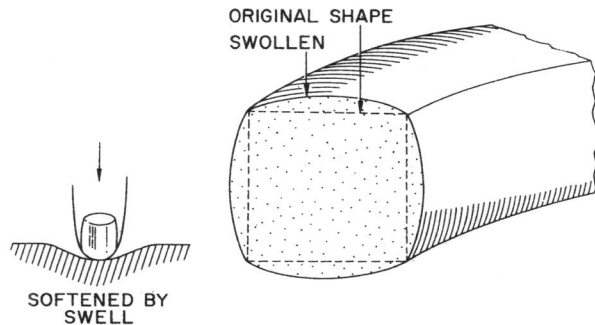

Figure 15.5 Rubber-fluid incompatibility

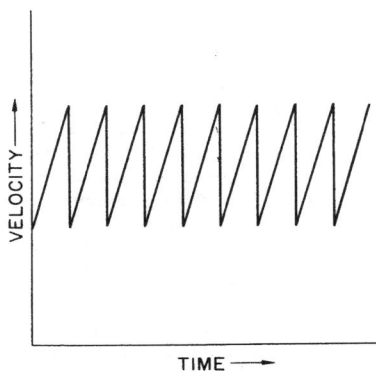

Figure 15.6 'Stick-slip'

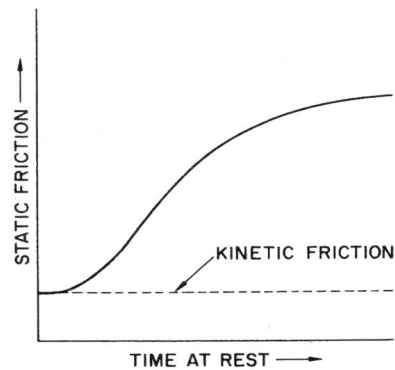

Figure 15.7 'Stiction'

O-Rings, Rectangular Rubber Rings, etc.

Table 15.3 Common failure mechanisms

Symptoms	Cause	Remedy
Fine circumferential cut set back slightly from sliding contact zone. Rubber 'nibbled' on one side. Ring completely ejected from its groove	Extrusion damage (Figure 15.8)	Reduce back clearance, check concentricity of parts; fit back-up ring; use reinforced seal; use harder rubber
Wear, not restricted to sliding contact zone. Partial or total fracture	Ring rolling or twisting in groove	Replace O-ring by rectangular section ring or a lobed type ring

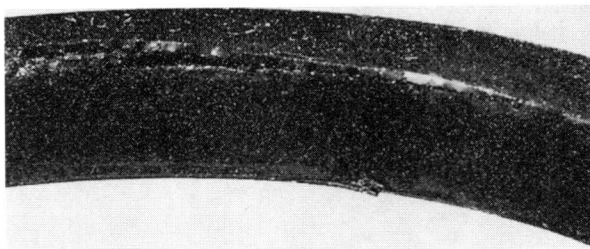

Figure 15.8 A rectangular-section rubber seal ring showing extrusion damage. Where damage is less severe (r.h.s.) only a knife-cut is visible, but material has been nibbled away where extrusion was severe. The circumferential variation of the damage indicates eccentricity of the sealed components

Figure 15.9 Wear failure of a rubberised-fabric square-back U-ring due to inadequate lubrication when sealing distilled water. Friction was also bad

Reciprocating Seals

Table 15.4 Common failure mechanisms

Symptoms	Cause	Remedy
Excessive wear and/or high friction	Poor lubrication. (Figure 15.9) Seal overloaded	If multiple seals are in use replace with single seal. Heavy duty seal may cure overloading. With aqueous fluids leather may be better than rubber
Non-uniform wear circumferentially	Dirt ingress Deposits on rod Side load	Fit wiper or scraper " Check bearings

Rotary Lip Seals

Table 15.5 Common failure mechanisms

Symptoms	Cause	Remedy
Excessive leakage	Damaged lip. Machining has left a spiral lead on shaft Unsuitable shaft surface	Check for damage and cause, e.g. contact with splines or other rough surface during assembly or careless handling/storage. Use mandrel for assembly. Eliminate shaft lead Finish the surface to R_A 0.1–0.5μm
Lip cracked in places	Excessive speed. Poor lubrication. Hot environment	Consider alternative rubber compounds. Improve lubrication. Reduce environmental temperature

PACKED GLANDS

Table 15.6 Common failure mechanisms of packed glands

Symptoms	Cause	Remedy
Packing extruded into clearance between shaft and housing or gland follower (Figure 15.10)	Designed clearance excessive or parts worn by abrasives or shaft bearings inadequate	Reduce clearances, check bearings
Leakage along outside of gland follower (Figure 15.11)	Packing improperly fitted or housing bore condition bad	Repack with care after checking bore condition
Used packing scored on outside surface, possibly leakage along outside of gland follower	Packing rotating with shaft due to being undersized	Check dimensions of housing and packing
Packing rings near gland follower very compressed (Figure 15.12) or rings embedded into each other	Packing incorrectly sized	Repack with accurately sized packing rings
Bore of used packing charred or blackened possibly shaft material adhering to packing	Lubrication failure	Change packing to one with more suitable lubricants or fit lantern ring with lubricant feed
Shaft badly worn along its length (Figure 15.13)	Lubrication failure or abrasive solids present	Change packing to one with more suitable lubricants or fit lantern ring with lubricant feed. Flush abrasives

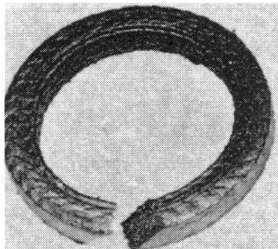

Figure 15.10 Soft packing rings after use.
Left: *normal appearance;* top: *scored ring due to rotation of the ring in its housing;* right: *extruded ring due to excessive clearance between housing and shaft*

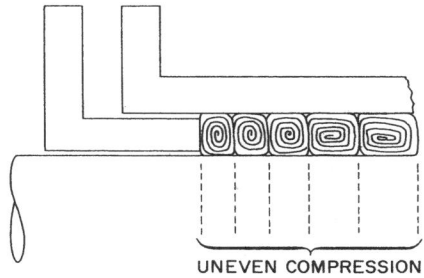

Figure 15.11 Packed gland showing abnormal leakage outside the gland follower

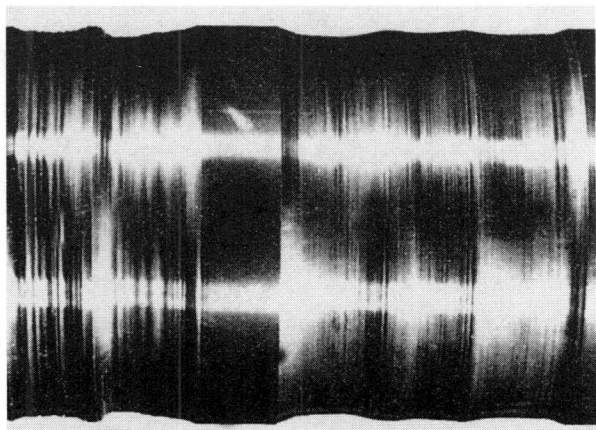

Figure 15.12 Packed gland showing uneven compression due to incorrect installation

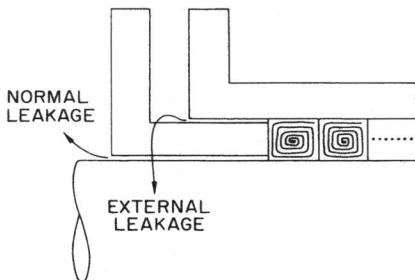

Figure 15.13 Severe wear of a bronze shaft caused by a soft packing with inadequate boundary lubricant

Some of the more common brake and clutch troubles are pictorially presented in subsequent sections; although these faults can affect performance and shorten the life of the components, only in exceptional circumstances do they result in complete failure.

BRAKING TROUBLES

Metal surface

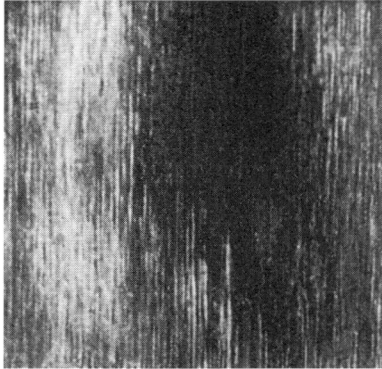

Heat spotting

Characteristics

Small isolated discoloured regions on the friction surface. Often cracks are formed in these regions owing to structural changes in the metal, and may penetrate into the component.

Causes

Friction material not sufficiently conformable to the metal member; or latter is distorted so that contact occurs only at small heavily loaded areas.

Crazing

Characteristics

Randomly orientated cracks on the rubbing surface of a mating component, with main cracks approximately perpendicular to the direction of rubbing. These can cause severe lining wear.

Causes

Overheating and repeated stress-cycling from compression to tension of the metal component as it is continually heated and cooled.

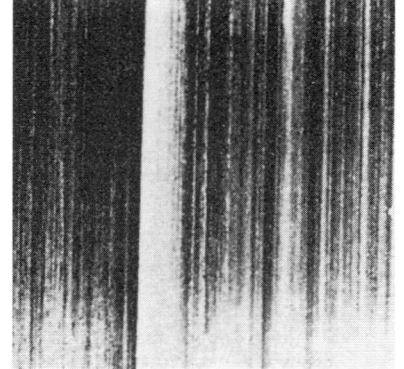

Scoring

Characteristics

Scratches on the rubbing path in the line of movement.

Causes

Metal too soft for the friction material; abrasive debris embedded in the lining material.

Friction material surface

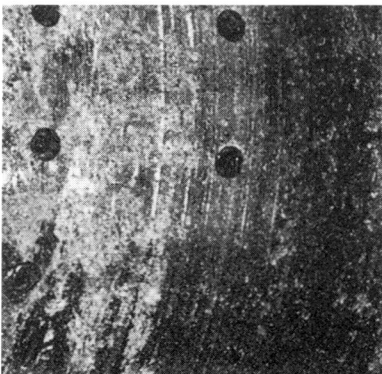

Heat spotting

Characteristics

Heavy gouging caused by hard proud spots on drum resulting in high localised work rates giving rise to rapid lining wear.

Causes

Material rubbing against a heat-spotted metal member.

Crazing

Characteristics

Randomly orientated cracks on the friction material, resulting in a high rate of wear.

Causes

Overheating of the braking surface from overloading or by the brakes dragging.

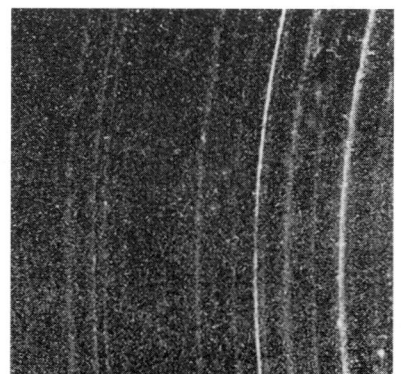

Scoring

Characteristics

Grooves formed on the friction material in the line of movement, resulting in a reduction of life.

Causes

As for metal surface or using new friction material against metal member which needs regrinding.

Fade

Characteristics

Material degrades at the friction surface, resulting in a decrease in μ and a loss in performance, which may recover.

Causes

Overheating caused by excessive braking, or by brakes dragging.

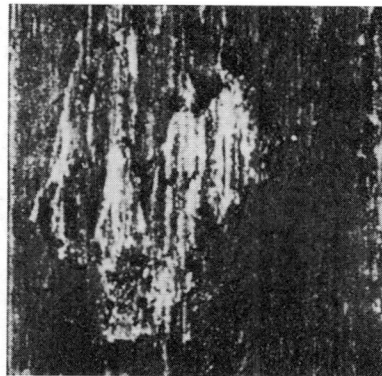

Metal pick-up

Characteristics

Metal plucked from the mating member and embedded in the lining.

Causes

Unsuitable combination of materials.

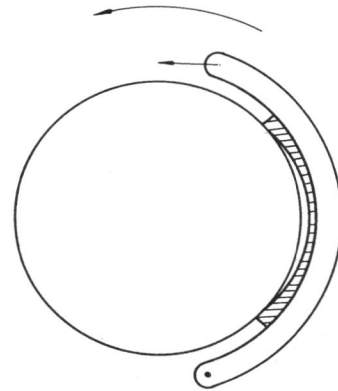

Grab

Characteristics

Linings contacting at ends only ('heel and toe' contact) giving high servo effect and erratic performance. The brake is often noisy.

Causes

Incorrect radiusing of lining.

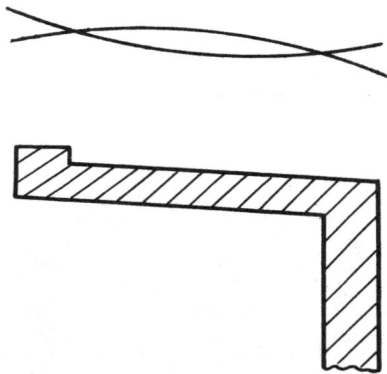

Strip braking

Characteristics

Braking over a small strip of the rubbing path giving localised heating and preferential wear at these areas.

Causes

Distortion of the brake path making it concave or convex to the lining, or by a drum bell mouthing.

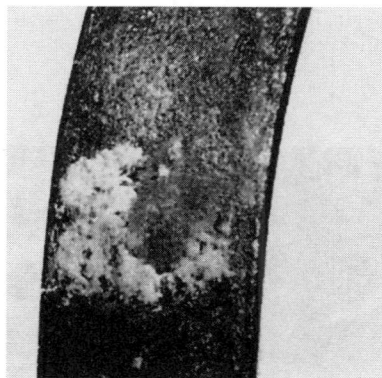

Neglect

Characteristics

Material completely worn off the shoe giving a reduced performance and producing severe scoring or damage to the mating component, and is very dangerous.

Causes

Failure to provide any maintenance.

Misalignment

Characteristics

Excessive grooving and wear at preferential areas of the lining surface, often resulting in damage to the metal member.

Causes

Slovenly workmanship in not fitting the lining correctly to the shoe platform, or fitting a twisted shoe or band.

CLUTCH TROUBLES

As with brakes, heat spotting, crazing and scoring can occur with clutches; other clutch troubles are shown below.

Dishing*

Characteristics

Clutch plates distorted into a conical shape. The plates then continually drag when the clutch is disengaged, and overheating occurs resulting in thermal damage and failure. More likely in multi-disc clutches.

Causes

Lack of conformability. The temperature of the outer region of the plate is higher than the inner region. On cooling the outside diameter shrinks and the inner area is forced outwards in an axial direction causing dishing.

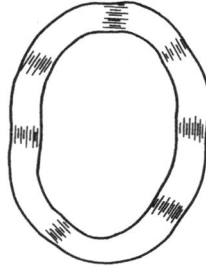

Waviness or buckling*

Characteristics

Clutch plates become buckled into a wavy pattern. Preferential heating then occurs giving rise to thermal damage and failure. More likely in multi-disc clutches.

Causes

Lack of conformability. The inner area is hotter than the outer area and on cooling the inner diameter contracts and compressive stresses occur in the outer area giving rise to buckling.

Band crushing*

Characteristics

Loss of friction material at the ends of a band in a band clutch. Usually results in grooving and excessive wear of the opposing member.

Causes

Crushing and excessive wear of the friction material owing to the high loads developed at the ends of a band of a positive servo band clutch.

Bond failure*

Characteristics

Material parting at the bond to the core plate causing loss of performance and damage to components.

Causes

Poor bonding or overheating, the high temperatures affecting bonding agent.

*These refer to oil immersed applications.

Material transfer

Characteristics

Friction material adhering to opposing plate, often giving rise to excessive wear.

Causes

Overheating and unsuitable friction material.

Burst failure

Characteristics

Material splitting and removed from the spinner plate.

Causes

High stresses on a facing when continually working at high rates of energy dissipation, and high speeds.

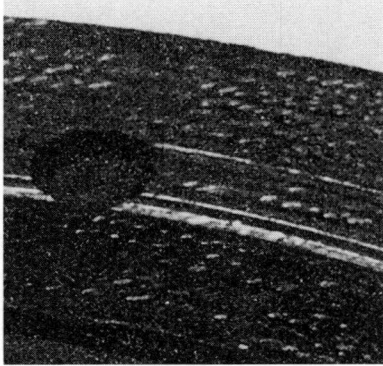

Grooving

Characteristics

Grooving of the facing material on the line of movement.

Causes

Material transfer to opposing plate.

Reduced performance

Characteristics

Decrease in coefficient of friction giving a permanent loss in performance in a dry clutch.

Causes

Excess oil or grease on friction material or on the opposing surface.

Distortion

Characteristics

Facings out of flatness after high operating temperatures giving rise to erratic clutch engagement.

Causes

Unsuitable friction material.

GENERAL NOTES

The action required to prevent these failures recurring is usually obvious when the causes, as listed in this section, are known.

Other difficulties can be experienced unless the correct choice of friction material is made for the operating conditions.

If the lining fitted has too low a coefficient of friction the friction device will suffer loss of effectiveness. Oil and grease deposited on dry linings and facings can have an even more marked reduction in performance by a factor of up to 3. If the μ is too high or if a badly matched set of linings are fitted, the brake may grab or squeal.

The torque developed by the brake is also influenced by the way the linings are bedded so that linings should be initially ground to the radius of the drum to ensure contact is made as far as possible over their complete length.

If after fitting, the brake is noisy the lining should be checked for correct seating and the rivets checked for tightness. All bolts should be tightened and checks made that the alignment is correct, that all shoes have been correctly adjusted and the linings are as fully bedded as possible. Similarly, a clutch can behave erratically or judder if the mechanism is not correctly aligned.

A wire rope is said to have failed when the condition of either the wire strands, core or termination has deteriorated to an unacceptable extent. Each application has to be considered individually in terms of the degree of degradation allowable; certain applications may allow for a greater degree of deterioration than others.

Complete wire rope failures rarely occur. The more common modes of failure/deterioration are described below.

DETERIORATION

Mechanical damage

Characteristics

Damage to exposed wires or complete strands, often associated with gross plastic deformation of the steel material. Damage may be localised or distributed along the length of the rope.

Inspection by visual means only.

Causes

There are many potential causes of mechanical damage, such as:

- rubbing against a static structure whilst under load
- impact or collision by a heavy object
- misuse or bad handling practices

External wear

Characteristics

Flattened areas formed on outer wires. Wear may be distributed over the entire surface or concentrated in narrow axial zones. Severe loss of worn wires under direct tension. Choice of rope construction can be significant in increasing wear resistance (e.g. Lang's lay ropes are usually superior to ordinary lay ropes).

Assess condition visually and also by measuring the reduction in rope diameter.

Causes

Abrasive wear between rope and pulleys, or between successive rope layers in multi-coiled applications, particularly in dirty or contaminated conditions (e.g. mining). Small oscillations, as a result of vibration, can cause localised wear at pulley positions.

Regular rope lubrication (dressings) can help to reduce this type of wear.

External fatigue

Characteristics

Transverse fractures of individual wires which may subsequently become worn. Fatigue failures of individual wires occur at the position of maximum rope diameter ('crown' fractures).

Condition is assessed by counting the number of broken wires over a given length of rope (e.g. one lay length, 10 diameters, 1 metre).

Causes

Fatigue failures of wires is caused by cyclic stresses induced by bending, often superimposed on the direct stress under tension. Tight bend radii on pulleys increases the stresses and hence the risk of fatigue. Localised Hertzian stresses resulting from ropes operating in oversize or undersize grooves can also promote premature fatigue failures.

Internal damage

Characteristics

Wear of internal wires generates debris which when oxidised may give the rope a rusty (or 'rouged') appearance, particularly noticeable in the valleys between strands.

Actual internal condition can only be inspected directly by unwinding the rope using clamps while under no load.

As well as a visual assessment of condition, a reduction in rope diameter can give an indication of rope deterioration.

Causes

Movement between strands within the rope due to bending or varying tension causes wear to the strand cross-over points (nicks). Failure at these positions due to fatigue or direct stress leads to fracture of individual wires. Gradual loss of lubricant in fibre core ropes accelerates this type of damage.

Regular application of rope dressings minimises the risk of this type of damage.

Corrosion

Characteristics

Degradation of steel wires evenly distributed over all exposed surfaces. Ropes constructed with galvanised wires can be used where there is a risk of severe corrosion.

Causes

Chemical attack of steel surface by corrosive environment e.g. seawater.

Regular application of rope dressings can be beneficial in protecting exposed surfaces.

Deterioration at rope terminations

Characteristics

Failure of wires in the region adjacent to the fitting. Under severe loading conditions, the fitting may also sustain damage.

Causes

Damage to the termination fitting or to the rope adjacent to the fitting can be caused by localised stresses resulting from sideways loads on the rope.

Overloading or shock loads can result in damage in the region of the termination.

Poor assembly techniques (e.g. incorrect mounting of termination fitting) can give rise to premature deterioration at the rope termination.

All photographs courtesy of Bridon Ropes Ltd., Doncaster

INSPECTION

To ensure safety and reliability of equipment using wire ropes, the condition of the ropes needs to be regularly assessed. High standards of maintenance generally result in increased rope lives, particularly where corrosion or fatigue are the main causes of deterioration.

The frequency of inspections may be determined by either the manufacturer's recommendations, or based on experience of the rate of rope deterioration for the equipment and the results from previous inspections. In situations where the usage is variable, this may be taken into consideration also.

Inspection of rope condition should address the following items:

- mechanical damage or rope distortions
- external wear
- internal wear and core condition
- broken wires (external and internal)
- corrosion
- rope terminations
- degree of lubrication
- equality of rope tension in multiple-rope installations
- condition of pulleys and sheaves

During inspection, particular attention should be paid to the following areas:

- point of attachment to the structure or drums
- the portions of the rope at the entry and exit positions on pulleys and sheaves
- lengths of rope subject to reverse or multiple bends

In order to inspect the internal condition of wire ropes, special tools may be required.

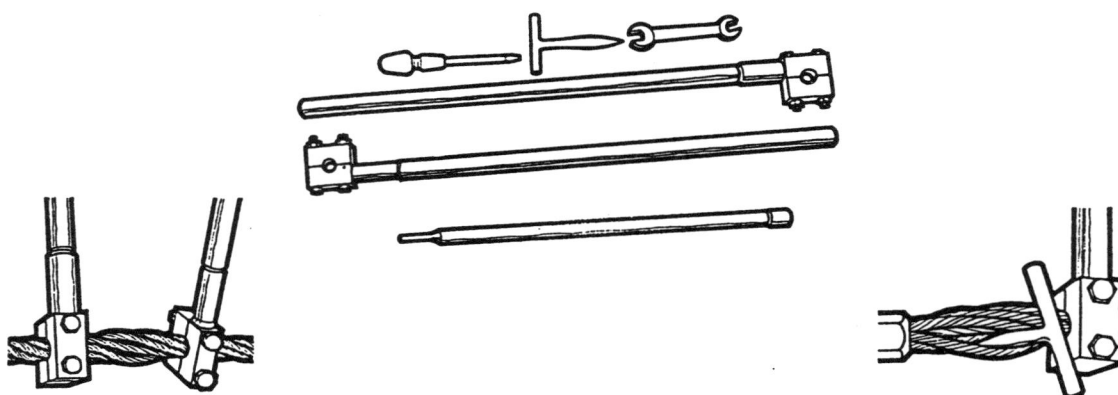

Figure 7.1 Special tools for internal examination of wire rope

MAINTENANCE

Maintenance of wire ropes is largely confined to the application of rope dressings, general cleaning, and the removal of occasional broken wires.

Wire rope dressings are usually based on mineral oils, and may contain anti-wear additives, corrosion inhibiting agents or tackiness additives. Solvents may be used as part of the overall formulation in order to improve the penetrability of the dressing into the core of the rope. Advice from rope manufacturers should be sought in order to ensure that selected dressings are compatible with the lubricant used during manufacture.

The frequency of rope lubrication depends on the rate of rope deterioration identified by regular inspection. Dressings should be applied at regular intervals and certainly before there are signs of corrosion or dryness.

Dressings can be applied by brushing, spraying, dripfeed, or by automatic applicators. For best results, the dressing should be applied at a position where the rope strands are opened up such as when the rope passes over a pulley.

When necessary and practicable ropes can be cleaned using a wire brush in order to remove any particles such as dirt, sand or grit.

Occasional broken wires should be removed by using a pair of pliers to bend the wire end backwards and forwards until it breaks at the strand cross-over point.

REPLACEMENT CRITERIA

Although the assessment of rope condition is mainly qualitative, it is possible to quantify particular modes of deterioration and apply a criterion for replacement. In particular the following parameters can be quantified:

- the number of wire breaks over a given length
- the change in rope diameter

Guidance for the acceptable density of broken wires in six and eight strand ropes is given below.

Table 17.1 Criterion for replacement based on the maximum number of distributed broken wires in six and eight strand ropes operating with metal sheaves

Total number of wires in outer strands (including filler wires)	Example of rope construction	Number of visible broken wires necessitating discard in a wire rope operating with metal sheaves when measured over a length of 10× nominal rope diameter	
		Factor of safety <5	Factor of safety >5
Less than 50	6 × 7 (6/1)	2	4
51–120	6 × 19 (9/9/1)	3–5	6–10
121–160	6 × 19F (12/6+6F/1)	6–7	12–14
161–220	8 × 19F (12/6+6F/1)	8–10	16–20
221–260	6 × 37 (18/12/6/1)	11–12	22–24

Rope manufacturers should be consulted regarding other types of rope construction.

Guidance for the allowable change in rope diameter is given below.

Table 17.2 Criterion for replacement based on the change in diameter of a wire rope

Rope construction	Replacement criteria related to rope diameter*
Six and eight strand ropes	Replacement necessary when the rope diameter is reduced to 90% of the nominal diameter at any position.
Multi-strand ropes	A more detailed rope examination is necessary when the rope diameter is either reduced to 97%, or has increased to 105%, of the nominal diameter.

*The diameter of the rope is measured across the tips of the strands (i.e. the maximum rope diameter).

BASIC MECHANISMS

Fretting occurs where two contacting surfaces, often nominally at rest, undergo minute oscillatory tangential relative motion, which is known as 'slip'. It may manifest itself by debris oozing from the contact, particularly if the contact is lubricated with oil.

    Colour of debris: red on iron and steel, black on aluminum and its alloys.

On inspection the fretted surfaces show shallow pits filled and surrounded with debris. Where the debris can escape from the contact, loss of fit may eventually result. If the debris is trapped, seizure can occur which is serious where the contact has to move occasionally, e.g. a machine governor.

    The movement may be caused by vibration, or very often it results from one of the contacting members undergoing cyclic stressing. In this case fatigue cracks may be observed in the fretted area. Fatigue cracks generated by fretting start at an oblique angle to the surface. When they pass out of the influence of the fretting they usually continue to propagate straight across the component. This means that where the component breaks, there is a small tongue of metal on one of the fracture surfaces corresponding to the growth of the initial part of the crack.

    Fretting can reduce the fatigue strength by 70–80%. It reaches a maximum at an amplitude of slip of about 8 μm. At higher amplitudes of slip the reduction is less as the amount of material abraded away increases.

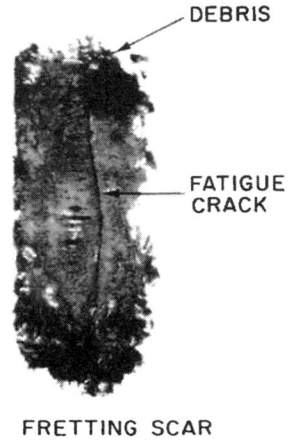

FRETTING SCAR

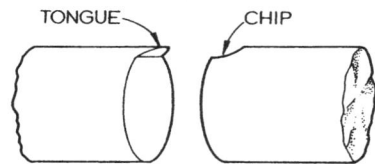

Figure 18.1 A typical fatigue fracture initiated by fretting

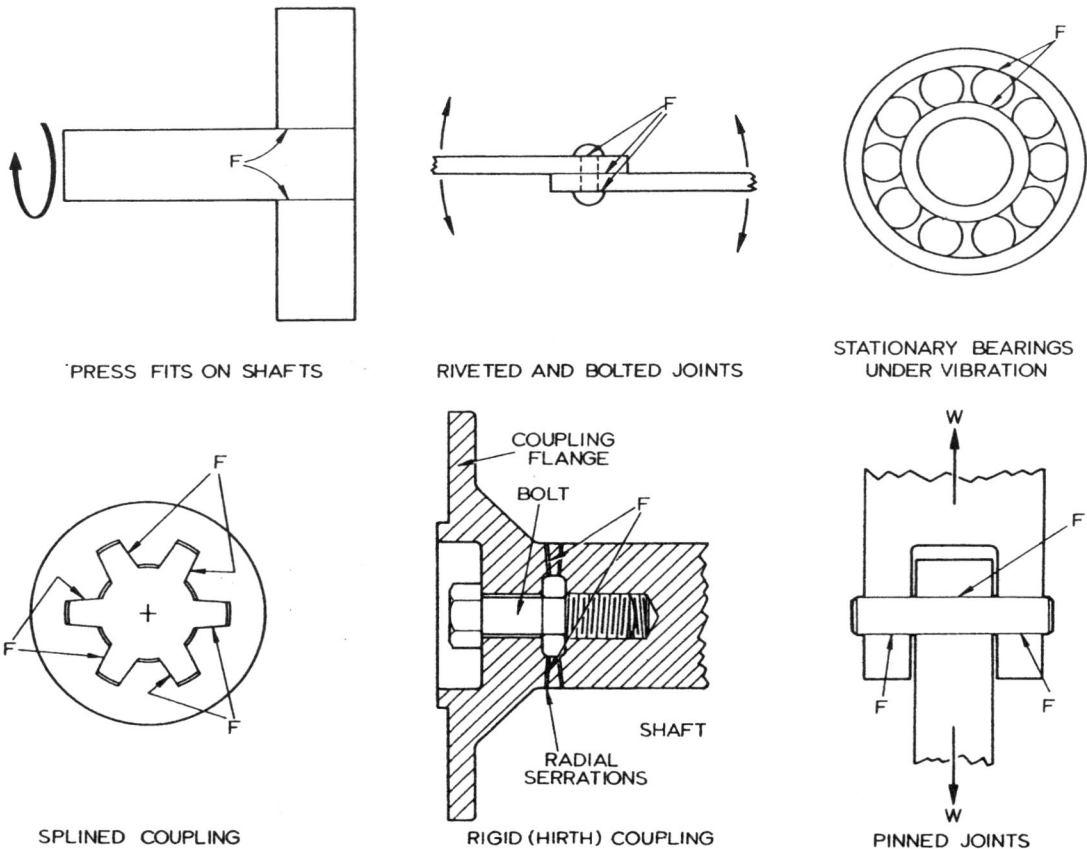

PRESS FITS ON SHAFTS   RIVETED AND BOLTED JOINTS   STATIONARY BEARINGS UNDER VIBRATION

SPLINED COUPLING   RIGID (HIRTH) COUPLING   PINNED JOINTS

Figure 18.2 Typical situations in which fretting occurs. Fretting sites are at points F.

Detailed mechanisms

Rupture of oxide films results in formation of local welds which are subjected to high strain fatigue. This results in the growth of fatigue cracks oblique to the surface. If they run together a loose particle is formed. One of the fatigue cracks may continue to propagate and lead to failure. Oxidation of the metallic particles forms hard oxide debris, i.e. Fe_2O_3 on steel, Al_2O_3 on aluminium. Spreading of this oxide debris causes further damage by abrasion. If the debris is compacted on the surfaces the damage rate becomes low.

Where the slip is forced, fretting wear damage increases roughly linearly with normal load, amplitude of slip, and number of cycles. Damage rate on mild steel – approx. 0.1 mg per 10^6 cycles, per MN/m^2 normal load, per μm amplitude of slip. Increasing the pressure can, in some instances, reduce or prevent slip and hence reduce fretting damage.

PREVENTION

Design

(*a*) elimination of stress concentrations which cause slip
(*b*) separating surfaces where fretting is occurring
(*c*) increasing pressure by reducing area of contact

Lubrication

Where the contact can be continuously fed with oil, the lubricant prevents access of oxygen which is advantageous in reducing the damage. Oxygen diffusion decreases as the viscosity increases. Therefore as high a viscosity as is compatible with adequate feeding is desirable. The flow of lubricant also carries away any debris which may be formed. In other situations greases must be used. Shear-susceptible greases with a worked penetration of 320 are recommended. E.P. additives and MoS_2 appear to have little further beneficial effect, but anti-oxidants may be of value. Baked-on MoS_2 films are initially effective but gradually wear away.

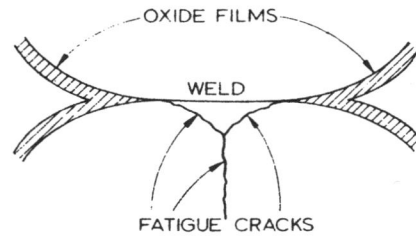

Figure 18.3 Oxide film rupture and the development of fatigue cracks

Non-metallic coatings

Phosphate and sulphidised coatings on steel and anodised coatings on aluminum prevent metal-to-metal contact. Their performance may be improved by impregnating them with lubricants, particularly oil-in-water emulsions.

Metallic coatings

Electrodeposited coatings of soft metals, e.g. Cu, Ag, Sn or In or sprayed coatings of Al allow the relative movement to be taken up within the coating. Chromium plating is generally not recommended.

Non-metallic inserts

Inserts of rubber, or PTFE can sometimes be used to separate the surfaces and take up the relative movement.

Choice of metal combinations

Unlike metals in contact are recommended – preferably a soft metal with low work hardenability and low recrystallisation temperature (such as Cu) in contact with a hard surface, e.g. carburised steel.

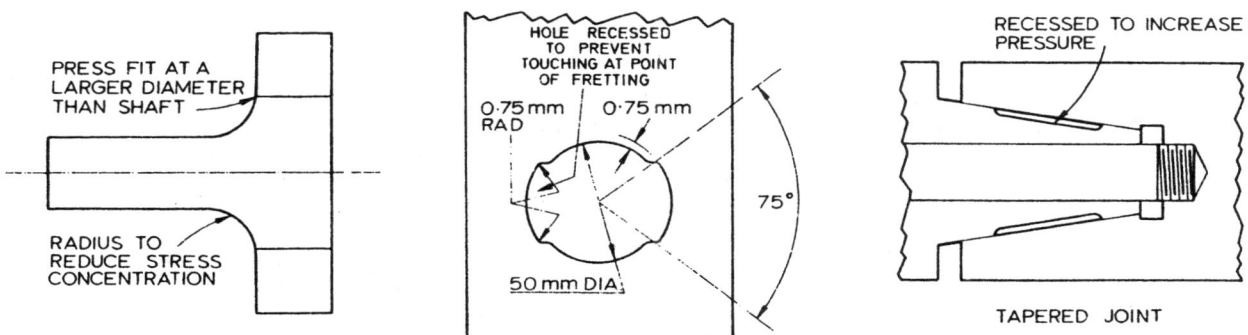

Figure 18.4 Design changes to reduce the risk of fretting

Wear can be defined as the progressive loss of substance resulting from mechanical interaction between two contacting surfaces. In general these surfaces will be in relative motion, either sliding or rolling, and under load. Wear occurs because of the local mechanical failure of highly stressed interfacial zones and the failure mode will often be influenced by environmental factors. Surface deterioration can lead to the production of wear particles by a series of events characterised by adhesion and particle transfer mechanisms or by a process of direct particle production akin to machining or, in certain cases, a surface fatigue form of failure. These three mechanisms are referred to as adhesive, abrasive and fatigue wear and are the three most important.

In all three cases stress transfer is principally via a solid–solid interface, but fluids can also impose or transfer high stresses when their impact velocity is high. Fluid erosion and cavitation are typical examples of fluid wear mechanisms. Chemical wear has been omitted from the list because environmental factors, such as chemical reaction, influence almost every aspect of tribology and it is difficult to place this subject in a special isolated category. Chemical reaction does not itself constitute a wear mechanism; it must always be accompanied by some mechanical action to remove the chemical products that have been formed. However, chemical effects rarely act in such a simple manner; usually they interact with an influence on a wear process, sometimes beneficially and sometimes adversely.

ADHESIVE WEAR

The terms cohesion and adhesion refer to the ability of atomic structures to hold themselves together and form surface bonds with other atoms or surfaces with which they are in intimate contact. Two clean surfaces of similar crystal structure will adhere strongly to one another simply by placing them in contact. No normal stress is theoretically required to ensure a complete bond. In practice a number of factors interfere with this state of affairs, particularly surface contamination, and measurable adhesion is only shown when the surfaces are loaded and translated with respect to each other causing the surface films to break up. Plastic deformation frequently occurs at the contacting areas because of the high loading of these regions, and this greatly assists with the disruption of oxide films.

Since the frictional force required to shear the bonded regions is proportional to their total area, and this area is proportional to the load under plastic contact conditions (also with multiple elastic contacts), a direct relationship exists between these two forces; the ratio being termed the coefficient of friction. However, it is important to realise that the coefficient of friction is not a fundamental property of a pair of materials, since strong frictional forces can be experienced without a normal load so long as the surfaces are clean and have an intrinsic adhesive capability.

Any factor which changes the area of intimate contact of two surfaces will influence the frictional force and the simple picture of plastic contact outlined above is only an approximation to the real behaviour of surfaces. Plasticity theory predicts that when a tangential traction is applied to a system already in a state of plastic contact, the junction area will grow as the two surfaces are slid against each other. The surfaces rarely weld completely because of the

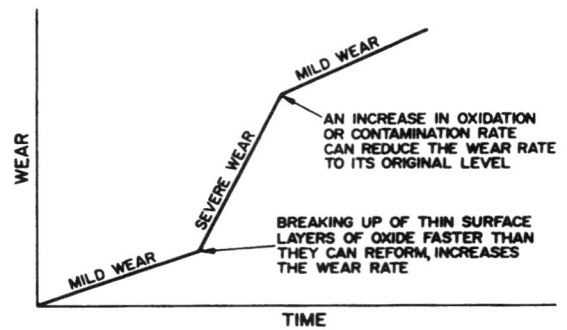

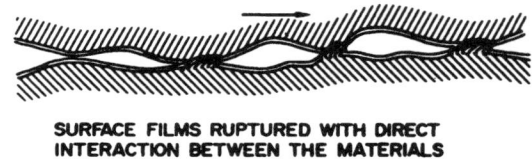

A simplified picture of adhesive wear

remarkable controlling influence exerted by the interfacial contaminating layers. Even a small degree of contamination can reduce the shear strength of the interface sufficiently to discourage continuous growth of the bonded area. Coefficients of friction therefore tend to remain finite. Control of the growth of contact regions can also be encouraged by using heterogeneous rather than homogeneous bearing surfaces, whilst the provision of a suitable finish can assist matters greatly. The direction of the finishing marks should be across the line of motion so that frequent interruptions occur.

The actual establishment of a bond, or cold weld as it is sometimes called, is only the first stage of a wear mechanism and does not lead directly to the loss of any material from the system. The bonded region may be strengthened by work hardening and shear may occur within the body of one of the bearing components, thus allowing a fragment of material to be transferred from one surface to another. Recent observations indicate that the bond plane may rotate as well as grow when a tangential traction is applied, the axis of rotation being such that the two surfaces appear to interlock and the deformation bulges formed on each surface act like prow waves to each other. If the result of a bond fracture is material transfer, then no wear occurs until some secondary mechanism encourages this particle to break away. Often transferred material resides on a surface and may even back transfer to the original surface. Quite frequently groups of particles are formed and they break away as a single entity. Numerous explanations have been put forward to explain this final stage of the wear process, but the stability of a group of particles will be affected by the environment. One view is that break-away occurs when the elastic energy just exceeds the surface energy; the latter being greatly reduced by environmental reaction.

It is useful to look upon the adhesive wear system as being in a state of dynamic equilibrium with its environment. Continuous sliding and the exposure of fresh surfaces cannot go on indefinitely and the situation is usually stabilised by the healing reaction of the air or other active components of the surrounding fluid. The balance between the rupturing and healing processes can be upset by changing the operating parameters, and surfaces may abruptly change from a low to a high wearing stage. Increasing the speed of sliding, for instance, reduces the time available for healing reactions to occur, but it also encourages higher surface temperatures which may accelerate chemical reactions or resorb weakly bound adsorbants. The particular course which any system will take will thus depend greatly upon the nature of the materials employed.

Many wear processes start off as adhesive mechanisms, but the fact that the wear process leads to the generation of debris inevitably means that there is always a possibility that it may change to one of abrasion. In most cases, wear debris becomes, or is formed, as oxide, and such products are invariably hard and hence abrasive. A typical situation where this can arise is when two contacting surfaces are subjected to very small oscillatory slip movements. This action is referred to as fretting and the small slip excursion allows the debris to build up rapidly between the surfaces. This debris is often in a highly oxidised condition. The actual rate of wear tends to slow down because the debris acts as a buffer between the two surfaces. Subsequent wear may occur by abrasion or by fatigue.

ABRASION

Wear caused by hard protrusions or particles is very similar to that which occurs during grinding and can be likened to a cutting or machining operation, though a very inefficient one by comparison. Most abrasive grits present negative rake angles to the rubbed material and the cutting operation is generally accompanied by a large amount of material deformation and displacement which does not directly lead to loose debris or chips. The cutting efficiency varies considerably from one grit to another and on average only a small amount, 15–20% of the groove volume is actually removed during a single passage.

During abrasion a metal undergoes extensive work hardening and for this reason initial hardness is not a particularly important factor so long as the hardness of the abrasive grit is always substantially greater than that of the metal surface. Under this special condition there is a relatively simple relationship between wear resistance and hardness. For instance, pure metals show an almost linear relationship between wear resistance and hardness in the annealed state.

When the hardness of a metal surface approaches that of the abrasive grains, blunting of the latter occurs and the wear resistance of the metal rises. The form of the relationship between wear resistance and the relative hardnesses of the metal and abrasive is of considerable technical importance. As an abrasive grain begins to blunt so the mode of wear changes from one of chip formation, perhaps aided by a plastic fatigue mechanism, to one which must be largely an adhesive-fatigue process. The change is quite rapid and is usually fully accomplished over a range of $H_{metal}/H_{abrasive}$ of 0.8 to 1.3, where H_{metal} is the actual surface hardness.

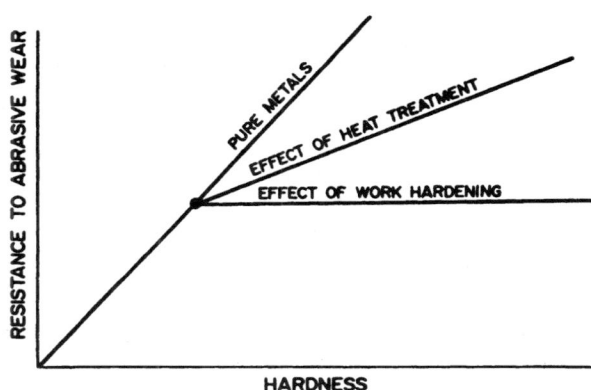

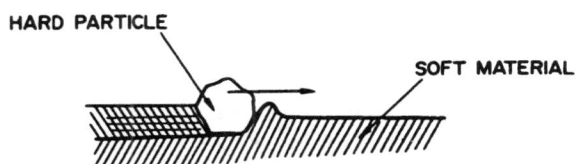

Some aspects of abrasive wear

Heterogeneous materials composed of phases with a considerable difference in hardness form a common and important class of wear resistant materials. When the abrasive is finely divided, the presence of relatively coarse, hard, material in these alloys increases the wear resistance considerably, but, when the abrasive size increases and becomes comparable with the scale of the heterogeneity of the structure, such alloys can prove disappointing. The reason seems to be that the coarse abrasive grits are able to gouge out the hard wear resistant material from the structure.

Brittle non metallic solids behave in a somewhat different way to the ductile metals. In general, the abrasion is marked by extensive fracture along the tracks and the wear rates can exceed those shown by metals of equivalent hardness by a factor of ten. With very fine abrasive material, brittle solids can exhibit a ductile form of abrasion. As in the case of metals, the effective wear resistance of a brittle material is a function of the relative hardnesses of the solid and the abrasive, but whereas with metals the effect is negligible until H_m/H_a reaches 0.8, with non metallic brittle solids blunting appears to take place at much lower values of this hardness ratio, indeed there seems to be no threshold. The wear resistance climbs slowly over a very wide range of $H_{brittle\ solid}/H_{abrasive}$.

Abrasion is usually caused either by particles which are embedded or attached to some opposing surface, or by particles which are free to slide and roll between two surfaces. The latter arrangement causing far less wear than the former. However, the abrasive grits may also be conveyed by a fluid stream and the impact of the abrasive laden fluid will give rise to erosive wear of any interposed surface. The magnitude and type of wear experienced now depends very much upon the impinging angle of the particles and the level of ductility, brittleness or elasticity of the surface. Many erosive wear mechanisms are similar to those encountered under sliding conditions, although they are modified by the ability of the particles to rebound and the fact that the energy available is limited to that of the kinetic energy given to them by the fluid stream. Rotation of the particles can also occur, but this is a feature of any loose abrasive action.

CONTACT FATIGUE

Although fatigue mechanisms can operate under sliding wear conditions, they tend to occupy a much more prominent position in rolling contact where the stresses are high and slip is small. Such contacts are also capable of effective elastohydrodynamic lubrication so that metal to metal contact and hence adhesive interaction is reduced or absent altogether. Ball and roller bearings, as well as gears and cams, are examples where a fatigue mechanism of wear is commonly observed and gives rise to pitting or spalling of the surfaces.

The mechanisms of rolling contact fatigue can be understood in terms of the elastic stress fields established within the surface material of the rolling elements. Elastic stress analysis indicates that the most probable critical stress in contact fatigue is the maximum cyclic orthogonal shear stress rather than the unidirectional shear stress which occurs at somewhat greater depths. The Hertzian stress distribution is adversely affected by numerous factors, including such features as impurity inclusions, surface flaws, general misalignment problems and other geometrical discontinuities, as well as the elastohydrodynamic pressure profile of the lubricant and the tangential traction.

Although the maximum cyclic stress occurs below the immediate surface, the presence of surface flaws may mean that surface crack nucleation will become competitive with those of sub-surface origin and hence a very wide range of surface spalls can arise. Furthermore, it is important to remember that if the surfaces are subjected to considerable tangential traction forces then the positions of the shear stress maxima slowly move towards the surface. This last condition is likely to arise under inferior conditions of lubrication, as when the elastohydrodynamic film thickness is unable to prevent asperity contact between the rolling elements. Pure sub-surface fatigue indicates good lubrication and smooth surfaces, or potent stress raising inclusions beneath the surface.

As in other aspects of fatigue, the environment can determine not only the stress required for surface crack nucleation, but more significantly the rate of crack propagation once a crack has reached the surface. The presence of even small amounts of water in a lubricant can have very serious consequences if suitable lubricant additives are not incorporated. It has also been suggested that a lubricant can accelerate crack propagation by the purely physical effect of becoming trapped and developing high fluid pressures in the wedge formed by the opening and closing crack.

FLUID AND CAVITATION EROSION

Both these wear mechanisms arise from essentially the same cause, namely the impact of fluids at high velocities. In the case of fluid erosion, the damage is caused by small drops of liquid, whilst in the case of cavitation, the impact arises from the collapse of vapour or gas bubbles formed in contact with a rapidly moving or vibrating surface.

Fluid erosion frequently occurs in steam turbines and fast flying aircraft through the impact of water droplets. The duration of impact is generally extremely small so that very sharp intense compression pulses are trans-

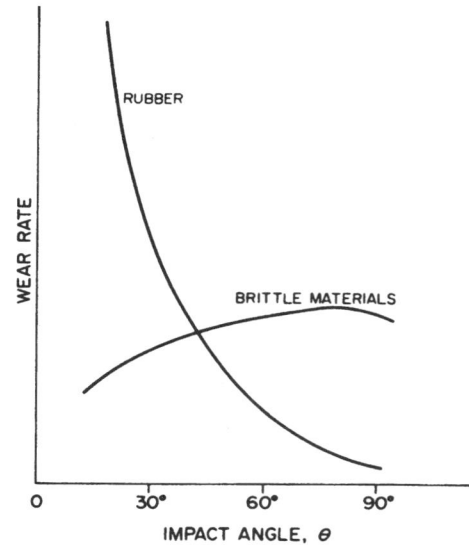

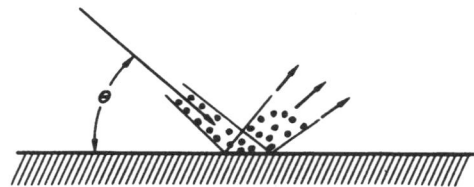

Wear by fluids containing abrasive particles

ferred to the surface material. This can generate ring cracks in the case of such brittle materials as perspex, or form plastic depressions in a surface. As the liquid flows away from the deformation zone, it can cause strong shear deformation in the peripheral areas. Repeated deformation of this nature gives rise to a fatigue form of damage and pitting or roughening of the surfaces soon becomes apparent.

With cavitation erosion, damage is caused by fluid cavities becoming unstable and collapsing in regions of high pressure. The cavities may be vaporous, or gaseous if the liquid contains a lot of gas. The damage caused by the latter will be less than the former. The physical instability of the bubbles is determined by the difference in pressure across the bubble interface so that factors such as surface tension and fluid vapour pressure become important. The surface energy of the bubble is a measure of the damage which is likely to occur, but other factors such as viscosity play a role. Surface tension depressants have been used successfully in the case of cavitation attack on Diesel engine cylinder liners. Liquid density and bulk modulus, as well as corrosion, may be significant in cavitation, but since many of these factors are interrelated it is difficult to assess their individual significance.

Attempts to correlate damage with material properties has led to the examination of the ultimate resilience characteristic of a material. This is essentially the energy that can be dissipated by a material before any appreciable deformation or cracking occurs and is measured by $\frac{1}{2}$ (tensile strength)^2/elastic modulus. Good correlation has been shown with many materials. The physical damage to metals is of a pitting nature and obviously has a fatigue origin.

The surfaces of most components which have been worn, corroded or mis-machined can be built up by depositing new material on the surface. The new material may be applied by many different processes which include weld deposition, thermal spraying and electroplating.

The choice of a suitable process depends on the base material and the final surface properties required. It will be influenced by the size and shape of the component, the degree of surface preparation and final finishing required, and by the availability of the appropriate equipment, materials and skills. The following tables give guidance on the selection of suitable methods of repair.

Table 20.1 Common requirements for all processes

Requirement	Characteristic
Re-surfacing should be cheaper than replacement with a new part	Larger components often tend to be the most suitable and usually can be repaired on site. Also, the design may be unique or changes may have rendered the part obsolete, making replacement impossible. If the worn area was not surfaced originally the replacement surface may give better performance than that obtained initially. Repair, and re-surfacing are conservationally desirable processes, saving raw materials and energy and often proving environmentally friendly.
Appropriate surface condition prior to coating	Whatever process is considered applicable the surfaces for treatment should be in sound metallurgical condition, i.e. free from scale, corrosion products and mechanically damaged or work-hardened metal. The part may have the remains of a previous coating and unless it is possible to apply the new coating on top of this the old coating must be removed, generally by machining or grinding, sometimes by grit-blasting, anodic dissolution or heat treatment. Surfaces which have been nitrided, carburised, etc., may possibly be modified by heat treatment, otherwise mechanical removal will be necessary.
Preparatory machining of the surface prior to coating	If the component has to be restored to its original dimensions it will need to be undercut to allow for the coating thickness. In many cases this undercutting should be no more than $250\,\mu\text{m}/500\,\mu\text{m}$ deeper than the maximum wear which can be tolerated on the new surface. With badly damaged parts it will be necessary to undercut to the extent of the damage and this may limit the choice of repair processes which could be used. For precision components which require finishing by machining it is important to preserve reference surfaces to ensure that the finished component is dimensionally correct.

Table 20.2 Factors affecting the choice of process

Factor	Effect
SIZE AND SHAPE Size and weight of component Thickness of deposit Weight to be applied Location of surfaces to be protected Accessibility	For small parts hand processes suitable. Spraying is a line-of-sight process limiting coating in bores, but some processes will operate in quite small diameters. Electroplating requires suitable anodes. Generally no limitation with weld deposition other than access of rod or torch in bores, however, some processes do not permit positional welding.
DESIGN OF ASSEMBLY Has brazing, welding, interference fits, riveting been used in the assembly? Are there temperature sensitive areas nearby? Have coaxial tolerances to be maintained?	Some processes put very little heat into an assembly so brazing filler metals would not be melted. Welds need careful cleaning and smoothing. Consideration must be given to the effect of heat on fitted parts. Suitable reference points are essential if tolerances are critical.
DISTORTION PROBLEMS What distortion is permissible? Is the part distorted as received? Will preparation cause distortion? Will surfacing cause distortion? Can distortion occur on cooling?	Eliminate distortion in part for repair. Remove cold-work stresses. Bond coats may replace grit-blasting. Some processes do not require surface roughening. Some processes provide more severe temperature gradients giving greater distortion. Allowance must be made for thermal expansion. Slow cooling must be available. In some cases pre-setting before coating may be used. Straightening or matching after coating may also be possible.
PROPERTIES OF BASE METAL Can chemical composition be determined? Have there been previous surface treatments/coating? What is surface condition? What is the surface hardness? What metallographic structure and base metal properties are ultimately required?	The depth of penetration and degree of surface melting can eliminate surface treatments, but previous removal may be necessary. If too hard for grit-blasting softening may be possible or a bond coat used. Suitable post-surfacing heat treatment may be needed.
DEPOSIT REQUIREMENTS Dimensional tolerances acceptable Surface roughness allowable/desirable Need for machining Further assembly to be carried out Additional surface treatments required, e.g. plating, sealing, painting, polishing Possible treatment of adjacent areas such as carburising Thermal treatments needed to restore base metal properties	Arc welding not suitable for thin coatings. To retain dimensions, a low or uniform heat input essential. If machining is needed a smooth deposit saves time and cost. Further assembly demands a dense deposit as do many surface treatments. Painting may benefit from some surface roughness, lubrication from porosity, some surfaces need to be smooth, low friction, others need frictional grip. These are some of the many factors needing consideration in selecting the process.

EFFECT ON BASE METAL PROPERTIES
Consideration must be given to the effect of the process on base metal strength. Some sprayed coatings can reduce base metal fatigue strength as can hard, low ductility fused deposits in some environments. The same applies to some electro-deposited coats. Such coatings on high UTS steels can produce hydrogen embrittlement. Applied coatings do not usually contribute to base metal strength and allowance must be made for this, particularly where wear damage is machined out, prior to reclamation of the part. Welding will produce a heat affected zone which must be assessed. Fusing sprayed deposits can cause metallurgical changes in the base metal and the possibility of restoring previous properties must be considered. Unless coatings can be used as deposited, finish machining must be employed. Some coatings, e.g. flame sprayed or electro-deposited coatings must be finish ground, many dense adherent coatings may be machined. If porosity undesirable the coating must be sealed or densified.

OTHER CONSIDERATIONS
The number of components to be treated at one time or the frequency of repairs required, influence the expenditure permissible on surfacing equipment, handling devices, manipulators and other equipment.

The re-surfacing of equipment in situ such as quarry equipment, steelmill plant, paper making machinery, military equipment and civil engineering plant may limit or dictate choice of process.

Table 20.3 The surfacing processes that are available

Type	Symbol	Process	Coating materials and their form	Energy source and other materials	Surface preparation and coating procedure	Coating thickness	Treatment after coating
Gas Welding	GW	Gas welding	Rods, wires, tungsten carbide containing rods	Oxygen, acetylene	Flux application may be necessary if substrate contains elements forming refractory oxides, e.g. Cr	Up to 3 mm	Slow cooling Machining Grinding Flux removal
	PW	Powder welding	Self-fluxing alloy powder	Oxygen, acetylene. Possibly flux	Flux application to high Cr steels. Pre-heating in some cases	Up to 5 mm	Slow cooling Machining Grinding Filing softer deposits
Arc Welding AC and/or DC	MMA	Arc welding Manual (stick) welding	Bare rods, pre-alloyed wire electrodes or with alloying in the coating or tungsten carbide containing tubes	AC and/or DC Flux	May need pre-setting and/or pre-heating. If wear excessive, preliminary build-up with low alloy steel economic	Thick coatings possible	Possibly slow cooling. If suitable heating treatment. If required straighten and/or machine. Generally flux/slag removal
	TIG	Tungsten inert gas	Bare rods or wire	Argon. Tungsten electrodes	May need pre-setting and/or pre-heating. May use manipulator	2 mm upwards	Possibly slow cooling/heat treatment. Straightening, stress relieving, machining
	MIG	Metal inert gas, gas shielded arc	Rods, wires or tubes	Carbon dioxide	As TIG	3 mm upwards	As TIG
	FCA	Flux-cored arc	Tubes, cored wires		As TIG	3 mm upwards	As TIG, flux removal may be necessary
	SA	Submerged arc	Wires, strips	Granulated flux	As TIG. Will need use of an automatic manipulator	3 mm upwards	As TIG. Flux removal needed
	PTA	Plasma transferred arc	Metal or alloy powder	Argon, hydrogen and/or nitrogen, helium	As TIG. Use of manipulator/robot general	2–5 mm	As TIG
Plating	EP	Electroplating	Ingots, plates, wires	Electric DC. Controlled chemical solutions. Cr, Ni possibly Cu anodes	Generally smooth turning or grinding. Protect areas where plating not desired. If steel has tensile strength exceeding $1000\,N/mm^2$ or fatigue strength is important shot peen	Wide range	Heat treat to reduce coating stresses Machine, grind

Table 20.3 (continued)

Type	Symbol	Process	Coating materials and their form	Energy source and other materials	Surface preparation and coating procedure	Coating thickness	Treatment after coating
Thermal spraying/Metallising	MP	Powder spraying	Metal or ceramic powder	Oxygen, acetylene sometimes hydrogen. Compressed air	Grit-blasting with or without grooving, rough threading or application of bond coat. Mask area not to be blasted or sprayed. Warm to prevent condensation from flame. Heat to reduce contraction stresses or improve adhesion	2 mm upwards	May machine, impregnate, seal, use as sprayed or densify mechanically or by HIPping
	MW	Wire spraying	Metal wire, ceramic rods, powder/plastic wire	Oxygen, acetylene sometimes hydrogen. Compressed air	As MP	75 μm upwards	As MP
	MA	Arc spraying	Solid or tubular wires	Electric power. Compressed air	Generally as MP. Al bronze an additional bond coat. Adjustment of spray conditions may provide coarse adherent deposit as initial bond coat. Mechanical handling often used. Warming/heating not used	2 mm upwards	Use as sprayed, machine, seal
	HVOF	High velocity oxy/flame spraying	Metal powder, sometimes ceramic powder	Oxygen, composite gas, acetylene, hydrogen. Carrier gas Ar, He or N$_2$	As MP	Wide range	As MP
	PF	Plasma flame spraying	Metal powder, perhaps wire or rod	Electric power. Argon, hydrogen nitrogen helium	As MP	Up to 2 mm	As MP
Sprayed and fused	SW	Sprayed and fused coatings	Self-fluxing alloy powders, occasionally bonded into wire	Oxygen, acetylene, (hydrogen), compressed air. Pre-heating torches/burners	Use only chilled iron or nickel alloy grit which must be clean and free from fines. Do not use bond coats. If needed pre-set/pre-heat. Mask areas not to be blasted/sprayed. Often lathe mounted and thermal expansion must be accommodated	0.4–2.0 mm	Fuse. Slow cool. Possibly heat treat. Use as fused or machine

Table 20.4 Characteristics of surfacing processes

Process	Advantages	Disadvantages
Gas welding	Small areas built up easily with thick deposits, grooves and recesses can be filled accurately. Thin, smooth coatings can be deposited. Minimum melting of the parent metal is possible with low dilution of the surfacing alloy. This is advantageous if using highly alloyed consumables or if a thin coating only desired. The process is under close control by the operator. Equipment is inexpensive and requires only fuel gases and possibly flux. A wide range of consumables available. There is minimum solution of carbon/carbide granules from tubular rods. Self-fluxing alloys can be deposited without flux, except on highly alloyed substrates, simplifying post-welding cleaning.	Process slow and not suitable for surfacing large areas. Build up of heat may overheat component and lead to distortion. Careful control of flame adjustment. As it is a manual process results are dependent on operator skill, fitness and degree of fatigue. Good technique is essential to ensure sound bond with the interface, especially if fusion with the substrate is not involved. There is a lack of NDT methods to check adhesion between coating and base metal.
Powder welding	Requires less skill than gas welding. Equipment not expensive but the self-fluxing powders are. The consumables have a wide temperature range between liquidus and solidus and between these temperatures the pasty consistency of the deposit enables thin, for sharp edges and worn corners to be built up. Once an initial, thin coating has been built up high deposition rates of subsequent coatings are easily maintained. The ability to put down a thin first coating makes it possible to coat small areas on large components. Smooth or contoured deposits are possible requiring little finish machining. Soft deposits may be finished by hand filing. The process leaves one hand free to manipulate the work.	As for gas welding. Additionally there is a limited range of suitable consumables. It is essential that the interface temperature reaches 1000°C to ensure that the coating is bonded to and not just cast on the base metal and that there is enough heat and time for the self-fluxing action of the coating alloy to clean the surface of interfering oxides.
Manual metal arc welding	Equipment cost low, requiring very little maintenance. It is adaptable to small or large complex parts, can be used with limited access and positional welding possible, i.e. vertical. A wide range of consumables available. Deposition rates up to 5 kghr^{-1}. Ideal for one off and small series work and is useful when only small quantities of hardfacing alloys are required.	A skilled operator is needed for high quality deposits and slag removal is necessary. Dilution tends to be high. Granular carbides in tubular electrodes are usually melted.
TIG welding	The process can be closely controlled by the operator using hand-held torch and hand-held filler rod but can be mechanised for special applications. Small areas can be surfaced e.g. small pores in hard surfacing deposits. A deposit thickness of 2 mm upwards is achievable and a deposition rate up to 2 kghr^{-1}. High quality deposits can be made.	The process is slow and unsuitable for surfacing large areas. A limited range of consumables is available. The equipment is expensive and not suitable for site work, needing a workshop in which the shielding gas can be protected from disturbing draughts.
MIG welding	A continuous process which is used semi-automatically with a hand-held gun or is wholly mechanised by traversing the gun and/or the workpiece. Slag removal is not needed. It provides a positional surfacing capability and guns are available for internal bore work. High deposition rates 3–8 kghr^{-1} with deposit thickness 3 mm upwards.	Equipment relatively expensive requiring regular maintenance. Use of shielding gas makes the process marginally less transportable than MMA and the gas must be selected to suit the surfacing alloy. Alloys for hard surfacing are not generally available in wire form so consumables restricted to mild steel (for build-up), stainless steels, aluminium bronze or tin bronze. High levels of UV radiation produced especially using high peak current pulse welding.
Flux-cored arc welding	Similar in principle to MIG surfacing but uses a tubular electrode containing a flux which decomposes to provide a shield to protect the molten pool. A separate shielding gas supply is not required. It is a continuous process used semi-automatically with a hand-held gun or is wholly mechanised. A wide range of consumables is available. High deposition rates up to 8 kghr^{-1} for CO_2 shielded process and 11 kghr^{-1} higher for self-shielded process.	With dilution of 15–30%, depending on technique, the process is not suitable for non-ferrous surfacing alloys. Equipment is fairly expensive and needs regular maintenance. Deposit quality may be lower than with MIG surfacing. Restrictions on use and transportation similar to the MIG process.

Table 20.4 (continued)

Process	Advantages	Disadvantages
Submerged arc	It is a fully automatic process providing high deposition rates 10 kghr^{-1} upwards on suitable workpieces. Deposit thickness 3 mm and over. A wide range of consumables available giving high quality deposits of excellent appearance requiring minimum finishing. Slag removal easy.	Intended primarily for workshop use with a fixed installation the equipment is very expensive and needs regular maintenance. Applications are limited. Generally, to large cylindrical or flat components; there is limited access to internal surfaces or larger bores.
PTA surfacing	A mechanised process giving close control of surface profile with minimum finishing required. Penetration and dilution low. Deposit thickness in range 2–5 mm. Deposition rate higher than with TIG process: 3.5 kghr^{-1}. Torches available for surfacing small bores down to about 25 mm diameter up to 400 mm deep. Very useful for surfacing internal valve seats. Balancing the power input between the arc plasma and the transferred arc enables deposits to be made on a wide range of components varying greatly in thickness and base metal composition.	Equipment expensive and not readily portable. Process costs high requiring a very skilled operator.
Electroplating	As operating temperatures never exceed 100°C work should not distort or suffer undesirable metallurgical changes. Coatings dense and adherent to substrate, molecular bonding may be as strong as 1000 N/mm^2. Plating conditions may be adjusted to modify hardness, internal stress and metallurgical characteristics of the deposits. No technical limits to thickness of deposits but most applications require thin coatings. Areas not requiring build-up can be masked. Brush plating can be used on localised areas, the equipment is portable and can be taken to the work. Deposition rates can exceed those of vat plating and may reach 200–400 mmhr^{-1}.	The thickness of deposit is proportional to current density and plating time, seldom exceeding 75 μmhr^{-1}. As current density over workpiece surface is seldom uniform coatings tend to be thicker at edges and corners and thinner in recesses and the centre of large flat areas.

Although application of coatings is not confined to line-of-sight the ability to plate round corners may be limited, however anode design and location may assist. The size of the vat limits dimensions of the work. Brush plating is labour intensive and requires considerable skill. The electrolytes are expensive but small volumes generally only required. |
| Powder spraying | Low cost equipment which can be used by semi-skilled operators. The most useful flame spray process for high alloy and self-fluxing surfacing materials continuously fed to the pistol. Coatings can be provided of materials which cannot be produced as rods or wires. Spray rates relatively high. Can provide coatings of uniform, controlled thickness, ideal for large cylindrical components but also possible on small and/or irregular parts. Low heat input to the base material minimises distortion and adverse metallurgical changes. Machining if needed can be minimal. | Most suitable for workshop use with adequate dust extraction available. Requires rough surface preparation. A line-of-sight process with limited access to bores and limitations on deposit thickness. Provides only a mechanical bond to the substrate. Deposit porous and may need sealing or densification. Needs compressed air as well as combustion gases. Not suitable for ceramic spraying. |
| Wire spraying | Generally as with powder spraying but much more suitable for site work. Wire reels can be at considerable distance from the spraying torch making it possible to work inside large constructions. Spray rates are fast and a wide range of surfacing materials can be sprayed. The spraying can be mechanised. Large areas easily sprayed given adequate design of work station. Can give useful porosity, thin coatings (75–150 μm) easily applied. For corrosion resistance sealed Zn, Al, or ZnAl coatings generally better than multi-layer paint systems. Ceramic spraying possible using rods. | As with powder spraying but does not require controlled mesh size of the consumables. |

Table 20.4 (continued)

Process	Advantages	Disadvantages
Arc spraying	Faster than gas spraying, providing extremely good bond and denser coatings. Either solid or tubular wires can be used giving a wide range of coating alloys. Using two different wires, composite or 'pseudo-alloy' coatings are possible with compositions/structures unobtainable by other means. Can be used for thick build-ups on cylindrical parts. Surface preparation not as critical as for flame spraying.	Equipment more expensive than for flame spraying. Weight of gun and great spraying rate makes mechanical handling desirable. Limited to consumables available in wire form. Density and adhesion lower than the plasma spraying or HVOF process.
High velocity oxyflame spraying	Produces dense, high bond strength coatings and thick, low stress coatings independent of coating hardness. Tight spray pattern allows accurate placement of the deposit. Low heat input to the component. Torches available for spraying internal surfaces down to 100 mm diameter. Some equipments can spray ceramics. Adhesion is high eliminating need for a bond coat.	More expensive than flame spraying the process needs careful control and monitoring. Not really suitable for manual spraying, manipulators should be used. Consumables used are in powder form.
Plasma spraying	The high temperature enables almost all materials to be sprayed giving high density coatings strongly bonded to the substrate, with low heat input to it. This can cause problems with differential contraction between coating and base metal. High density, high adhesion, low oxide coatings with properties reproducable to close limits are possible by spraying in a vacuum chamber back-filled with argon. Heat input to the component is then much greater.	Capital cost higher than gas arc spraying. A spray booth is desirable. Vacuum spraying requires pumping equipment with a suitable manipulator. The size of the chamber limits work dimensions. Process time slow. High heat input to the workpiece may cause distortion or metallurgical changes in the substrate.
Spray, fused coatings	The coating is dense, non-porous and metallurgically bonded to the substrate. Spraying thickness and uniformity easily monitored and controlled. Although restricted to Ni- and Co-base self-fluxing alloys a large range of these available with a wide spread of hardness and other properties. Often used 'as fused' the smooth accurate surface facilitates machining with minimum wastage of the surface alloy. Tungsten carbide particles often incorporated in the coating to enhance wear resistance. Torches may be held manually or mechanically manipulated. Coatings, generally 0.4–2 mm thick can be put on components varying greatly in size and shape. Fusing can be carried out manually, in furnaces (generally vacuum), by induction heating or using lasers.	A lot of heat is introduced into the part. The interface must reach about 1000°C to ensure metallurgical bonding at the interface. This can cause distortion and metallurgical changes in the base material. Cooling from above 1000°C can provide problems due to the differential thermal contaction of coating and substrate metal. Coatings on transformable steels can provide cracking problems.

Table 20.5 General guidance on the choice of process

Process	Cost of equipment	Ease of operation	Operator skill	Applicability of the process to:											Heat input on surfacing			Suitable surfacing alloys
				Large areas	Small areas	Thick deposits	Thin deposits	Thick sections	Thin sections	Massive work	Small parts	Edge build-up	Site work	Machined components	Before	During	After	
Gas welding	Low	Easy	Medium	3	1	1	2	3	1	3	1	2	1	2	Fair	High diffuse	May be desirable	Many. Excellent for WC containing rods.
Powder welding	Low	Easy	Low	3	1	2	1	2	1	2	1	1	1	1	Low	Fairly high	None	Restricted to self-fluxing alloys.
Manual metal arc welding	Fairly low	Fairly easy	Fairly high	2	2	1	3	1	3	1	3	3	1	3	Fair	Very high, steep gradients	May be desirable	Wide range available.
TIG welding	Medium high	Moderately easy	Fairly high	3	1	2	2	1	2	2	2	2	2	2	Low	Very high, fairly confined	May be desirable	Wide range available.
MIG welding	Medium	Moderately easy	Medium	2	3	2	3	1	3	2	3	3	2	3	Low	Very high, steep gradients	May be desirable	Restricted to alloys available as wires.
Flux-cored arc welding	Medium	Moderately easy	Medium	1	3	2	3	1	3	1	4	3	2	3	Low	Very high, steep gradients	May be desirable	Needs cored wires.
Submerged arc welding	Very high	Moderately hard	Fairly high	1	4	1	4	1	4	1	4	4	4	4	Low	Very high, diffuse	Some	Needs suitable coating materials or flux additions.
PTA surfacing	High	Hard	High	4	1	3	2	3	1	3	2	3	4	1	Low	High, moderate gradient	May be desirable	Many alloys possible.
Electroplating	High	Fairly easy	Medium	1	1	2	1	1	1	2	1	3	4	1	Low	Low	Low	Generally Cr, Ni and Cu for resurfacing.
Powder spraying	Fairly low	Easy	Low	2	2	4	1	1	1	2	2	4	1	1	Very low	Low	None	Limited, other than self-fluxing alloys.
Wire spraying	Fairly low	Easy	Low	1	2	4	1	1	1	2	2	4	1	1	Very low	Low	None	Considerable range including bond coats. Some ceramic rods.
Arc spraying	High	Easy	Fairly low	1	3	2	2	1	1	1	2	4	1	1	Very low	Fairly low	None	Limited range of wires.
HVOF spraying	Fairly high	Moderately easy	Medium	2	2	2	1	1	1	2	2	3	2	1	Very low	Fairly low	None	Considerable range of metallic powders. Some equipments spray plastics.
Plasma spraying	High	Fairly hard	High	3	1	3	1	2	1	1	1	3	4	1	Low	Fairly low	None	Many alloys and ceramics, usually powder
Sprayed and fused coatings	Fairly low	Fairly easy	Medium	2	2	3	1	3	1	3	2	2	3	1	Low	Fairly low	Fairly high, uniform	Fairly high, Needs self-fluxing alloys.

1, Very applicable; 2, Not very applicable; 3, Not really suitable; 4, Unsuitable.

Table 20.6 Available coating materials

Group	Sub-group	Alloy system	Important properties	GW	PW	MMA	TIG	MIG	FCA	SA	PTA	EP	MP	MW	MA	HVOF	PF	SW	Typical applications
STEEL (up to 1.7% C)	Pearlitic	Low carbon	Crack resistant, low cost, good base for hard-surfacing	1		X								1	1				Build up to restore dimensions. Track links, rollers, idlers
	Martensitic	Low, medium or high carbon. Up to 9% alloying elements	Abrasion resistance increases with carbon content, resistance to impact decreases. Economical	1		X		X	X	1				1	X				Bulldozer blades, excavator teeth, bucket lips, impellers, conveyor screws, tractor sprockets, steel mill wobblers, etc
	High speed	Complex alloy	Heat treatable to high hardness	X		1													Working and conveying equipment
	Semi-austenitic	Manganese chromium	Tough crack resistant. Air and work hardenable	X		1			1										Mining equipment, especially softer rocks
	Austenitic	Manganese	Work hardening	X		X	1	1	1										Rock crushing equipment
		Alloyed manganese	Work hardening, less susceptible to thermal embrittlement. Useful build-up	X		X	1	1	1										Build up normal manganese steel prior to application of other hard-surfacing alloys
		Chromium nickel	Stainless, tough, high temperature and corrosion resistant (low carbon)	1		1	X	X	X	X				1	X				Furnace parts, chemical plant
IRON (above 1.7% C)	High chromium	Martensitic	Show improved hot hardness and increased abrasion resistance	1		X	1	1	1	1									Steelworks equipment, scraper blades, bucket teeth
		Multiple alloy	Hardenable. Can anneal for machining and re-harden. Good hot hardness	1		X	1	1	1	1									Mining equipment, dredger parts
		Austenitic	Wide plastic range, can be hot shaped, brittle. Oxidation resistant	1		X	1	1	1	1									Low stress abrasion and metal-to-metal wear. Agricultural equipment
	Martensitic alloy	Chromium tungsten / Chromium molybdenum / Nickel chromium	Very good abrasion resistance, very high compressive strength so can resist light impact. Can be heat-treated. Considerable variation in properties between gas and arcweld deposits	X		X	1	1	1	1									Cutting tools, shear blades, rolls for cold rolling
	Austenitic alloy	Chromium molybdenum / Nickel chromium	Lower compressive strength and abrasion resistance. Less susceptible to cracking. Will work harden	X		X	1	1	1	1	1								Mixing and steelworks equipment, agricultural implements

Table 20.6 (continued)

Group	Sub-group	Alloy system	Important properties	GW	PW	MMA	TIG	MIG	FCA	SA	PTA	EP	MP	MW	MA	HVOF	PF	SW	Typical applications
CARBIDE	Iron base	Tungsten carbides in steel matrix	Resistant to severe abrasive wear. Care required in selection and application	X		X	1	1	1	1									Rock drill bits. Earth handling and digging equipment. Extruder screw augers
	Cobalt base	Tungsten carbides in cobalt alloy matrix	Matrix gives improved high temperature properties and corrosion resistance	X	X	1	1	1	1		1							X	Oil refinery components, etc
	Nickel base	Tungsten carbides in nickel alloy matrix	Matrix gives improved corrosion resistance		X						X							X	Screws, pump sleeves etc. in corrosive environments
	Copper base	Tungsten carbides in copper alloy matrix	Often larger carbide particles to give cutting and sizing properties	X															Oil field equipment
NICKEL BASE	Nickel		High corrosion resistance	1		1	1					X	1	1					Chemical plant. Bond costs for ceramics
	Nickel Chromium Boron	With Fe, Si and C; W or Mo may be added	Self-fluxing alloys available in wide range of hardnesses. Abrasion, corrosion, oxidation resistant. Can be applied as thin, dense, impervious layers. Metallurgical bond to substrate	X	X	1	1				X	X	1	1		1	1	X	Glass mould equipment, engineering components, chemical and petrochemical industries
	Nickel Chromium	With possibly C, Mo or W to improve hardness and hot strength. Fe modifies thermal expansion improves creep resistance	Relatively soft and ductile. Good hot gas corrosion resistance. Very good corrosion resistance	X		X	1		1		X								High temperature engineering applications. Chemical industry. I.C. engine valves
	Nickel Iron molybdenum	60% Ni, 20% Mo, 20% Fe or 65% Ni, 30% Mo, 5% Fe	Resistant to HCl, also sulphuric, formic and acetic acids	1		1	1	X	X		1								Chemical plant
	Nickel Copper	Monel	Corrosion resistant	1		1	1	1	X					1	1				Chemical plant
COBALT	Cobalt Chromium Tungsten	About 30% Cr, increasing W and C increases hardness	Superlative high temperature properties. Wear, oxidation and corrosion resistant	X	X	X	X	X	1		X						1		Oilwell equipment, steelworks, chemical engineering plant, textile machinery
	Self-fluxing	With B, Si, Ni	Modified to provide self-fluxing properties. Good abrasion, corrosion resistance	1	X	1	1				1	X	X	1		1	1	X	Chemical and petrochemical industries. Extrusion screws

Table 20.6 (continued)

Group	Sub-group	Alloy system	Important properties	GW	PW	MMA	TIG	MIG	FCA	SA	PTA	EP	MP	MW	MA	HVOF	PF	SW	Typical applications
COPPER BASE	Copper		Electrical conductivity	1		1						X		X	1				Electrical equipment, paper-working machinery
	Bronze	Aluminium manganese, tobin, phosphor, commercial	Resistance to frictional wear and some chemical corrosion. Al-bronze excellent bond coat arc metallising	1		1								X	X				Bearing shells, shafts, slides, valves, propellers, etc
	Brass		Bearing properties, decorative finishes	1										X					Water tight seals. Electric discharge machining electrodes
CHROMIUM			Wear, corrosion resistant									X							Engineering components
MOLYBDENUM			High adhesion to base metal											X	1				Bond coats. Engineering parts reclamation
ALUMINIUM			Corrosion and heat resistance										1	1	X				Steel structures. Furnace parts
ZINC			Corrosion resistance											X	1				Steel structures, gas cylinders, tanks, etc. HF shielding
ZINC/ALUMINIUM ALLOYS			Corrosion resistance											X	X				Steel structures
LEAD BASE	Lead		Resistance to chemical attack	1								1		X	1				Resistance to sulphuric acid. Radiation shielding
	Solder	Often 60/40% Pb/Sn	Joining	1										X					Tinning surfaces for subsequent joining
TIN BASE	Tin		High corrosion resistance	1								X		X	1				Electrical contacts. Food industry plant
	Babbitt	Sn, Sb, Cu	Bearing alloys	1										X	1				Bearing shells
OXIDES		Principally of Al, Zn, Cr, Ti and mixtures	High temperature oxidation resistance. Wear resistance	1										X	1	X			Pump sleeves, aerospace parts
REFRACTORY METALS		W, Ti, Ta, Cr	Good, high temperature properties. Often develop stable, protective oxide films										1	1	1	X			Electrical contacts. High density areas, corrosion protection
CARBIDES/BORIDES		Cr. W. B. possibly + Co or Ni	Very high wear resistance	1		X	1							1	X	X			Thin cutting edges. Wear resistant areas
COMPLEX CERAMICS		Silicides, titanates, zirconates, etc	Wear, oxidation, and erosion resistance											1	X	X			Thermal barriers, coating equipment handling molten metal and glass

X frequently used, 1 sometimes used.

Table 20.7 Factors affecting choice of coating material

Factor	Effect
PROPERTIES REQUIRED OF DEPOSIT Function of surface. Nature of adjacent surface or rubbing materials. What is the service temperature? What is the working environment (corrosive, oxidising, abrasive, etc.) What coefficient of friction required?	Is porosity desirable or to be avoided? Sprayed coatings are porous; fused or welded coatings are impervious. Porosity can be advantageous if lubrication required. What compressive stresses are involved? A porous deposit can be deformed and detached from the substrate. What degree of adhesion to the substrate is necessary? In many applications corrosion protection is satisfactory with mechanical adhesion – in some cases a metallurgical, bond will be required. How important is macro-hardness, metallographic structure – this often affects abrasion resistance. Is there danger of electro-chemical corrosion? If abrasion resistance is required, is it high or low stress? Many different alloys are available and needed to meet the many wear conditions which may be encountered. Are similar applications known?
PROPERTIES NEEDED IN ALLOY Physical Chemical Metallurgical Mechanical	Thermal expansion, or contraction, compared with the base metal, affects distortion. Chemical composition relates to corrosion, erosion and oxidisation resistance. The metallurgical structure is very important, influencing properties such as abrasion resistance and frictional properties. Mechanical properties can be critical. It is necessary to assess the relative importance of properties needed such as resistance to abrasion, corrosion, oxidation, erosion, seizure, impact and to what extent machinability, ductility, thermal conductivity or resistance, electrical conductivity or resistance may be required.
PROCESS CONSIDERATIONS Surfacing processes available. Surface preparation methods available and feasible; auxiliary services which can be used. Location and size of work. Thickness of deposit required.	Some consumables can be produced only as wires – or powders – or cast rods. Choice could be limited by the equipment in use and operator experience. Lack of specific surface preparation methods could limit choice of process and this, in turn, prohibits use of certain materials. Similar limitations could arise from lack of necessary pre-heating or post-heat treatment plant. Can resurfacing be carried out without dismantling the assembly? Can work be carried out on site? Are only small areas on a large part to be repaired?
ECONOMICS How much cost will the repair bear?	Often much more important than a direct cost comparison – cost of the repair compared to the purchase of a new part – are many other features, e.g. saving in subsequent maintenance labour and material costs, value of the lost production which is avoided, lower scrap rate during subsequent processing, time saved in associated production units, improved quality of component and product, increased production rates, reduction in consumption of raw materials.

Table 20.8 Methods of machining electroplated coatings

Deposited metal	Machining with cutting tools	Grinding
Chromium	Chromium is too hard	Grinding or related procedures are the only suitable process. Soft or medium wheels should be used at the highest speed consistent with the limits of safety. Coolant must be continuous and copious. Light cuts only—preferably not exceeding 0.0075 mm should be taken. Heavy cuts can cause cracking or splintering of the deposit
Nickel	High-speed steel is, in general, the most satisfactory material for cutting nickel. Tipped tools are not recommended. The shape of the tools should be similar to those used for steel but with somewhat increased rake and clearance. Nickel easily work hardens therefore tools must be kept sharp and well supported to ensure that the cut is continuous	To avoid glazing, use an open textured wheel with a peripheral speed of 25–32 m/s. Coolant as for grinding steel

Table 20.9 Bearing materials compatible with electroplated coatings

Material	Excellent	Good	Avoid
Chromium	Copper–lead, lead–bronze, white metal, Fine grain cast iron	Rubber or plastic, water-lubricated. Soft or medium hard steel with good lubrication and low speed. Brass, gun metal	Hard steel, phosphor bronze,* light alloys*
Nickel		Bronze, brass, gun metal, white metal	Ferrous metals, phosphor bronze

*May be satisfactory in some conditions.

When using deposited metals in sliding or rotating contact with other metals, adequate lubrication must be assured at all times.

Table 20.10 *Examples of successful repairs*

Application	Coating material	Process used
Cast iron glass container moulds	Ni base self fluxing alloy in rod form Ni base alloy	Gas welding Powder welding
Steel plungers on molten glass pumps	Ni base alloy	Metal spraying
Excavator teeth	Bulk welding all over with chromium carbide alloy	Bulk welding
Excavator tooth tips	Hard alloy	Flux cored welding
Metering pump pistons	Aluminium oxide titanium oxide composite	Plasma flame spraying
Fire pump shafts	Stainless steel	Plasma wire spraying
Printing press rolls	Stainless steel on a nickel alumnide bond coat	Arc spraying
Large hydraulic press ram	Martensitic steel on a bond coat and sealed with a vinyl sealer	Wire spraying
Shafts of pumps handling slurries	Ni and Co base alloys with tungsten carbide	Plasma spraying
Screw conveyors	Ni base alloy rods	Gas welding
Small valves less than 12 cm diameter	Co–Cr–Ni–W alloys	Gas welding
Medium size valves	Co–Cr–Ni–W alloys	Powder welding
Very large valves	Co–Cr–Ni–W alloys	Submerged arc welding
Paper mill drums	Stainless steel wire	Arc spraying
Ammonia compressor pistons	Whitemetal	Wire spraying
Hydraulic pump plungers	Ni base alloy	Sprayed and fused
Guides on steel mills	Ni base alloy	Gas welding

ABRASIVE WEAR

Abrasive wear is the loss of material from a surface that results from the motion of a hard material across this surface.

There are several types of abrasive wear. Since the properties required of a wear-resistant material will depend on the type of wear the material has to withstand, a brief mention of these types of wear may be useful.

There are three main types of wear generally considered: gouging abrasion (impact), Figure 21.1; high-stress abrasion (crushing), Figure 21.2; and low-stress abrasion (sliding), Figure 21.3. This classification is made more on the basis of operating stresses than on the actual abrading action.

Gouging abrasion

This is wear that occurs when coarse material tears off sizeable particles from wearing surfaces. This normally involves high imposed stresses and is most often encountered when handling large lumps.

High-stress abrasion

This is encountered when two working surfaces rub together to crush granular abrasive materials. Gross loads may be low, while localised stresses are high. Moderate metal toughness is required; medium abrasion resistance is attainable.

Rubber now competes with metals as rod and ball mill linings with some success. Main advantages claimed are longer life at a given cost, with no reduction in throughput, lower noise level, reduced driving power consumption, less load on mill bearings and more uniform wear on rods.

Low-stress abrasion

This occurs mainly where an abrasive material slides freely over a surface, such as in chutes, bunkers, hoppers, skip cars, or in erosive conditions. Toughness requirements are low, and the attainable abrasion resistance is high.

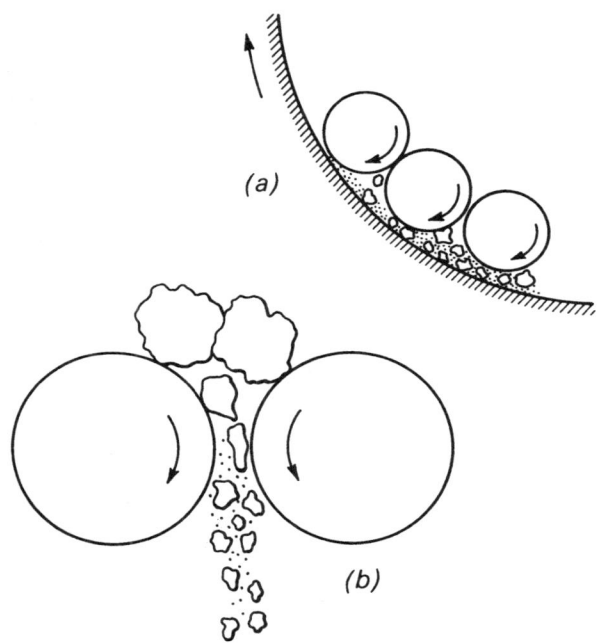

Figure 21.2 Types of high-stress abrasion: (a) rod and ball mills; (b) roll crushing

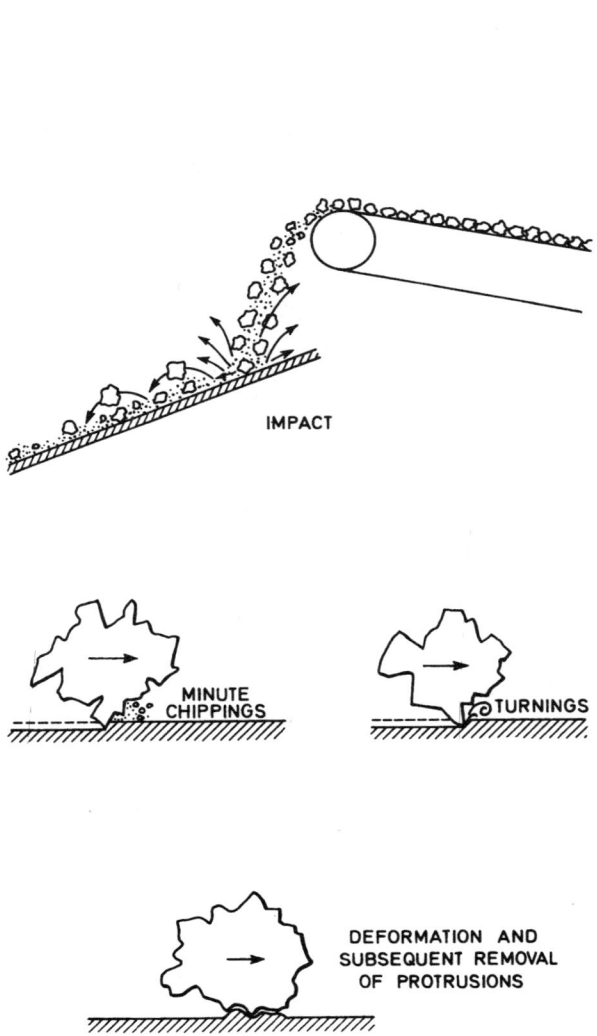

Figure 21.1 Types of gouging abrasion

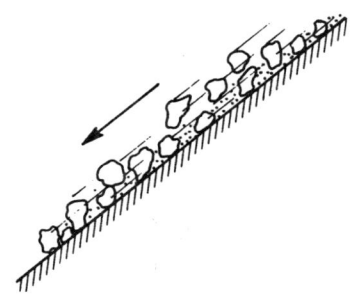

Figure 21.3 Low-stress abrasion

MATERIAL SELECTION

Very generally speaking the property required of a wear-resistant material is the right combination of hardness and toughness. Since these are often conflicting requirements, the selection of the best material will always be a compromise. Apart from the two properties mentioned above, there are few *general* properties. Usually the right material for a given wear-resistant application can only be selected after taking into consideration other factors that determine the rate of wear. Of these the most important are:

Ambient temperature, or temperature of material in contact with the wear surface.

Size distribution of particles flowing over the wear surface.

Abrasiveness of these particles.

Type of wear to which wear surface is subjected (i.e. gouging, sliding, impact, etc.).

Velocity of flow of material in contact with wear surface.

Moisture content or level of corrosive conditions.

General conditions (e.g. design of equipment, headroom available, accessibility, acceptable periods of non-availability of equipment).

Tables 21.1 and 21.2 give some general guidance on material selection and methods of attaching replaceable components.

Table 21.3 gives examples of actual wear rates of various materials when handling abrasive materials.

The subsequent tables give more detailed information on the various wear resistant materials.

Table 21.1 Suggested materials for various operating conditions

Operating conditions	Properties required	Material
High stress, impact	Great toughness; work-hardening properties	Austenitic manganese steel, rubber of adequate thickness
Low stress, sliding	1, Great hardness; 2, toughness less important; 3, quick replacement	Hardened and/or heat-treated metals, hardfacing, ceramics
	1, Cheapness of basic material; 2, replacing time less important	Ceramics, quarry tiles, concretes
	1, Maximum wear resistance; cost is immaterial	Tungsten carbide
Gouging wear	High toughness	Usually metals, i.e. irons and steels, hardfacing
Wet and corrosive conditions	Corrosion resistance	Stainless metals, ceramics, rubbers, plastics
Low stress; contact of fine particles; low abrasiveness	Low coefficient of friction	Polyurethane, PTFE, smooth metal surfaces
High temperature	Resistance to cracking, spalling, thermal shocks; general resistance to elevated temperatures	Chromium-containing alloys of iron and steel; some ceramics
Minimum periods of shut-down of plant	Ease of replacement	Any material that can be bolted in position and/or does not require curing
Curved, non-uniform irregular surface and shapes	Any one or a combination of the above properties	Hardfacing weld metal; most trowellable materials
Arduous and hot conditions		Hardfacing weld metal

Table 21.2 Methods of attachment of replaceable wear-resistant components

Method of fixing		Suitable for:
1	Bolting, nuts, or nuts and bolts	Metals, ceramics, rubbers, plastics
2	Sticking, adhesives or cement mortar	Ceramics, concretes, plastics, rubbers
3	Filled fabricated metal trays, provided with studs, then fixed as 1 above	Concretes, pastes, poured plastics
4	Cast-in bolts or studs	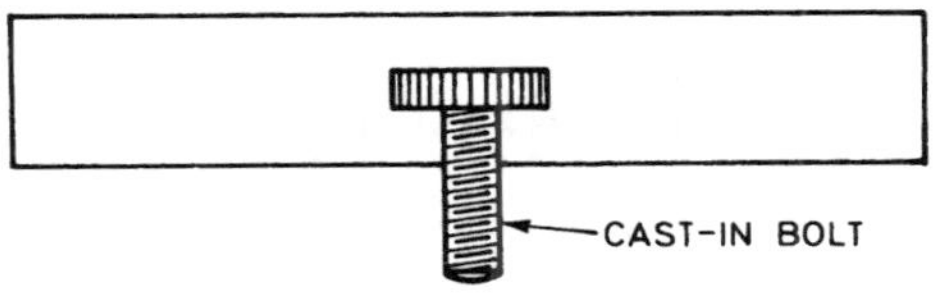 Cast irons, ceramics, concretes
5	Fabricated panels	Ceramics
6	'T' bars	Rubbers, plastics
7	Welded studs	Metal plates previously plasticoated or coated with weld or spray metal
8	Tack-welding	Mainly for steel or steel-based components

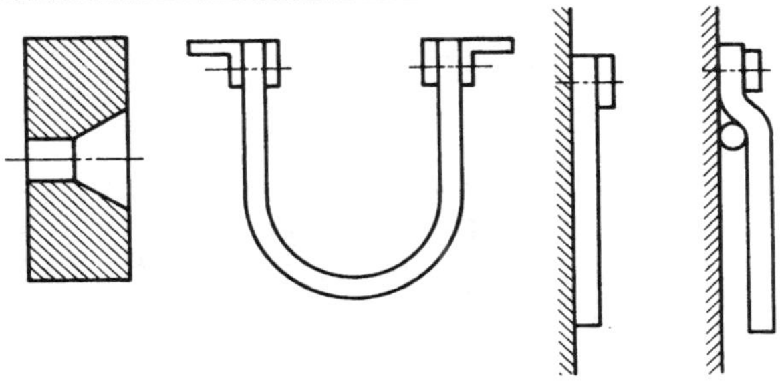

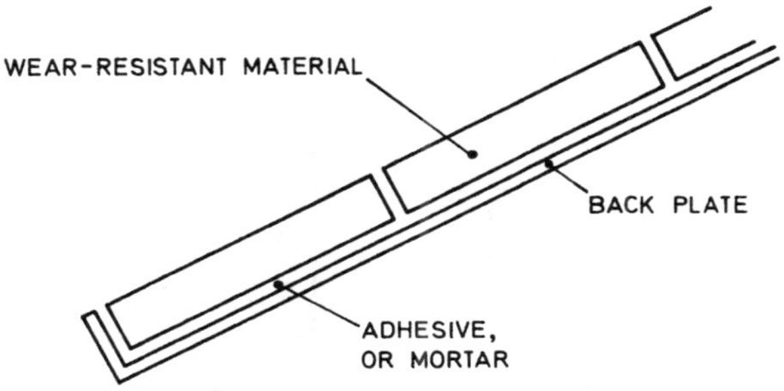

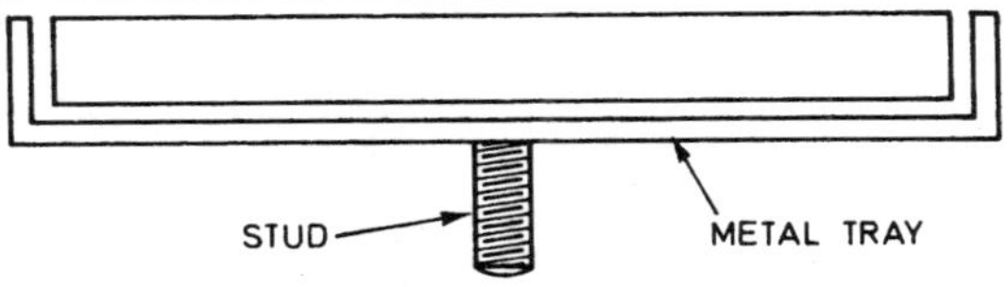

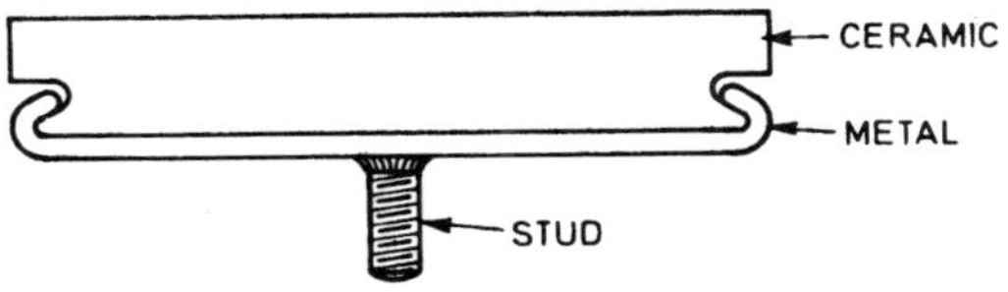

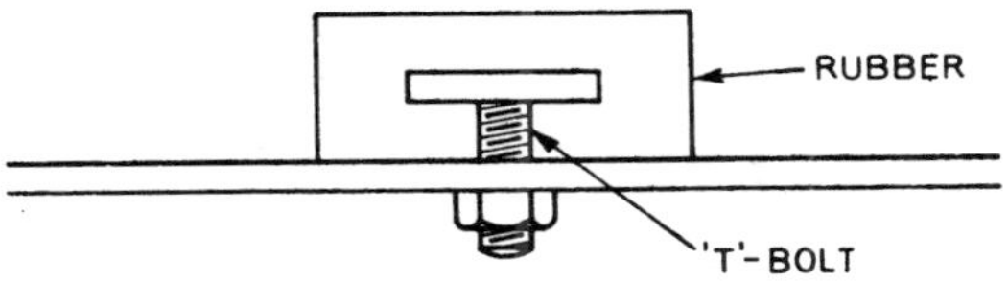

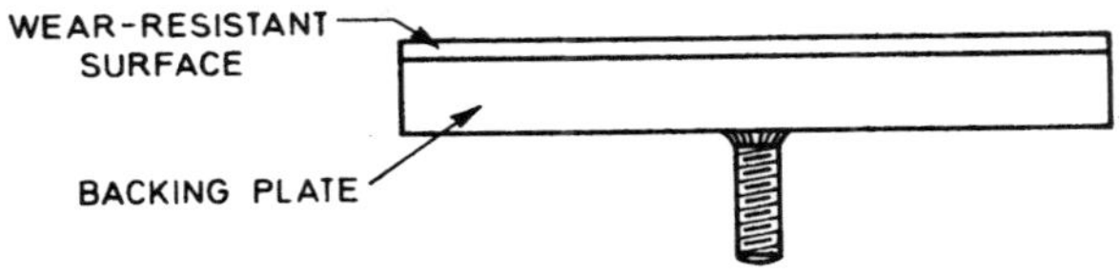

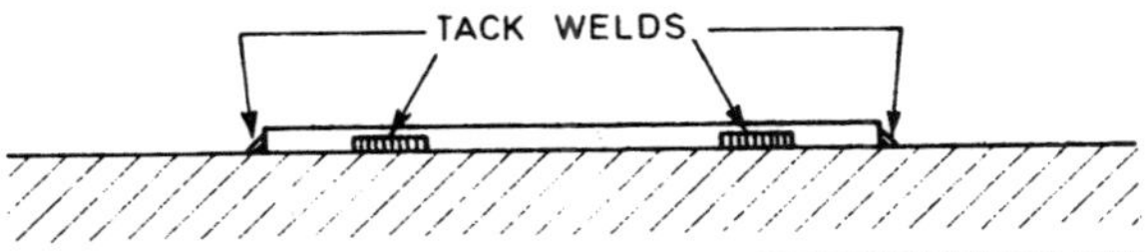

WEAR-RESISTANT MATERIAL

BACK PLATE

ADHESIVE, OR MORTAR

STUD

METAL TRAY

CAST-IN BOLT

CERAMIC

METAL

STUD

RUBBER

'T'-BOLT

WEAR-RESISTANT SURFACE

BACKING PLATE

TACK WELDS

Table 21.3 Typical performance of some wear-resistant materials as a guide to selection

Type	Some typical materials	Sliding wear-rate* by coke	by sinter	Temperature limitations	Ease and convenience of replacement	General comments
Cast irons	Ni-hard type martensitic white iron	0.11	0.06		Yes	The most versatile of the materials which, now, by varying alloying elements, method of manufacture and application are able to give a wide range of properties. Their main advantage is the obtainable combination of strength, i.e. toughness and hardness, which accommodates a certain amount of abuse. Other products are sintered metal and metal coatings
	High chrome martensitic white irons	0.12	0.11			
	Spheroidal graphite-based cast iron	0.22	0.09			
	High phosphorus pig iron	0.32	0.91			
	Low alloy cast iron	–	0.44			
Cast steels	3¼ Cr–Mo cast steel	0.17	–			
	13 Mn austenitic cast steel	0.22	–			
	1½ Cr–Mo cast steel	0.43	–			
Rolled steels	Armour plate	0.12	–			
	Work-hardened Mn steel	0.13	–			
	Low alloy steel plate, quenched and tempered	0.31	0.30–0.84			
	EN8 steel	0.43	0.63			
Hard facings	High chrome hardfacing welds, various	0.09–0.16	0.05–0.14	No	Could be difficult if applied *in situ*	
Ceramics	Fusion-cast alumina-zirconia-silica	0.05	0.11–0.14		Yes, if bolted. Not so convenient if fixed by adhesive or cement mortar, as long curing times may be unacceptable	Great range of hardness. Most suitable for low-stress abrasion by low-density materials, and powders. Disadvantage: brittleness
	Slageram	0.15	0.33			
	Fusion-cast basalt	0.17	0.53			
	Acid-resisting ceramic tile	0.19	1.27			
	Plate glass	0.81	–			
	Quarry floor tiles	2.2–3.4	–			
Concretes	Aluminous cement concrete.	0.42	4.0–4.4		Could be messy. Might be difficult under dirty conditions	Advantages: cheapness, castability. Disadvantage: long curing or drying-out times
	Quartz–granite aggregate-based concrete	0.87	6.5			
Rubbers	Rubbers, various	2.1–3.2	–	Yes	Bonded and bolted. Stuck with adhesive, could be difficult under dirty conditions	Main advantage is resilience and low density, with a corresponding loss in bulk hardness. The most useful materials where full advantage at the design stage can be taken of their resilience and anti-sticking properties
Rubber-like plastics	Polyurethanes, various	2.3–5.4	2.3			
Other plastics	High-density polyethylene	6.4	–		In sheet form it is difficult to stick	Low coefficient of friction, good antisticking properties. Best for low-stress abrasion by fine particles
	Polytetrafluoroethylene (PTFE)	8.2	–			
Resin-bonded compounds	Resin-bonded calcined bauxite	2.3	–		Trowelled; could be messy. Difficult in dirty and inaccessible situations	These materials are only as strong as their bonding matrix and therefore find more application where low-stress wear by powders or small particles (grain, rice) takes place

*Wear rate is expressed in in³ of material worn away per 1000 tons of the given bulk materials per ft² of area in contact with the abrading material. The results were obtained from field trials in a chute feeding a conveyor belt.

This data is provided as examples of the relative wear rates of the various materials when handling abrasive bulk materials.

The following tables give more detailed information on the materials listed in Table 21.3 with examples of some typical applications in which they have been used successfully.

When selecting the materials for other applications, it is important to identify the wear mechanism involved as this is a major factor in the choice of an optimum material. Further guidance on this is given in Table 21.1.

Table 21.4 Cast irons

Type	Nominal composition	Hardness Brinell	Characteristics	Typical application
Grey irons BS 1452 ASTM A48	Various	150–300	Graphite gives lubrication	Brake blocks and drums, pumps
Spheroidal graphite	Meehanite WSH2	Up to 650	Heat-treatable. Can be lined with glass, rubber, enamels	Many engineering parts, crusher cones, gears, wear plates
High phosphorus	3.5%C 2.0%P	Up to 650	Brittle, can be reinforced with steel mesh.	Sliding wear
Low alloy cast iron	3%C 2%Cr 1%Ni	250–700		Sliding wear, grate bars, cement handling plant, heat-treatment
NiCr Martenstic irons Typical examples: NiHard, BF 954	2.8–3.5%C, 1.5–10% Cr 3–6% Ni	470–650		High abrasion Ore handling, sand and gravel
Ni Hard 4	7–9% Cr, 5–6.5% Ni, 4–7% Mn			More toughness. Resists fracture and corrosion
CrMoNi Martensitic irons Typical examples: Paraboloy	14–22% Cr, 1.5% Ni, 3.0% Mo	500–850		Ball and rod mills, wear plates for fans, chutes, etc.
High Chromium irons Typical examples: BF 253 HC 250	22–28% Cr	425–800	Cast as austenite Heat-treated to martensite	Crushing and grinding Plant Ball and Rod mills Shot blast equipment, pumps

Table 21.5 Cast steels

Type	Nominal composition	Hardness	Characteristics	Typical applications
Carbon steel BS 3100 Grade A		Up to 250		Use as backing for coatings
Low alloy steels BS 3100 Grade B	Additions of Ni, Cr, Mo up to 5%	370–550		For engineering 'lubricated' wear conditions
Austenitic BS 3100 BW10	11% Mn min.	200 soft Up to 600 when work-hardened		For heavy impact wear, Jaw and Cone crushers, Hammer mills
High alloy steels BS 3100 Grade C	30% Cr 65% Ni + Mo, Nb etc.	500		Special alloys for wear at high temperature and corrosive media
Tool steels Many individual specifications	17% Cr, 4% Ni, 9% Mo, 22% W, 10% Co	Up to 1000		Very special applications, usually as brazed-on plates.

Table 21.6 Rolled steels

Type	Nominal composition	Hardness	Characteristics	Typical application
Carbon steels BS 1449 Part 1 Typical examples: BS 1449 Grade 40 (En8 plate) Abrazo 60	.06/1.0% C, 1.7% Mn	160–260	Higher carbon for low/medium wear	For use as backing for hard coatings
Low alloy steels Many commercial specifications Typical examples: ARQ Grades, Tenbor 25 30, Wp 300 and 500, Creusabro Grades, Abro 321 and 500, OXAR 320 and 450, Red Diamond 20 & 21, Compass B555	Up to 3.5% Cr, 4% Ni, 1% Mo	250–500	Quenched and tempered. Are weldable with care	Use for hopper liners, chutes, etc.
Austenitic manganese steel Typical examples: Cyclops 11/14 Mn, Red Diamond 14	11/16% Mn	200 in soft condition 600 skin hardened by rolling		
High alloy and stainless steel BS 1449 Part 2	Up to 10% Mn, 26% Cr, 22% Ni, + Nb, Ti	Up to 600	Heat and corrosion resistant	Stainless steels

Table 21.7 Wear resistant coatings for steel

	Method	Technique	Materials	Characteristics and applications
Weld applied surfaces	Gas welding	Manual	Rods of wide composition. Mainly alloys of Ni, Cr, Co, W etc.	For severe wear, on small areas. Thickness up to 3 mm
	Arc welding	Manual	Coated tubular electrodes. Specifications as above	Wear, corrosion and impact resistant. Up to 6 mm thickness.
		Semi- or full automatic	Solid or flux-cored wire, or by bulk-weld Tapco process. Wide range of materials	As above, but use for high production heavy overlays 10–15 mm.
	Fused paste	Paste spread onto surface, then fused with oxy-fuel flame or carbon-arc.	Chromium boride in paste mix	Excellent wear resistance. Thin (1 mm) coat. Useful for thin fabrications: fans, chutes, pump impellors, screw conveyors, agricultural implements
Flame spray processes	Oxy-fuel	Consumable in form of powder, wire, cord or rod, fed through oxy-fuel gun. Deposit may be 'as-sprayed' or afterwards fused to give greater adhesion	Materials very varied, formulated for service duty	Wear, corrosion heat, galling, and impact resistance. Thicknesses vary depending on material. Best on cylindrical parts or plates
	Arc spray	Wire consumable fed through electric arc with air jet to propel molten metal	Only those which can be drawn into wire	High deposition rate, avoid dew-point problems, therefore suited to larger components
Plasma spray processes	Non-transferred	Similar to flame spray, but plasma generated by arc discharge in gun	Materials as for flame-spray, but refractory metals, ceramics and cermets in addition, due to very high temperatures developed	High density coatings. Very wide choice of materials. Application for high temperature resistance and chemical inertness
	Transferred	As above but part of plasma passes through the deposit causing fusion	Mainly metal alloys	High adhesion Low dilution Extremely good for valve seats
	Detonation gun	Patent process of Union Carbide Corp. Powder in special gun, propelled explosively at work	Mainly hard carbides and oxides	Very high density. Requires special facilities
	High velocity oxy-fuel	Development of flame gun, gives deposit of comparable quality to plasma spray	Similar to plasma spray	More economic than plasma
Others	Hard chromium plating	Electrode position	Hard chrome Up to 950 HV (70 Rc)	Wear, corrosion and sticking resistant
	Electroless nickel	Chemical immersion	Nickel phosphide 850 HV after heat treatment	Similar to hard chrome
	Putty or paint	Applied with spatula or brush	Epoxy or polyester resins, or self-curing plastics, filled with wear resistant materials	Wear and chemical resistant

Table 21.8 Some typical wear resistant hardfacing rods and electrodes

Material type	Name	Typical application
Low alloy steels	Vodex 6013, Fortrex 7018, Saffire Range. Tenosoudo 50, Tenosoudo 75, Eutectic 2010	Build-up, and alternate layering in laminated surfaces
Low alloy steels	Brinal Dymal range. Deloro Multipass range. EASB Chromtrode and Hardmat. Metrode Met-Hard 250, 350, 450. Eutectic N6200, N6256. Murex Hardex 350, 450, 650, Bostrand S3Mo. Filarc 350, Filarc PZ6152/PZ6352. Suodometal Soudokay 242-0, 252-0, 258-0, Tenosoudo 105, Soudodur 400/600, Abrasoduril. Welding Alloys WAF50 range Welding Rods Hardrod 250, 350, 650	Punches, dies, gear teeth, railway points
Martensitic chromium steels	Brinal Chromal 3, ESAB Wearod, Metrode Met-Hard 650. Murex Muraloy S13Cr. Filarc PZ6162. Oelikon Citochrom 11/13. Soudometal Soudokay 420, Welding Rods Serno 420FM. Welding Alloys WAF420	Metal to metal wear at up to 600C. High C types for shear blades, hot work dies and punches, etc.
High speed steels	Brinal Dyma H. ESAB OK Harmet HS. Metrode Methard 750TS. Murex-Hardex 800, Oerlikon Fontargen 715. Soudometal Duroterm 8, 12, 20, Soudostel 1, 12, 21. Soudodur MR	Hot work dies, punches, shear blades, ingot tongs
Austenitic stainless steels	Murex Nicrex E316, Hardex MnP, Duroid 11, Bostrand 309. Metrode Met-Max 20.9.3, Met-Max 307, Met-Max 29.9 Soudometal Soudocrom D	Ductile buttering layer for High Mn steels on to carbon steel base. Furnace parts, chemical plant
Austenitic manganese steels	Brinal Mangal 2. Murex Hardex MnNi Metrode Workhard 13 Mn, Workhard 17 MnMo, Workhard 12MnCrMo. Soudometal Soudomanganese, Filarc PZ6358	Hammer and cone crushers, railway points and crossings
Austenitic chromium manganese steels	Metrode Workhard 11Cr9Mn, Workhard 14Cr14Mn. Soudometal Comet MC, Comet 624S	As above but can be deposited on to carbon steels. More abrasion resistant than Mn steels
Austenitic irons	Soudometal Abrasodur 44. Deloro Stellite Delcrome 11	Buttering layer on chrome irons, crushing equipment, pump casings and impellors
Martensitic irons	Murex Hardex 800. Soudometal Abrasodur 16. Eutectic Eutectdur N700	For adhesive wear, forming tools, scrapers, cutting tools
High chromium austenitic irons	Murex Cobalarc 1A, Soudometal Abrasodur 35, 38. Oerlikon Hardfacing 100, Wear Resistance WRC. Deloro Stellite Delcrome 91	Shovel teeth, screen plate, grizzly bars, bucket tips
High chromium martensitic irons	Metrode Met-hard 850, Deloro Stellite Delcrome 90	Ball mill liners, scrapers, screens, impellors
High complex irons	Brinal Niobal. Metrode Met-hard 950, Met-Hard 1050. Soudometal Abrasodur 40, 43, 45, 46	Hot wear applications, sinter breakers and screens
Nickel alloys	Metrode 14.75Nb, Soudonel BS, Incoloy 600. Metrode 14.75MnNb, Soudonel C, Incoloy 800. Metrode HAS C, Comet 95, 97, Hastelloy types	Valve seats, pump shafts, chemical plant
Cobalt alloys	EutecTrode 90, EutecRod 91	Involving hot hardness requirement: Valve seats, hot shear blades
Copper alloys	Saffire Al Bronze 90/10, Citobronze, Soudobronze	Bearings, slideways, shafts, propellers
Tungsten carbide	Cobalarc 4, Diadur range	Extreme abrasion: fan impellors, scrapers

Table 21.9 Wear resistant non-metallic materials

	Type	Nominal composition	Hardness	Characteristics	Typical application
Ceramic materials	**Extruded ceramic** Indusco Vesuvius	High density Alumina	9 Moh	Process limits size to $100 \times 300 \times 50$ mm. Low stress wear, also at high temperature	
	Ceramic plates Hexagon-shaped, cast Indusco				Suitable for lining curved surfaces
	Sintered Alumina Alumina 1542	$96\%Al_2O_3$ $2,4\%SiO_2$	9 Moh	Low stress wear also at high temperatures	
	Isoden 90	$90\%Al_2O_3$			
	Isoden 95	$95\%Al_2O_3$			
	Fusion-cast Alumina Zac 1681	$50\%Al_2O_3$ $32.5\%ZrO_2$ $16\%SiO_2$		Can be produced in thick blocks to any shape. Low stress and medium impact, also at high temperatures	
	Concrete Alag Ciment Fondu	Mainly calcium silico-aluminates $40\%Al_2O_3$		Low cost wear-resistant material. High heat and chemical resistance	Floors, coke wharves, slurry conveyors, chutes
	Cast Basalt Heat-treated	Remelted natural basalt	7–8 Moh	Low stress abrasion. Brittle	Floors, coke chutes, bunker, pipe linings, usually 50 mm thick minimum. Therefore needs strong support
	Plate Glass		Glass	Very brittle	Best suited for fine powders, grain, rice etc.
Rubber and plastics	**Rubber** Trellex Skega Linatex	Various grades 95% Natural rubber	Various Fairly soft	Resilient, flexible	Particularly suitable for round particles, water borne flow of materials
	Ceramic ballsheet Hoverdale	Rubber filled with ceramic balls		Enhanced wear resistance	
	Plastics Duthane Flexane Tivarthane (Polyhi-Solidor) Scandurathane (Scandura) Supron (Slater)	Polyurethane based, rubber-like materials		Low stress abrasion applications	Floors, chute liners screens for fine materials
	Duplex PTFE	Polytetra-fluorethylene		Low coefficient of friction	For fine powders light, small particles
	Resins Belzona Devcon Greenbank AD1 Thortex Systems Nordbak	Resin-based materials with various wear-resistant aggregates		Can be trowelled. Specially suitable for curved and awkward surfaces but not for lumpy materials	Floors, walls, chutes, vessels. In-situ repairs

In general, the repair of bearings by relining is confined to the low melting-point whitemetals, as the high pouring temperatures necessary with the copper or aluminium based alloys may cause damage or distortion of the bearing housing or insert liner. However, certain specialist bearing manufacturers claim that relining with high melting-point copper base alloys, such as lead bronze, is practicable, and these claims merit investigation in appropriate cases.

For the relining and repair of whitemetal-lined bearings three methods are available:

(1) Static or hand pouring.
(2) Centrifugal lining.
(3) Local repair by patching or spraying.

Table 22.1 Guidance on choice of lining method

Type of bearing	Relining method	Field of application
Direct lined housings	Static pouring or centrifugal lining	Massive housings. To achieve dynamic balance during rotation, parts of irregular shape are often 'paired' for the lining operation, e.g. two cap half marine type big-end bearings lined together, ditto the rod halves
INSERT LINERS *'Solid inserts'*	Not applicable	New machined castings or pressings required
Lined inserts Thick walled Medium walled	Static pouring or centrifugal lining	Method adopted depends on size and thickness of liner, and upon quantities required and facilities available
Thin walled	Not recommended	Relining not recommended owing to risk of distortion and loss of peripheral length of backing. If relining essential (e.g. shortage of supplies) special lining jigs and protective measures essential

1 PREPARATION FOR RELINING

(a) Degrease surface with trichlorethylene or similar solvent degreaser. If size permits, degrease in solvent tank, otherwise swab contaminated surfaces thoroughly.
(b) Melt off old whitemetal with blowpipe, or by immersion in melting-off pot containing old whitemetal from previous bearings, if size permits.
(c) Burn out oil with blowpipe if surface heavily contaminated even after above treatment.
(d) File or grind any portions of bearing surface which remain contaminated or highly polished by movement of broken whitemetal.
(e) Protect parts which are not to be lined by coating with whitewash or washable distemper, and drying. Plug bolt holes, water jacket apertures, etc., with asbestos cement or similar filler, and dry.

2 TINNING

Use pure tin for tinning steel and cast iron surfaces; use 50% tin, 50% lead solder for tinning bronze, gunmetal or brass surfaces.

Flux surfaces to be tinned by swabbing with 'killed spirit' (saturated solution of zinc in concentrated commercial hydrochloric acid, with addition of about 5% free acid), or suitable proprietary flux.

Tinning cast iron presents particular difficulty due to the presence of graphite and, in the case of used bearings, absorption of oil. It may be necessary to burn off the oil, scratch brush, and flux repeatedly, to tin satisfactorily. Modern methods of manufacture embodying molten salt bath treatment to eliminate surface graphite enable good tinning to be achieved, and such bearings may be retinned several times without difficulty.

Tin bath

(i) Where size of bearing permits, a bath of pure tin held at a temperature of 280°–300°C or of solder at 270°–300°C should be used.
(ii) Flux and skim surface of tinning metal and immerse bearing only long enough to attain temperature of bath. Prolonged immersion will impair bond strength of lining and cause contamination of bath, especially with copper base alloy housings or shells.
(iii) Flux and skim surface of bath to remove dross, etc., before removing bearing.
(iv) Examine tinned bearing surface. Wire brush any areas which have not tinned completely, reflux and re-immerse.

Stick tinning

(*i*) If bearing is too large, or tin bath is not available, the bearing or shell should be heated by blowpipes or over a gas flame as uniformly as possible.

(*ii*) A stick of pure tin, or of 50/50 solder is dipped in flux and applied to the surface to be lined. The tin or solder should melt readily, but excessively high shell or bearing temperatures should be avoided, as this will cause oxidation and discoloration of the tinned surfaces, and impairment of bond.

(*iii*) If any areas have not tinned completely, reheat locally, rub areas with sal-ammoniac (ammonium chloride) powder, reflux with killed spirit, and retin.

3 LINING METHODS

(a) Static lining

(i) Direct lined bearings

The lining set-up depends upon the type of bearing. Massive housings may have to be relined *in situ*, after preheating and tinning as described in sections (1) and (2). In some cases the actual journal is used as the mandrel (see Figures 22.1 and 22.2).

Journal or mandrel should be given a coating of graphite to prevent adhesion of the whitemetal, and should be preheated before assembly.

Sealing is effected by asbestos cement or similar sealing compounds.

(ii) Lined shells

The size and thickness of shell will determine the type of lining fixture used. A typical fixture, comprising face plate and mandrel, with clamps to hold shell, is shown in Figure 22.5 while Figure 22.6 shows the pouring operation.

Figure 22.1 Location of mandrel in end face of direct lined housing

Figure 22.3 Direct lined housing. Pouring of whitemetal

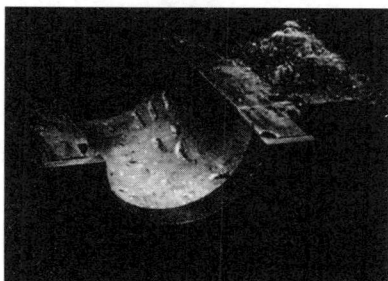

Figure 22.2 Outside register plate, and inside plate machined to form radius

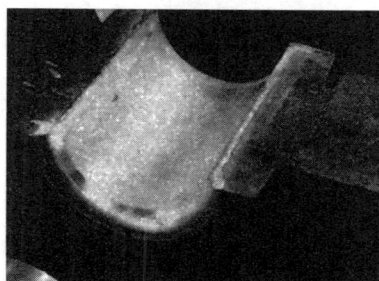

Figure 22.4 Direct lined housing, as lined

(b) Centrifugal lining

This method is to be preferred if size and shape of bearing are suitable, and if economic quantities require relining.

(i) Centrifugal lining equipment

For small bearings a lathe bed may be adapted if suitable speed control is provided. For larger bearings, or if production quantities merit, special machines with variable speed control and cooling facilities, are built by specialists in the manufacture or repair of bearings.

(ii) Speed and temperature control

Rotational speed and pouring temperature must be related to bearing bore diameter, to minimise segregation and eliminate shrinkage porosity.

Rotational speed must be determined by experiment on the actual equipment used. It should be sufficient to prevent 'raining' (i.e. dropping) of the molten metal during rotation, but not excessive, as this increases segregation. Pouring temperatures are dealt with in a subsequent section.

(iii) Cooling facilities

Water or air–water sprays must be provided to effect directional cooling from the outside as soon as pouring is complete.

(iv) Control of volume of metal poured

This is related to size of bearing, and may vary from a few grams for small bearings to many kilograms for large bearings.

The quantity of metal poured should be such that the bore will clean up satisfactorily, without leaving dross or surface porosity after final machining.

Excessively thick metal wastes fuel for melting, and increases segregation.

(v) Advantages

Excellent bonding of whitemetal to shell or housing.
Freedom from porosity and dross.
Economy in quantity of metal poured.
Directional cooling.
Control of metal structure.

(vi) Precautions

High degree of metallurgical control of pouring temperatures and shell temperatures required.

Close control of rotational speed essential to minimise segregation.

Measurement or control of quantity of metal poured necessary.

Control of timing and method of cooling important.

Figure 22.5 Lining fixture for relining of shell type bearing

Figure 22.6 Pouring operation in relining of shell type bearing

Figure 22.7 Purpose-build centrifugal lining machine for large bearings

Figure 22.8 Assembling a stem tube bush 680 mm bore by 2150 mm long into a centrifugal lining machine

4 POURING TEMPERATURES

(a) Objective

In general the minimum pouring temperature should be not less than about 80°C above the liquidus temperature of the whitemetal, i.e. that temperature at which the whitemetal becomes completely molten, but small and thin 'as cast' linings may require higher pouring temperatures than thick linings in massive direct lined housings or large and thick bearing shells.

The objective is to pour at the minimum temperature consistent with adequate 'feeding' of the lining, in order to minimise shrinkage porosity and segregation during the long freezing range characteristic of many whitemetals. Table 22.2 gives the freezing range (liquidus and solidus temperatures) and recommended minimum pouring temperatures of a selection of typical tin-base and lead-base whitemetals. However, the recommendations of manufacturers of proprietary brands of whitemetal should be followed.

(b) Pouring

The whitemetal heated to the recommended pouring temperature in the whitemetal bath, should be thoroughly mixed by stirring, without undue agitation. The surface should be fluxed and cleared of dross immediately before ladling or tapping. Pouring should be carried out as soon as possible after assembly of the preheated shell and jig.

(c) Puddling

In the case of large statically lined bearings or housings, puddling of the molten metal with an iron rod to assist the escape of entrapped air, and to prevent the formation of contraction cavities, may be necessary. Puddling

must be carried out with great care, to avoid disturbance of the structure of the freezing whitemetal. Freezing should commence at the bottom and proceed gradually upwards, and the progress of solidification may be felt by the puddler. When freezing has nearly reached the top of the assembly, fresh molten metal should be added to compensate for thermal contraction during solidification, and any leakage which may have occurred from the assembly.

(d) Cooling

Careful cooling from the back and bottom of the shell or housing, by means of air–water spray or the application of damp cloths, promotes directional solidification, minimises shrinkage porosity, and improves adhesion.

5 BOND TESTING

The quality of the bond between lining and shell or housing is of paramount importance in bearing performance. Non destructive methods of bond testing include:

(a) Ringing test

This is particularly applicable to insert or shell bearings. The shell is struck by a small hammer and should give a clear ringing sound if the adhesion of the lining is good. A 'cracked' note indicates poor bonding.

(b) Oil test

The bearing is immersed in oil, and on removal is wiped clean. The lining is then pressed by hand on to the shell or housing adjacent to the joint faces or split of the bearing. If oil exudes from the bond line, the bonding is imperfect.

Table 22.2 Whitemetals, solidification range and pouring temperatures

| Specification | Nominal composition % | | | | | Solidus temp °C | Liquidus temp °C | Min. pouring temp °C |
	Antimony	Copper	Other	Tin	Lead			
ISO 4381	12	6	0	Remainder	2	183	400	480
Tin base	7	3	0	Remainder	0.4	233	360	440
Alloys	7	3	1.0 Cd	Remainder	0.4	233	360	440
ISO 4381	14	0.7	1.0 As	1.3	Remainder	240	350	450
Lead base	15	0.7	0.7 Cd 0.6 As	10	Remainder	240	380	480
Alloys	14	1.1	0.5 Cd 0.6 As	9	Remainder	240	400	480
	10	0.7	0.25 As	6	Remainder	240	380	480

(c) Ultrasonic test

This requires specialised equipment. A probe is held against the lined surface of the bearing, and the echo pattern resulting from ultrasonic vibration of the probe is observed on a cathode ray tube. If the bond is satisfactory the echo occurs from the back of the shell or housing, and its position is noted on the C.R.T. If the bond is imperfect, i.e. discontinuous, the echo occurs at the interface between lining and backing, and the different position on the C.R.T. is clearly observable. This is a very searching method on linings of appropriate thickness, and will detect small local areas of poor bonding. However, training of the operator in the use of the equipment, and advice regarding suitable bearing sizes and lining thicknesses, must be obtained from the equipment manufacturers.

This method of test which is applicable to steel backed bearings is described in ISO 4386-1 (BS 7585 Pt 1). It is not very suitable for cast iron backed bearings because the cast iron dissipates the signal rather than reflecting it. For this material it is better to use a gamma ray source calibrated by the use of step wedges.

(d) Galvanometer method

An electric current is passed through the lining by probes pressed against the lining bore, and the resistance between intermediate probes is measured on an ohm-meter. Discontinuities at the bond line cause a change of resistance. Again, specialised equipment and operator training and advice are required, but the method is searching and rapid within the scope laid down by the equipment manufacturers.

6   LOCAL REPAIR BY PATCHING OR SPRAYING

In the case of large bearings, localised repair of small areas of whitemetal, which have cracked or broken out, may be carried out by patching using stick whitemetal and a blowpipe, or by spraying whitemetal into the cavity and remelting with a blowpipe. In both cases great care must be taken to avoid disruption of the bond in the vicinity of the affected area, while ensuring that fusion of the deposited metal to the adjacent lining is achieved.

The surface to be repaired should be tinned as described in section (2) prior to deposition of the patching metal. Entrapment of flux must be avoided.

The whitemetal used for patching should, if possible, be of the same composition as the original lining.

Patching of areas situated in the positions of peak loadings of heavy duty bearings, such as main propulsion diesel engine big-end bearings, is not recommended. For such cases complete relining by one of the methods described previously is to be preferred.

THE PRINCIPLE OF REPLACEMENT BEARING SHELLS

Replacement bearing shells, usually steel-backed, and lined with whitemetal (tin or lead-base), copper lead, lead bronze, or aluminium alloy, are precision components, finish machined on the backs and joint faces to close tolerances such that they may be fitted directly into appropriate housings machined to specified dimensions.

The bores of the shells may also be finish machined, in which case they are called 'prefinished bearings' ready for assembly with shafts or journals of specified dimensions to provide the appropriate running clearance for the given application.

In cases where it is desired to bore *in situ*, to compensate for misalignment or housing distortion, the shells may be provided with a boring allowance and are then known as 'reboreable' liners or shells.

The advantages of replacement bearing shells may be summarised as follows:

1 Elimination of hand fitting during assembly with consequent labour saving, and greater precision of bearing contour.
2 Close control of interference fit and running clearance.
3 Easy replacement.
4 Elimination of necessity for provision of relining and machining facilities.
5 Spares may be carried, with saving of bulk and weight.
6 Lower ultimate cost than that of direct lined housings or rods.

Special note

'Prefinished' bearing shells must not be rebored *in situ* unless specifically stated in the maker's catalogue, as many modern bearings have very thin linings to enhance load carrying capacity, or may be of the overlay plated type. In the first case reboring could result in complete removal of the lining, while reboring of overlay-plated bearings would remove the overlay and change the characteristics of the bearing.

Linings are attached to their shoes by riveting or bonding, or by using metal-backed segments which can be bolted or locked on to the shoes. Riveting is normally used on clutch facings and is still widely used on car drum brake linings and on some industrial disc brake pads. Bonding is used on automotive disc brake pads, on lined drum shoes in passenger car sizes and also on light industrial equipment.

For larger assemblies it is more economical to use bolted-on or locked-on segments and these are widely used on heavy industrial equipment. Some guidance on the selection of the most appropriate method, and of the precautions to be taken during relining, are given in the following tables.

Table 23.1 Ways of attaching friction material

	Riveting	Bonding	Bolted-on-segments	Locked-on-segments
Shoe relining	De-riveting old linings and riveting on new linings can be done on site. General guidance is given in BS 3575 (1981) SAE J660	Shoes must be returned to factory. Cannot be done on site	Can be relined on site without dismantling brake assembly. Bolts have to be removed	Can be relined on site without dismantling brake assembly by slackening off bolts
Plate clutch relining	As above	As above	Not applicable	Not applicable
Use of replacement shoes already lined	Quick. Old shoes returned in part exchange	Quick. Old shoes returned in part exchange	Quick	Quick
Use of replacement clutch plates already lined	As above	As above	Not applicable	Not applicable
Friction surface	Reduced by rivet holes	Complete unbroken surface, giving full friction lining area	As for bonding apart from bolt slots in side of lining	As for bonding
Life	Amount of wear governed by depth of rivet head from working surface. If linings are worn to less than 0.8 mm (0.031 in) above rivets they should be replaced	Not affected by rivets or rivet holes. Can be worn right down. If linings are worn to less than 1.6 mm (0.062 in) above shoe they should be replaced	Governed by thickness of tie plates. Advantage over rivets, or countersunk screws. If linings are worn to less than 0.8 mm (0.031 in) above tie plates they should be replaced	Governed by depth of keeper plates. Comparable with use of rivets or countersunk screws. If linings are worn to less than 0.8 mm (0.031 in) above countersunk screws they should be replaced

Table 23.1 (continued)

	Riveting	Bonding	Bolted-on segments	Locked-on segments
Spares	Good. Linings drilled or undrilled can be supplied ex-stock together with rivets. Small space required for stocks	Bulky. Complete shoe or plate with lining attached required. Where large metal shoes or plates are involved there is a high cost outlay and extra storage space	Good. Only linings with tie plates bonded into them required	Good. Only linings required suitably grooved
Limitations	Less suitable for low-speed, high-torque applications	Less suitable for high ambient temperatures, corrosive atmospheres, or where bonding to alloy with copper content of over 0.4%	12 mm ($\frac{1}{2}$ in) thick or over, up to 610 mm (24 in) long	12 mm ($\frac{1}{2}$ in) thick or over, up to 610 mm (24 in) long
Suitable applications	General automotive and industrial	Large production runs. Attachment of thin linings. Attachment to shoes where other methods are not practicable	For attachment to shoes which are not readily dismantled such as large winding engines and excavating machinery brakes, also high torque applications	As for bolted-on segments

Table 23.2 Practical techniques and precautions during relining

	Riveting	Bonding	Bolted-on segments	Locked-on segments
Removing lining or facings	Best to strip old linings and facings by drilling out rivets, taking care to avoid damage to the rivet holes and shoe platform	Best done as a factory job	Slacken off brake adjustment. Slacken off nuts on main side. Remove nuts and bolts from outer side and slide linings across the slots in side of the lining	Slacken off brake adjustment. Slacken off bolts sufficiently to allow linings to slide along the keeper plates. If necessary tap the linings with a wooden drift to assist removal
Replacing lining or facings	Clean shoes and spinner plate, replace if distorted or damaged. If new linings or facings are drilled clamp to new shoe or pressure plate, insert rivets, clench lightly. Insert all rivets before securely fastening. If undrilled, clamp to shoe or spinner plate, locate in correct position and drill holes using the drilled metal part as a template. Counterbore on opposite side. Use same procedure as for drilled linings or drilled facings to complete the riveting. During relining particular attention should be given to rivet and hole size and also to the clench length	See above	Replace by the reverse procedure using the slots to locate the linings. Tighten up all bolts and readjust the brake	Replace by the reverse procedure. Afterwards readjust the brake

Table 23.2 (continued)

	Riveting	Bonding	Bolted-on segments	Locked-on segments
Precautions	Use brass or brass-coated steel rivets to avoid corrosion problems. Copper rivets may be used for passenger car linings and also in light industrial applications. There must be good support for the rivet and the correct punch must be used. Allow one third the thickness of lining material under rivet head	Best done as a factory job	Best to use high-tensile steel socket-head type of bolt. Can be applied to all linings over 12 mm ($\frac{1}{2}$ in) thick	Avoid over-tightening the bolts so as not to distort the heads or crack the linings. Generally linings must be 12 mm ($\frac{1}{2}$ in) thick or over for rigid material and at least 25 mm (1 in) thick for flexible material

Care must be taken during relining to avoid the lining becoming soaked or contaminated by oil or grease, as it will be necessary to replace the lining, or segment of lining, for if not its performance will be reduced throughout its life

Table 23.3 Methods of working the lining and finishing the mating surfaces

Material	Cutting	Drilling	Surface finishing	Handling
Woven materials	Hacksaw or bandsaw. Grinding is not recommended as it gives a scuffed surface and possible fire hazard	HSS tools are suitable. A burnishing tool is necessary to remove ragged edges	HSS tools are suitable for turning or boring of facings	Can be more easily bent to radius by heating to around 60°C. When machining and handling asbestos based materials, work must be carried out within the relevant asbestos dust regulations.
Moulded materials	Hacksaw, bandsaw or abrasive wheels	Tungsten carbide (WC) tipped tools are usually needed	Tungsten carbide (WC) tipped tools are usually needed for turning or boring of facings	Care must be taken to give adequate support during machining because of their brittle nature
Mating members	Fine/medium ground to 0.63–1.52 μm (25–60 μin) cla surface finish is best. Avoid chatter marks, keep drum ovality to within 0.127 mm (0.005 in) and discs parallel to within 0.076 mm (0.0003 in). Surface should be cleaned up if rust, heat damage or deep scoring is evident but shallow scoring can be tolerated. If possible the job should be done in situ or with discs or drums mounted on hubs or mandrels. The total amount removed by griding from the disc thickness or the drum bore diameter should not exceed: 1.27 mm (0.05 in) on passenger cars 2.54 mm (0.1 in) on commercial vehicles. If these values are exceeded a replacement part should be fitted. When components have been ground, a thicker lining should be fitted to compensate for the loss of metal. With manual clutches the metal face can be skimmed by amounts up to 0.25 mm (0.01 in). For guidance on the reconditioning of vehicle disc and drum brakes, reference should be made to the vehicle manufacturer's handbook.			

DEFINITION OF VISCOSITY

Viscosity is a measure of the internal friction of a fluid. It is the most important physical property of a fluid in the context of lubrication. The viscosity of a lubricant varies with temperature and pressure and, in some cases, with the rate at which it is sheared.

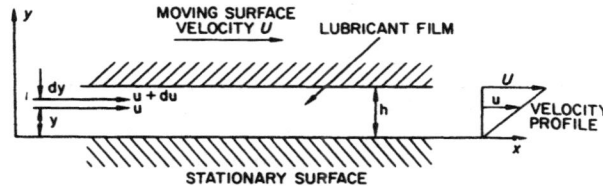

Figure 1.1 Lubricant film between parallel plates

Dynamic viscosity

Dynamic viscosity is the lubricant property involved in tribological calculations. It provides a relationship between the shear stress and the rate of shear which may be expressed as:

Shear stress = Coefficient of Dynamic Viscosity × Rate of Shear

or
$$\tau = \eta \frac{\partial u}{\partial y} = \eta D,$$

where τ = shear stress,
       η = dynamic viscosity,

$$\frac{\partial u}{\partial y} = D = \text{rate of shear.}$$

For the parallel-plate situation illustrated in Fig. 6.1.

$$\frac{\partial u}{\partial y} = \frac{U}{h}$$

and $\tau = \eta \dfrac{U}{h}$

If τ is expressed in N/m^2 and $\dfrac{\partial u}{\partial y}$ in s^{-1}

then η is expressed in Ns/m^2, i.e. viscosity in SI units.

The unit of dynamic viscosity in the metric system is the poise $\left(\dfrac{g}{cm\ s}\right)$:

$$1 \frac{Ns}{m^2} = 10 \text{ poise.}$$

Kinematic viscosity

Kinematic viscosity is defined as

$$v = \frac{\eta}{\rho}$$

where ρ is the density of the liquid.

If ρ is expressed in kg/m^3, then v is expressed in m^2/s, i.e. in SI units.

The unit of kinematic viscosity in the metric system is the stoke $\left(\dfrac{cm^2}{s}\right)$.

$$1 \frac{m^2}{s} = 10^4 \text{ stokes.}$$

Table 1.1 gives the factors for converting from SI to other units.

Table 1.1 Viscosity conversion factors

Dynamic viscosity $\left(\text{SI unit is } \dfrac{Ns}{m^2}\right)$	
$\dfrac{Ns}{m^2} \times 10$	$= \text{poise}\left(\dfrac{g}{cm.s}\right)$
poise $\times 0.1$	$= \dfrac{Ns}{m^2}$

Kinetic viscosity $\left(\text{SI unit is } \dfrac{m^2}{s}\right)$	
$\dfrac{m^2}{s} \times 10^4$	$= \text{stokes}\left(\dfrac{cm^2}{s}\right)$
stokes $\times 0.1^{-4}$	$= \dfrac{m^2}{s}$

ANALYTICAL REPRESENTATION OF VISCOSITY

The viscosities of most liquids decrease with increasing temperature and increase with increasing pressure. In most lubricants, e.g. mineral oils and most synthetic oils, these changes are large. Effects of temperature and pressure on the viscosities of typical lubricants are shown in Figs 1.2 and 1.3. Numerous expressions are available which describe these effects mathematically with varying degrees of accuracy. In general, the more tractable the mathematical expression the less accurate is the description. The simplest expression is:

$$\eta = \eta_o \exp(yp - \beta t)$$

where η_o = viscosity at some reference temperature and pressure, p = pressure, t = temperature, and y and β are constants determined from measured viscosity data. A more accurate representation is obtained from the expression:

$$\eta = \eta_o \exp\left[\frac{A + Bp}{t + t_o}\right]$$

where A and B are constants.

Numerical methods can be employed to give a greater degree of accuracy.

A useful expression for the variation of density with temperature used in the calculation of kinematic viscosities is:

$$\rho_t = \rho_s - a(t - t_s) + b(t - t_s)^2$$

where ρ_s is the density at temperature t_s, and a and b are constants.

The change in density with pressure may be estimated from the equation:

$$\frac{V_o P}{V_o - V} = K_o + mp$$

where V_o is the initial volume, V is the volume at pressure p, and K_o and m are constants.

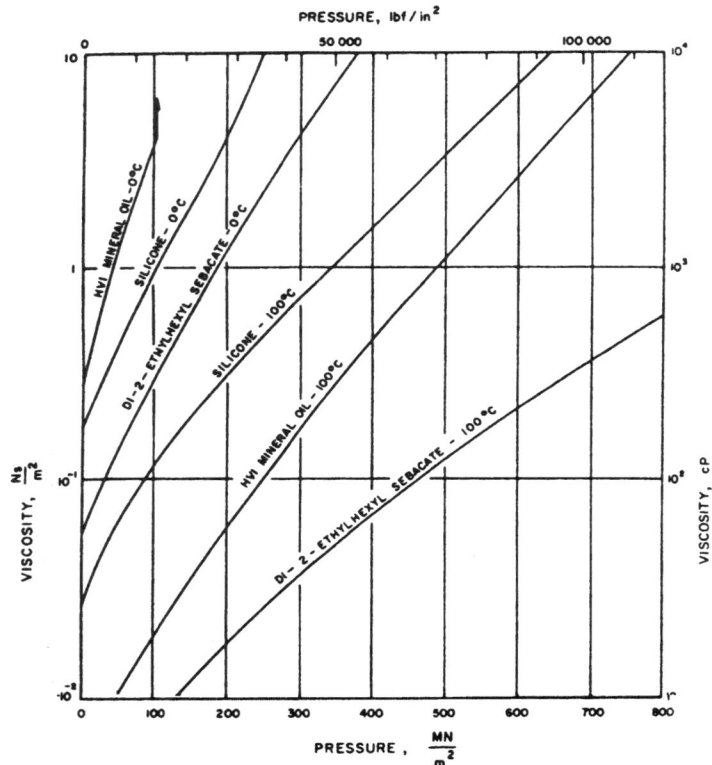

Figure 1.2 The variation of viscosity with pressure for some mineral and synthetic oils

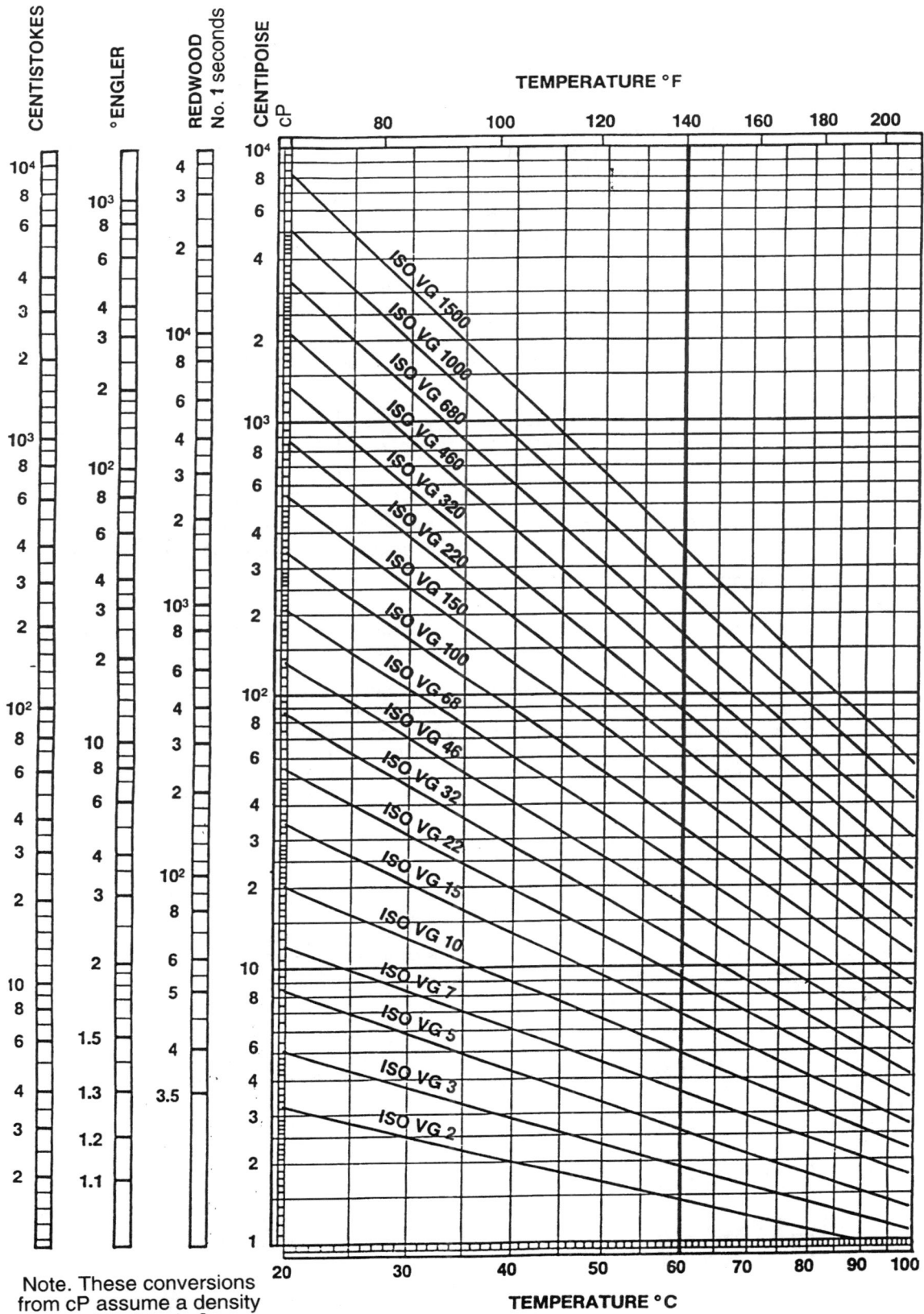

Figure 1.3 The viscosity of lubricating oils to ISO 3448 at atmospheric pressure

Note. These conversions from cP assume a density of 900 kg/m³

VISCOSITY OF NON-NEWTONIAN LUBRICANTS

If the viscosity of a fluid is independent of its rate of shear, the fluid is said to be Newtonian. Mineral lubricating oils and synthetic oils of low molecular weight are Newtonian under almost all practical working conditions.

Polymeric liquids of high molecular weight (e.g. silicones, molten plastics, etc.) and liquids containing such polymers may exhibit non Newtonian behaviour at relatively low rates of shear. This behaviour is shown diagrammatically in Fig. 1.4. Liquids that behave in this way may often be described approximately in the non linear region by a power-law relationship of the kind:

$$\tau = (\phi s)^n$$

where ϕ and n are constants. For a Newtonian liquid $n = 1$ and $\phi \equiv \eta$, and typically for a silicone, $n \approx 0.95$.

Greases are non-Newtonian in the above sense but, in addition, they exhibit a yield stress the magnitude of which depends on their constitution. The stress/strain rate characteristics for a typical grease is also indicated in Fig 1.4. This characteristic may be represented approximately in the non linear region by an expression of the form,

$$\tau = \tau_l + (\phi s)^n$$

where τ_l is the yield stress and ϕ and n are constants.

MEASUREMENT OF VISCOSITY

Viscosity is now almost universally measured by standard methods that use a suspended-level capillary viscometer. Several types of viscometer are available and typical examples are shown in Fig. 1.5. Such instruments measure the kinematic viscosity of the liquid. If the dynamic viscosity is required the density must also be measured, both kinematic viscosity and density measurements being made at the same temperature.

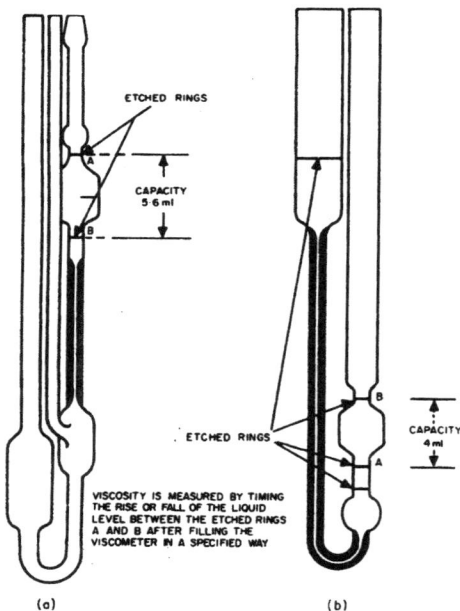

Figure 1.4 Shear stress/viscosity/shear rate characteristics of non-Newtonian liquids

If the viscosity/rate-of-shear characteristics of a liquid are required a variable-shear-rate instrument must be used. The cone-and-plate viscometer is the one most frequently employed in practice. The viscosity of the liquid contained in the gap between the cone and the plate is obtained by measuring the torque required to rotate the cone at a given speed. The geometry, illustrated in Fig. 1.6, ensures that the liquid sample is exposed to a uniform shear rate given:

$$D = \frac{U}{h} = \frac{r\omega}{r\alpha} = \frac{\omega}{\alpha}$$

where r = cone radius, ω = angular velocity and α = angle of gap.

From the torque, M, on the rotating cone the viscosity is then calculated from the expression:

$$\eta = \frac{3M\alpha}{2\pi r^3 \omega}$$

This instrument is thus an absolute viscometer measuring dynamic viscosity directly.

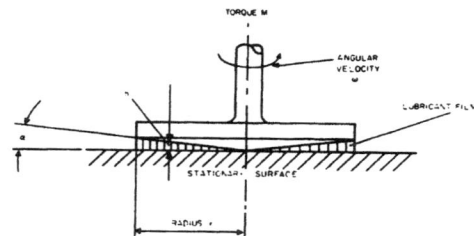

Figure 1.5 Typical glass suspended-level viscometers

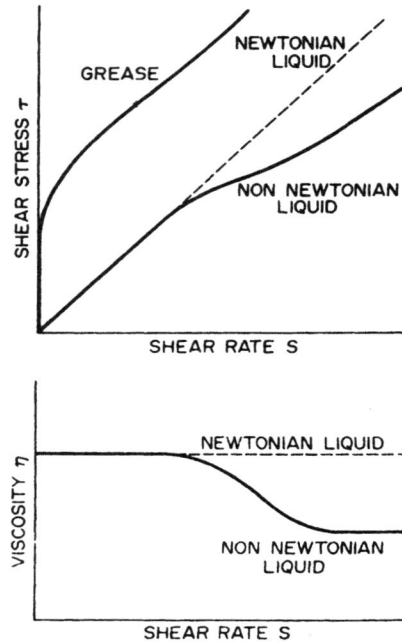

Figure 1.6 Cone-and-plate viscometer

INTRODUCTION

The hardness of the surface of components is an important property affecting their tribological performance. For components with non conformal contacts such as rolling bearings and gears, the hardness, and the corresponding compressive strength, of the surface material must be above a critical value. For components with conformal contacts such as plain bearings, the two sides of the contact require a hardness difference typically with a hardness ratio of 3:1 and ideally with 5:1. The component with the surface, which extends outside the close contact area, needs to be the hardest of the two, in order to avoid any incipient indentation at the edge of the contact. Shafts and thrust collars must therefore generally be harder than their associated support bearings.

HARDNESS MEASUREMENT

The hardness of component surfaces is measured by indenting the surface with a small indenter made from a harder material.

The hardness can then be inferred from the width or area of the indentation or from its depth.

The Brinell hardness test generally uses a steel ball 10 mm diameter which is pressed into the surface under a load of 30 kN. In the Vickers hardness test, a pyramid shaped indenter is pressed into the surface, usually under a load of 500 N. In both cases the hardness is then inferred from a comparison of the load and the dimensions of the indentation.

The Rockwell test infers the hardness from the depth of penetration and thus enables a direct reading of hardness to be obtained from the instrument. Hard materials are measured on the Rockwell C scale using a diamond rounded tip cone indenter and a load of 1.5 kN. Softer metals are measured on the Rockwell B scale using a steel ball of about 1.5 mm diameter and a load of 1 kN.

Table 2.1 gives a comparison of the various scales of hardness measurement, for the convenience of conversion from one scale to another. The values are reasonable for most metals but conversion errors can occur if the material is prone to work hardening.

Table 2.1 Approximate comparison of scales of hardness

Brinell	Vickers	Rockwell B	Rockwell C
120	120	67	—
150	150	88	—
200	200	92	—
240	250	100	22
300	320	—	32
350	370	—	38
400	420	—	43
450	475	—	47
500	530	—	51
600	680	—	59
650	750	—	62

INTRODUCTION

Manufacturing processes tend to leave on the surface of the workpiece characteristic patterns of hills and valleys known as the texture. The texture produced by stock removal processes is deemed to have components of roughness and waviness. These may be superimposed on further deviations from the intended geometrical form, for example, those of flatness, roundness, cylindricity, etc.

Functional considerations generally involve not only the topographic features of the surface, each having its own effect, but also such factors as the properties especially of the outer layers of the workpiece material, the operating conditions, and often the characteristics of a second surface with which contact is made.

While the properties of the outer layers may not differ from those of the material in bulk, significant changes can result from the high temperatures and stresses often associated particularly with the cutting and abrasive processes.

Optimised surface specification thus becomes a highly complex matter that often calls for experiment and research, and may sometimes involve details of the process of manufacture.

Surface profiles

The hills and valleys, although very small in size, can be visualised in the same way as can those on the surface of the earth. They have height, shape and spacing from one peak to the next. They can be portrayed in various ways.

An ordinary microscope will give useful information about their direction (the lay) and their spacing, but little or none about their height. The scanning electron microscope can give vivid monoscopic or stereoscopic information about important details of topographic structure, but is generally limited to small specimens. Optical interference methods are used to show contours and cross-sections of fine surfaces. The stylus method, which has a wide range of application, uses a sharply pointed diamond stylus to trace the profile of a cross-section of the surface.

The peak-to-valley heights of the roughness component of the texture may range from around $0.05\,\mu m$ for fine lapped, through $1\,\mu m$ to $10\,\mu m$ for ground, and up to $50\,\mu m$ for rough machined surfaces, with peak spacings along the surface ranging from $0.5\,\mu m$ to $5\,mm$. The height of the associated waviness component, resulting for example from machine vibration, should be less than that of the roughness when good machines in good order are used, but the peak spacing is generally much greater.

of the flanks, is shown in Fig. 3.1. The horizontal compression must always be remembered.

The principle of the stylus method is basically the same as that of the telescopic level and staff used by the terrestrial surveyor, and sketched in Fig. 3.2(a). In Fig. 3.2(b), the stylus T is equivalent to the staff and the smooth datum surface P is equivalent to the axis of the telescope. The vertical displacements of the stylus are usually determined by some form of electric transducer and amplifying system.

For convenience, the datum surface P of Fig. 3.2(b) is often replaced by another form of datum provided

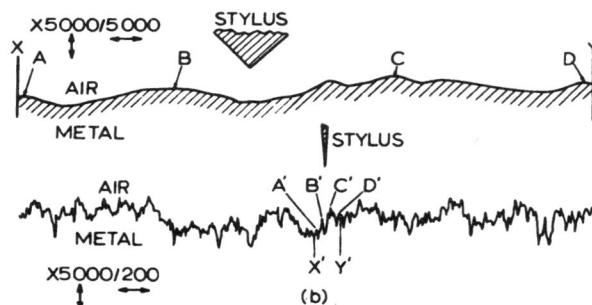

Figure 3.1 Effect of horizontal compression

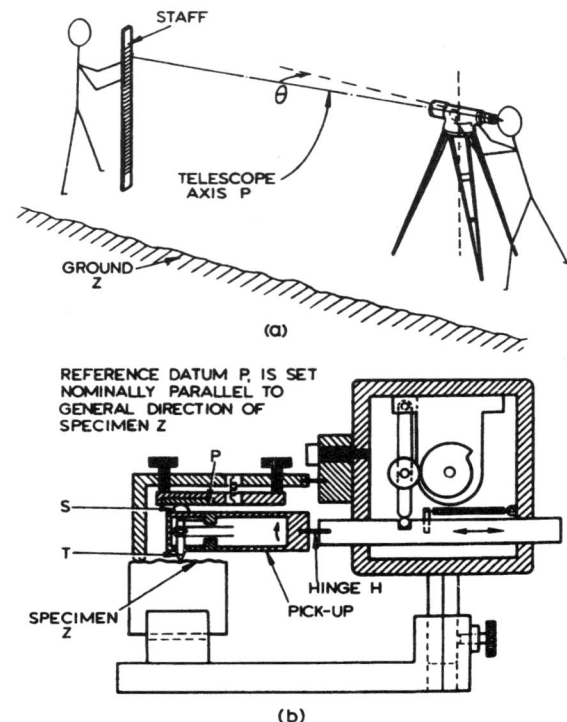

Figure 3.2 The surveyor takes lines of sight in many directions to plot contours. The engineer plots one or more continuous, but generally unrelated cross-sections (a) Telescope axis usually set tangential to mean sea level by use of bubble in telescope (b) Skid S slides along the reference surface. Stylus T is carried on flexure links or a hinge

Because of the need for portraying on a profile graph a sufficient length of surface to form a representative sample, and the small height of the texture compared with its spacing, it is generally necessary to use far greater vertical than horizontal magnification. The effect of this on the appearance of the graph, especially on the slopes

by the surface of the workpiece itself (Fig. 3.3), over which slides a rounded skid. This form of datum may be quicker to use but it is only approximate, as the unwanted vertical excursions of the skid combine in various indeterminate ways with the wanted excursions of the stylus, and in practice the combination can be accepted only when the asperities are deep enough and close enough together for the excursions of the skid of given radius to be small compared with those of the stylus.

The diamond stylus may have the form of a 90° cone or 4-sided pyramid, with standardised equivalent tip radii and operative forces of $2\,\mu m$ $(0.7\,mN)$ or $10\,\mu m$ $(16\,mN)$. While these tips are sharp enough for the general run of engineering surfaces, values down to $0.1\,\mu m$ $(10\,\mu N)$ can be used in specialised instruments to give better resolution of the finest textures (e.g. those of gauge blocks).

Some typical profiles of roughness textures, horizontally compressed in the usual way are shown in Fig. 3.4.

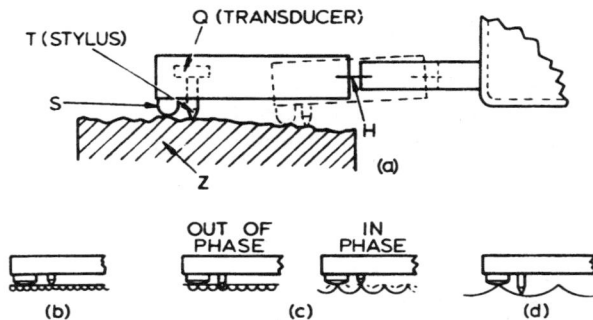

Figure 3.3 (a) Skid S slides over crests of specimen (b) acceptable approximation to independent datum (c) significant skin error, negative or positive according to whether the skid and stylus move in or out of phase (d) acceptable mechanical filtration of wave-lengths which are long compared with separation of skid and stylus

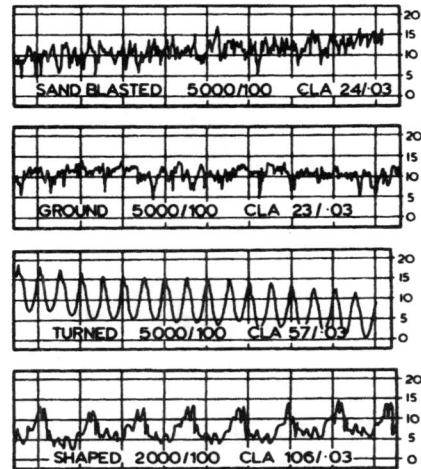

Figure 3.4 Typical profiles. Magnification is shown thus: vertical/horizontal

NUMERICAL ASSESSMENT

Significance and preparation

For purposes of communication, especially on drawings, it is necessary to describe surface texture numerically, and many approaches to this have been considered. A numerical evaluation of some aspect of the texture is often referred to as a 'parameter'. The height, spacing, slope, crest curvatures of the asperities, and various distributions and correlation factors of the roughness and waviness can all be significant and contribute to the sum total of information that may be required; but no single parameter dependent on a single variable can completely describe the surface, because surfaces having quite different profiles can be numerically equal with respect to one such parameter while being unequal with respect to others. Economic considerations dictate that the number admitted to workshop use should be minimised.

The profile found by the pick-up may exhibit (a) tilt relative to the instrument datum, (b) general curvature, (c) long wavelengths classed as 'waviness' and (d) the shorter wavelengths classed as roughness, shown collectively by the profile in Fig. 3.5(c). The first two of these being irrelevant to the third and fourth, some preparation of the profile is required before useful measurement can begin. Preparation involves recognition and isolation of the irregularities to be measured,

and the establishment of a suitable reference line from which to measure those selected.

Recognition of the different kinds of texture and of their boundaries may involve some degree of judgement and experience. Broadly, roughness is deemed to include all those irregularities normally produced by the process, these being identified primarily on the basis of their peak spacing. Cutting processes generally leave feed marks of which the spacing can be recognised immediately. Abrasive textures are more difficult because of their random nature, but experience has shown that the peak spacings of the finer ones can generally be assumed to be less than 0.8 mm, while those of the coarser ones, which may be greater, tend to become reasonably visible.

Isolation is effected on a wavelength basis by some form of filtering process which has the effect of ironing out (i.e. attenuating) the longer wavelengths that do not form part of the roughness texture.

The reference line used for the assessment of parameters is generally not the instrument datum of Figs 3.1 and 3.2, but a reference line derived from the profile itself, this line taking the form of a mean line passing through those irregularities of the profile that have been isolated by the filtering process.

The profile can be filtered graphically by restricting the measurement to a succession of very short sampling lengths, as shown in Fig. 3.5(a), through each of which is drawn a straight mean line parallel to its general direction. The length of each sample must be not less than the dominant spacing of the texture to be measured. If the individual samples are redrawn with their mean lines in line, the filtering effect becomes immediately apparent (Fig. 3.5(b)).

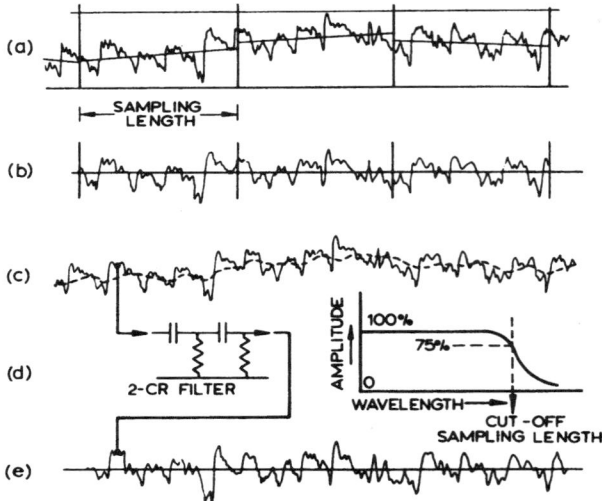

Figure 3.5 Comparative behaviour of graphical and electrical methods of filtering. Residual differences between (b) and (e) are referred to as Method Divergence

In the case of meter instruments, the alternating electric current representing the whole profile (Fig. 3.5(c)) is passed through an electric wave filter which transmits the shorter but attenuates the longer wavelengths. The standardised 2-CR filter network, its rate of attenuation and the position of the long wavelength cut-off accepted by convention (which is known as the *meter* cut-off) are shown in Fig. 3.5(d). When the meter cut-off is made equal to the graphical sampling length, the two methods are found to give, on average, equal numerical assessments.

The electric wave filter determines automatically a mean line which weaves its way through the input profile according to the way in which the filter reacts with the rates of change of the profile. This wave filter mean line is shown dotted in Fig. 3.5(c), where the profile is a repetition of Fig. 3.5(a). Relative to the output from the filter, the mean line becomes a straight line representing zero current (Fig. 3.5(e), cf. Fig. 3.5(b)). It is from this

line that the meter operates and about which the filtered profile would be displayed on a recorder having a sufficient frequency response. Sampling lengths and filter characteristics to suit the whole range of textures are standardised in British, US, ISO and other Standards the usual values being 0.25 mm, 0.8 mm and 2.5 mm.

Several graphical samples are usually taken consecutively to provide a good statistical basis. Meter instruments do the equivalent of this automatically; but care must be taken not to confuse the total length of traverse with the much shorter meter cut-off, for it is the latter that decides the greatest spacing on the surface to which the meter reading refers.

The 2-CR filter shown in Fig. 3.5(d) lends itself to simple instrumentation, but can distort the residual waveform that is transmitted for measurement. The amount of distortion is generally not sufficient to affect seriously the numerical assessments that are made.

An important point is that the signal fed to a recorder is generally *not* filtered, so that the graph can show as true a cross-section as the stylus, transducer and datum permit.

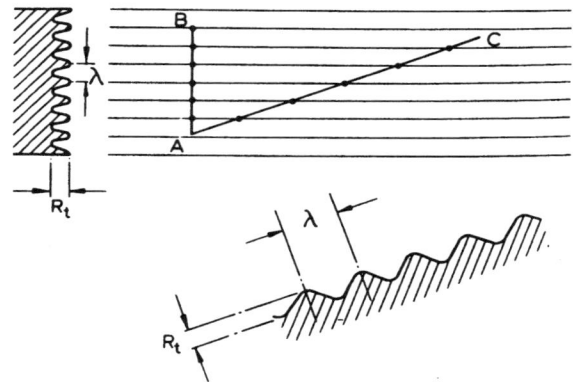

Figure 3.6 Oblique traverse increases λ but not R_t.

It is generally best to measure in the direction in which a maximum of information can be collected from the shortest possible traverse. In Fig. 3.6, representing a surface finished by a cutting process, that would be in the direction *AB* lying across the lay. In the oblique direction *AC* the peaks are farther apart, the slopes are less and the radii of curvature of the peaks are greater, but the *height* is the same. Thus, if a meter reading of the height is taken transversely with a just sufficient meter cut-off, and then obliquely without corresponding increase in the meter cut-off, a low reading is likely to result, even though the total traverse is increased. This would also apply to ground surfaces. In the case of textures having a random lay (e.g. shot blast or lapped with a criss-cross motion) the spacing, and hence the minimum meter cut-off, may be much the same in all directions.

Parameters

Except for the fairly periodic textures sometimes produced by cutting processes, surface textures tend to vary randomly in height and spacing. The problem of describing the different kinds of texture offers full scope for the devices of the statistician, from the simplest forms of averaging to the complexities of correlation functions.

Height

The most commonly used and easily measured height parameter is the average departure from the mean line of the filtered profile. It is known in the USA as the AA value. In Great Britain it was known as the cla value, but it is now known as the R_a value to line up with ISO terminology.

The rms value has been used in approximate form, obtained by multiplying the AA value by 1.11 which is the conversion factor for sine waves.

The height from the highest peak to deepest valley found anywhere along a selected part of the profile, known as R_t, is used especially in Europe, but it is unsatisfactory because the single extremes are too greatly dependent on the chance position selected.

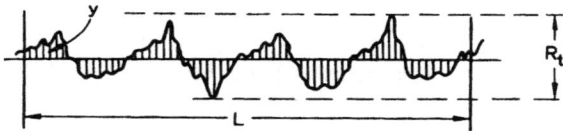

Figure 3.7 $cla = \dfrac{1}{L} \displaystyle\int_0^L |y|\, dL, \ rms = \left(\dfrac{1}{L} \int_0^L y^2\, dL\right)^{1/2}$

An averaged overall height is more representative than R_t. The 'ten point height R_z' is obtained by averaging the five highest peaks and five deepest valleys in the total traverse. In a German variant, known as R_{tm}, the average of the R_t values in five consecutive samples of equal length is taken.

Although the ratio of one height measure to another varies with the shape of the profile, some degree of conversion is generally possible as shown below:

Surfaces	rms / cla	10pt / cla	Peak-to-valley / cla
Turned*	1.1 to 1.15	4 to 5	4 to 5
Ground	1.18 to 1.30	5 to 7	7 to 14
Lapped	1.3 to 1.5	—	7 to 14
Statistically random	1.25	—	8.0

*With clean cut from round-nose tool.

Spacing

The spacing of the more significant peaks along the surface, may be functionally important.

Peaks per unit length have been counted, a peak being rated as such only if the adjacent valley exceeds a given depth. Bearing intercepts per unit length at a given level have also been counted.

Bearing area

A concept frequently encountered is that of 'bearing area', shown in abstract principle in Fig. 3.8, the level of the intersecting line being expressed either as a depth below the highest peak (an uncertain reference point) or as an offset from the mean line.

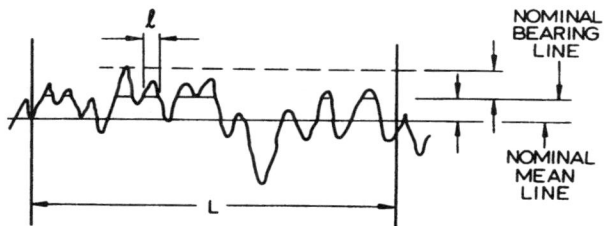

Figure 3.8 Nominal bearing area of sample

$$\% = \frac{\Sigma l}{L} \times 100$$

It must be remembered that this measure is confined to a small sample of the surface and does not represent the overall bearing area taking waviness and errors of form into account, nor does it allow for elastic deformation of the peaks under load. These and other considerations limit its value.

STRAIGHTNESS, FLATNESS, ROUNDNESS, CYLINDRICITY AND ALIGNMENT

Although these aspects are generally considered separately from surface texture, they can be highly significant to the functioning of surfaces, and must therefore receive at least some mention in the present section.

These aspects are generally measured with a blunt stylus that traces only the crests of the roughness, and does not appreciably enter the narrower scratch marks.

Straightness and flatness

Straightness and flatness are measured with the same basic type of apparatus as for texture in Fig. 3.1, but with the instrument datum made much longer. Horizontal magnifications are generally lower and compression ratios often higher than for texture measurement.

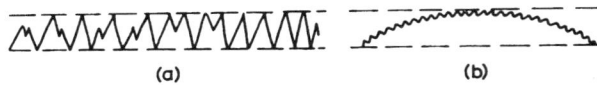

Figure 3.9. (a) Rough but straight (b) Smooth but curved

The normal engineering way of describing the deviations is in terms of the separation of two parallel lines or planes between which all deviations are contained. This is a maximum peak-to-valley measure that would not distinguish between the two surfaces in Fig. 3.9. A distinction can be made by measuring over the whole and also over a fraction of the length. This resembles the sampling length procedure used for surface texture.

Roundness

Roundness is generally measured by rotation of the pick-up or workpiece round a precisely generated axis (Fig. 3.10). Variations in the radius of the workpiece are

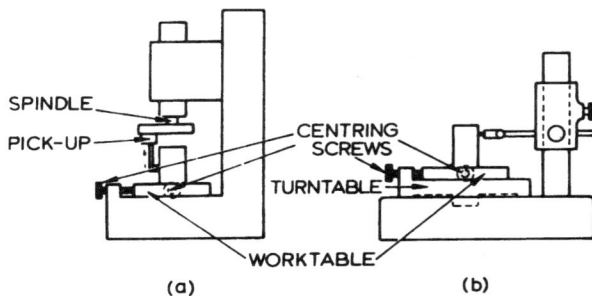

Figure 3.10 Principle of roundness instruments (a) rotating pick-up (b) rotating workpiece

plotted on a polar chart on which can be superposed a least-squares reference circle from which the radial deviations are determined. They are expressed in terms of the separation of two circles, drawn from a specified centre, that just contain the undulations. Four centres are possible, the two standardised being (a) the centre giving the minimum separation and (b) the least-squares centre (Fig. 3.11). The measure is again a maximum peak-to-valley value. For control of vibration in rolling bearings, the amplitude is sometimes assessed in three or four sharply defined wavebands.

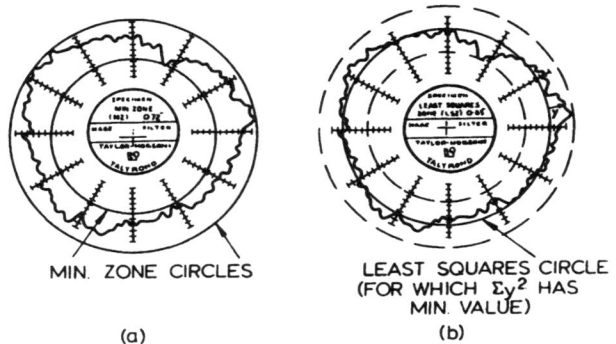

Figure 3.11 Polar graphs showing methods of assessing radial variations

Since the angular magnification is always unity, the effective compression ratio is generally very much greater than for texture measurement, and the resulting forms of distortion must be fully understood.

Cylindricity

The expression of cylindricity (Fig. 3.12) requires suitable instrumentation and display. The magnitude of the error is generally conceived in the same way as for straightness and roundness, but in this case as lying between two co-axial cylinders (or cones).

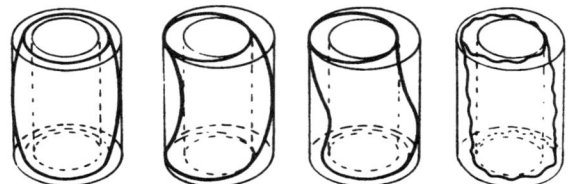

Figure 3.12 Errors in cylindricity

USE AND INTERPRETATION OF SURFACE MEASUREMENTS

The industrial requirement is (1) To investigate the characteristics of surfaces and identify the types that are functionally acceptable. (2) To specify the dominant requirements on drawings. (3) To control manufacture.

Ideally, the most economical surface or surfaces for a given application should be determined through the medium of experimental models closely allied to the intended method of manufacture. Surface texture tends to change as the surfaces are run in, and initially smooth surfaces are not necessarily best. The peaks can get smoothed down even after the first pass, but the run-in surface may retain traces of the original valleys (which may assist the distribution of lubricant) throughout the life of the machine. A recitation of the machining data used for the most successful model, in conjunction with the simplest topographic data (e.g. R_a) may provide a serviceable basis for manufacture, reflecting not only the topographic, but also the probable physical characteristics of the surface. If it is subsequently found expedient to change the process of manufacture, it may be advisable to reconsider both the functional consequences and the R_a value.

When the cost of fully experimental evolution cannot be accepted, it may suffice to rely on general experience, in which case the problem may arise of knowing what surface texture values would be descriptive of surfaces already familiar by sight and touch. Standard roughness comparison specimens can then be helpful. Electro-formed reproductions are available which show progressive grades of roughness of the usual machining processes, each grade having twice the R_a value of the previous one. This is about the smallest increment that can readily be detected by touch, and is often the smallest that matters functionally.

Sets of Roughness Comparison Specimens generally give a fair idea of the range of R_a values that can be achieved by each process, though it is as well to remember that as the fine end is approached, the cost of component production may rise rapidly, and have to be offset against other benefits that may accrue, for example in assembly or performance. The best surface of which any process is capable will involve many factors such as the stiffness of the workpiece, the material being worked, the condition of the machine and tool or wheel, the uniformity of the preceding process, the amount of stock to be removed, the time allowed for appropriately gentle cutting or grinding, and the care that is given to every detail of the process. The last two factors may determine the economic limit.

Eventually it may be possible to select optimum characteristics from tables, as is done for dimensional tolerances, but so many factors are involved that this seems a long way off.

The simple R_a parameter must be interpreted with full awareness of what it can and cannot tell about surface texture. It is best regarded as a practical index for comparing the heights of similar profiles on a linear basis – twice the index, twice the height. To this extent it has proved generally serviceable for process control. Its ability to compare dissimilar profiles is more limited. It corresponds with the sensory impression of roughness only for similar textures and over a limited range of heights and peak spacings, as examination of sets of roughness comparison specimens will show. It cannot provide a direct measure of the functional quality of a surface, because this aspect depends on many factors and often involves those of a second surface.

Statements of parameter values

A statement of height will have little meaning if it is not accompanied by a statement of the maximum peak spacing (generally expressed by the meter cut-off) to which it refers. The standard British way is to recite the height value and the parameter, followed by the meter cut-off in brackets. Thus, in metric units using micrometres for height and millimetres for the cut-off, an example would be $0.2\,\mu m\ R_a$ (2.5). A standard cut-off value often found suitable for the finer surfaces is 0.8 mm, and if this value was used or is to be used, the standards allow it to be assumed and direct statement omitted; but this does not mean that its significance can be ignored.

For fully co-ordinated control, the cut-off indicated on a drawing should be taken not only as the value to be used for inspection but also as the maximum significant spacing (e.g. traverse feed) that may result from production.

A further point is that the R_a value given on a drawing is often taken not as a target figure but as an upper limit, anything smoother being acceptable unless a lower limit is also given, so that manufacture must aim at something less if half the product is not to be rejected. On the other hand the R_a value marked on a roughness comparison specimen is the nominal value of the specimen. This difference in usage must be allowed for when choosing the required texture from a set of specimens, and indicating the choice on a drawing.

It is not possible to quote definite values for the tolerances of geometry maintainable by manufacturing process: values for any process can vary, not only from workshop to workshop, but within a workshop. The type of shop, the rate and quantity of production, the expected quality of work, the type of labour, the sequence of operations, the equipment available (and its condition) are among factors which must be taken into account. It is very difficult to apply values to these influencing factors, but as general guidance it might be expected to halve or double the maintainable tolerances by either better or worse practice. There are obviously no hard and fast rules.

The working values tabulated below are in good general agreement with modern practice. Achievable values can be better: they can be worse.

EXPLANATION OF TOLERANCES

Feature	Symbol*	The tolerance zone is limited by:	Diagrammatically:
Flatness	▱	Two parallel planes distance t apart	
Straightness	—	A cylinder of diameter t	
Parallelism	//	Two parallel straight lines/planes distance t apart and parallel to the datum line/plane	
Perpendicularity	⊥	Two parallel lines/planes distance t apart and perpendicular to the datum line/plane	
Concentricity	◎	A circle of diameter t the centre of which coincides with the datum point	
Circularity, roundness	⊙	Two concentric circles distance t apart	

TYPICAL TOLERANCE VALUE/SIZE RELATIONSHIPS

		Geometric tolerance feature			
Flatness of surface	Parallelism of cylinders or taper of cones on diameter (cylindricity, or conicity)	Straightness of cylinders or cones	Parallelism and/or squareness of flat surfaces	Parallelism or squareness of cylinders and flats	Roundness

Manufacturing process	Orders of tolerance t in mm/mm, or inches/inch length in surface or cylinder					
Turn, bore	0.00005	0.0001	0.0001	0.0001	0.0001	0.00004
	50×10^{-6}	100×10^{-6}	100×10^{-6}	100×10^{-6}	100×10^{-6}	40×10^{-6}
Fine turn, fine bore	0.00003	0.00004	0.00004	0.00005	0.00005	0.00003
	30×10^{-6}	40×10^{-6}	40×10^{-6}	50×10^{-6}	50×10^{-6}	30×10^{-6}
Cylindrical grind	0.00003	0.00005	0.00005	0.00005	0.00005	0.00002
	30×10^{-6}	50×10^{-6}	50×10^{-6}	50×10^{-6}	50×10^{-6}	20×10^{-6}
Fine cylindrical grind	0.00002	0.00002	0.00002	0.00003	0.00002	0.00001
	20×10^{-6}	20×10^{-6}	20×10^{-6}	30×10^{-6}	20×10^{-6}	10×10^{-6}

Achievable tolerance values

These can be obtained from the above table by multiplying the tolerance t in mm/mm or inches/inch by the size of the feature, bearing in mind that there will be a reasonable *minimum value*. These minimum values usually correspond to the values obtained by applying the above rules using a feature size of 25 mm or 1 inch except for *roundness*, where 50 mm or 2 inches gives more satisfactory values.

Examples

1 The minimum maintainable roundness tolerance for the diametral size of a turned bore would be

$$0.00004 \times 50 = 0.0020 \text{ mm}$$

2 The minimum straightness tolerance for a cylindrically-ground bore would be

$$0.00005 \times 25 = 0.0013 \text{ mm}$$

3 The tolerance on diametral parallelism appropriate to a cylindrically-ground parallel bore 200 mm long would be

$$\text{proportional value of } t \times \text{length} = 0.00005 \times 200$$
$$= 0.010 \text{ mm}$$

C5   SI units and conversion factors

The International System of Units (SI – Système International d'Unites) is used as a common system throughout this handbook. The International System of Units is based on the following seven basic units.

Physical quantity	SI unit	Symbol for the unit
length	metre	m
mass	kilogram	kg
time	second	s
electric current	ampere	A
thermodynamic temperature	kelvin	K
luminous intensity	candela	cd
amount of substance	mole	mol

The remaining mechanical engineering units are derived from these, and the most important derived unit is the unit of force. This is called the newton, and is the force required to accelerate a mass of 1 kilogram at 1 metre/second^2. The acceleration due to gravity does not come into the basic unit system, and any engineering formulae in SI units no longer need g correction factors. The whole system of units is consistent, so that it is no longer necessary to have conversion factors between, for example, the various forms of energy such as mechanical, electrical, potential, kinetic or heat energy. These are all measured in joules in the SI system.

Other SI units frequently used in mechanical engineering have the names and symbols given in the following table.

Physical quantity	SI unit	Symbol for the unit
force	newton	$N = kg\,m/s^2$
work, energy or quantity of heat	joule	$J = N\,m$
power	watt	$W = J/s$
velocity	metre/second	m/s
angular velocity	radian/second	rad/s
acceleration	metre/second^2	m/s^2
density	kilogramme/metre^3	kg/m^3
absolute or dynamic viscosity	†newton second/metre^2	N s/m^2
kinematic viscosity	†metre^2/second	m^2/s
volumetric flow rate	metre^3/second	m^3/s
pressure	newton/metre^2	n/m^2
torque	newton metre	N m

† The centipoise and centistokes are also acceptable as units with the SI system.

In many cases the basic SI unit for a physical quantity will be found to be an unsatisfactory size and multiples of the units are therefore used as follows:

10^{12}	tera	T	10^{-3}	milli	m
10^{9}	giga	G	10^{-6}	micro	μ
10^{6}	mega	M	10^{-9}	nano	n
10^{3}	kilo	k	10^{-12}	pico	p

A typical example is the watt, which for mechanical engineering is too small as a unit of power. For most purposes the kilowatt (kW i.e. 10^3W) is used, while for really large powers the megawatt (MW i.e. 10^6W) is more convenient.

It should be noted that the prefix symbol denoting a multiple of the basic SI unit is placed immediately to the left of the basic unit symbol without any intervening space or mark. The multiple unit is treated as a single entity, e.g. mm^2 means (mm)^2 i.e. $(10^{-3} \times m)^2$ or $10^{-6} \times m^2$ and *not* m(m)^2 i.e. *not* $10^3 \times m^2$.

SI units and conversion factors

The following table of conversion factors is arranged in a form that provides a simple means for converting a quantity of SI units into a quantity of the previous British units.

This table also makes it possible to get a feel for the size of the SI units, e.g. that one newton is just less than a quarter of a pound (about the weight of an apple).

Physical quantity	Symbol for SI unit	Conversion factor*	Symbol for familiar British units
acceleration	m/s^2	3.28	ft/s^2
angular acceleration	rad/s^2	57.3	deg/s^2
angular velocity	rad/s	57.3	deg/s
area	m^2	10.8	ft^2
coefficient of heat transfer	$W/m^2 K$	0.176	$Btu/h\,ft^2\,°F$
coefficient of linear expansion	$1/K$	0.556	$1/°F$
density	kg/m^3	6.24×10^{-2}	lb/ft^3
dynamic viscosity	$N\,s/m^2$	10^3	cP
energy, work	J	0.737	ft lbf
		2.78×10^{-7}	kW h
force	N	0.225	lbf
heat capacity	J/K	5.27×10^{-4}	$Btu/°F$ or $CHU/°C$
heat flow rate	W	3.41	Btu/h
heat flux	W/m^2	0.317	$Btu/h\,ft^2$
heat quantity	J	9.48×10^{-4}	Btu
		5.27×10^{-4}	CHU
kinematic viscosity	m^2/s	10^6	cSt
		10.8	ft^2/s
length	m	3.28	ft
mass	kg	2.20	lb
moment, torque	N m	0.738	lbf ft
moment of inertia	$kg\,m^2$	23.7	$lb\,ft^2$
power	W	1.34×10^{-3}	hp
pressure	N/m^2	1.45×10^{-4}	lbf/in^2
second moment of area	m^4	2.40×10^6	in^4
specific heat capacity	$J/kg\,K$	2.39×10^{-4}	$Btu/lb°F$
specific heat/unit volume	$J/m^3\,K$	1.49×10^{-5}	$Btu/ft^3°F$
stress	N/m^2	1.45×10^{-4}	lbf/in^2
surface tension	N/m	6.85×10^{-2}	lbf/ft
thermal conductivity	$W/m\,K$	0.578	$Btu/h\,ft°F$
velocity	m/s	3.28	ft/s
volume	m^3	35.3	ft^3
volumetric flow rate	m^3/s	1.32×10^4	Imp. gall/min

* Multiply the number of SI units by this conversion factor to obtain the number of familiar British units.

e.g. 100 metres $= 100 \times 3.28 = 328$ ft

The conversion factors in this table are correct to only three significant figures.

Index

Index

Index

Index